S0-CDW-989

POLYAMINES IN NORMAL AND NEOPLASTIC GROWTH

Polyamines in Normal and Neoplastic Growth

Proceedings of a Symposium of the National Cancer Institute, U.S.A.

Edited by

Diane H. Russell, Ph.D.
National Cancer Institute
Baltimore Cancer Research Center
Baltimore, Maryland, U.S.A.

Raven Press, Publishers • New York

Made in the United States of America

International Standard Book Number 0–911216–44–8
Library of Congress Catalog Card Number 72–96336

Contents

List of Contributors

Juhani Ahonen
Department of Medical Chemistry, University of Turku, Turku, Finland

Uriel Bachrach
Department of Molecular Biology, Hebrew University, Hadassah Medical School, Jerusalem, Israel

William T. Beck
Department of Pharmacology, Yale University School of Medicine, New Haven, Connecticut 06510

Miriam Ben-Joseph
Department of Molecular Biology, Hebrew University, Hadassah Medical School, Jerusalem, Israel

Anne Blackledge
Biochemistry Group, University of Sussex, School of Biological Sciences, Falmer, Brighton, Sussex, England

Craig V. Byus
Department of Biochemistry, University of New Hampshire, Durham, New Hampshire 03824

E. S. Canellakis
Department of Pharmacology, Yale University School of Medicine, New Haven, Connecticut 06510

Seymour S. Cohen
Department of Microbiology, University of Colorado Medical Center, Denver, Colorado 80220

Gordon L. Coppoc
Department of Veterinary Physiology and Biochemistry, Purdue University, Lafayette, Indiana 47907

E. Delain
Laboratoire de Microscopie Electronique, Institut Gustave Roussy, 94, Villejuif, France

M. Drue Denton
Department of Internal Medicine, University of Cincinnati College of Medicine, Cincinnati, Ohio 45229

Arnold S. Dion
Molecular Biology Section, Institute for Medical Research, Camden, New Jersey 08103

Robert H. Fillingame
Department of Biochemistry, University of Washington, School of Medicine, Seattle, Washington 98195

Robert C. Gallo
Laboratory of Tumor Cell Biology, National Cancer Institute, National Institutes of Health, Bethesda, Maryland 20014

Charles W. Gehrke
Department of Agricultural Chemistry, University of Missouri, Columbia, Missouri 65201

Eduard Gfeller
Department of Anatomy, Johns Hopkins University School of Medicine, Baltimore, Maryland 21205

Wade Gibson
The University of Chicago, Chicago, Illinois 60637

Helen S. Glazer
Department of Internal Medicine, University of Cincinnati College of Medicine, Cincinnati, Ohio 45229

Kenneth D. Graziano
Section of Immunology and Cell Biology, Baltimore Cancer Research Center, Baltimore, Maryland 21211

Luce Gresland
Unité de Biologie Moléculaire, Groupe de Recherche No. 8 du C.N.R.S., Institut Gustave Roussy, 94, Villejuif, France

Bruce Hacker
Division of Oncology, Albany Medical College, Albany, New York 12208

Pekka Hannonen
Department of Medical Chemistry, University of Helsinki, Helsinki, Finland

Sami I. Harik
Department of Pharmacology, Johns Hopkins University School of Medicine, Baltimore, Maryland 21205

Inez A. Hawk
Section on Enzymology and Drug Metabolism, Laboratory of Pharmacology, Baltimore Cancer Research Center, National Cancer Institute, Baltimore, Maryland 21211

Olle Heby
Baltimore Cancer Research Center, National Cancer Institute, Baltimore, Maryland 21211

Edward J. Herbst
Department of Biochemistry, University of New Hampshire, Durham, New Hampshire 03824

Brigid L. M. Hogan
Biochemistry Group, School of Biological Sciences, University of Sussex, Falmer, Brighton, Sussex, England

Erkki Hölttä
Department of Medical Chemistry, University of Helsinki, Helsinki, Finland

J. Huppert
Unité de Biologie Moléculaire, Groupe de Recherche No. 8 du C.N.R.S., Institut Gustave Roussy, 94, Villejuif, France

Juhani Jänne
Department of Medical Chemistry, University of Helsinki, Helsinki, Finland

Leon T. Kremzner
Department of Neurology, College of Physicians & Surgeons of Columbia University, New York, New York 10032

Kenneth C. Kuo
Department of Agricultural Chemistry, University of Missouri, Columbia, Missouri 65201

Phoebe S. Leboy
Department of Biochemistry, School of Dental Medicine, University of Pennsylvania, Philadelphia, Pennsylvania 19104

Carl C. Levy
Laboratory of Pharmacology, Baltimore Cancer Research Center, National Institutes of Health, Baltimore, Maryland 21211

Carol-Ann Manen
Department of Zoology, University of Maine, Orono, Maine 04473

Michael R. Mardiney, Jr.
Section of Immunology and Cell Biology, Baltimore Cancer Research Center, Baltimore, Maryland 21211

Laurence J. Marton
Laboratory of Pharmacology, Baltimore Cancer Research Center, Baltimore, Maryland 21211

William E. Mitch
Laboratory of Pharmacology, Baltimore Cancer Research Center, Baltimore, Maryland 21211

Dan H. Moore
Department of Cytological Biophysics, Institute for Medical Research, Camden, New Jersey 08103

David R. Morris
Department of Biochemistry, University of Washington, School of Medicine, Seattle, Washington 98195

Susan Murden
Biochemistry Group, University of Sussex, School of Biological Sciences, Falmer, Brighton, Sussex, England

Donald L. Nuss
Department of Biochemistry, University of New Hampshire, Durham, New Hampshire 03824

Gavril W. Pasternak
Department of Pharmacology, Johns

Hopkins University School of Medicine, Baltimore, Maryland 21205

Pamela Piester
Department of Biochemistry, School of Dental Medicine, University of Pennsylvania, Philadelphia, Pennsylvania 19104

Gérard A. Quash
Biochemistry Department, University of the West Indies, Kingston, Jamaica

Aarne Raina
Department of Biochemistry, University of Kuopio, Kuopio, Finland

Bernard Roizman
Departments of Microbiology and Biophysics, The University of Chicago, Chicago, Illinois 60637

Diane H. Russell
Laboratory of Pharmacology, Baltimore Cancer Research Center, National Cancer Institute, Baltimore, Maryland 21211

Ted T. Sakai
Department of Microbiology, University of Colorado Medical Center, Denver, Colorado 80220

Amelia Schenone
Ben May Laboratory, The University of Chicago, Chicago, Illinois 60637

Stephen C. Schimpff
Medicine Section, Baltimore Cancer Research Center, National Cancer Institute, Baltimore, Maryland 21211

Morton Schmukler
Laboratory of Pharmacology, Baltimore Cancer Research Center, Baltimore, Maryland 21211

Nikolaus Seiler
Max-Planck-Institut für Hirnforschung, Arbeitsgruppe Neurochemie, 6000 Frankfurt, Germany

Edward G. Shaskan
Section of Neurosciences, Division of Biological and Medical Sciences, Brown University, Providence, Rhode Island 02912

Richard M. Simon
Division of Cancer Treatment, National Cancer Institute, Bethesda, Maryland 20014

Frank G. Smith
Department of Internal Medicine, University of Cincinnati College of Medicine, Cincinnati, Ohio 45229

Solomon H. Snyder
Departments of Pharmacology and Psychiatry, Johns Hopkins University School of Medicine, Baltimore, Maryland 21205

Ching-Hsiang Su
Department of Microbiology, University of Pennsylvania, Philadelphia, Pennsylvania 19104

James G. Vaughn
Amino Acid Analyzer Applications Division, Beckman Instruments, Incorporated, Palo Alto, California

T. Phillip Waalkes
Department of Health, Education and Welfare, National Institutes of Health, National Cancer Institute, Bethesda, Maryland 20014

Thomas Walle
Department of Pharmacology, Medical University of South Carolina, Charleston, South Carolina 29401

George Weber
Department of Pharmacology, Indiana University School of Medicine, Indianapolis, Indiana 46202

H. G. Williams-Ashman
Ben May Laboratory, The University of Chicago, Chicago, Illinois 60637

Kwang B. Woo
Division of Cancer Treatment, National Cancer Institute, Bethesda, Maryland 20014

Christopher C. Wylie
Department of Anatomy, University College London, London, England

David C. Zellner
Department of Internal Medicine, University of Cincinnati College of Medicine, Cincinnati, Ohio 45229

Robert W. Zumwalt
Department of Agricultural Chemistry, University of Missouri, Columbia, Missouri 65201

Polyamines in Normal and Neoplastic Growth. Edited by
D. H. Russell. Raven Press, New York © 1973.

Polyamines in Growth — Normal and Neoplastic

Diane H. Russell

*Laboratory of Pharmacology, Baltimore Cancer Research Center,
National Cancer Institute, Baltimore, Maryland 21211*

It is indeed the fulfillment of much hope and effort that brings us together today. We are beginning a two-day program concerning polyamines, small organic cations which are prime candidates for many regulatory roles in the control of the growth process. This is a rather lofty position for the polyamines. These compounds are the most maligned of amines, as they bear names such as putrescine, spermidine, and spermine. These symbols immediately invoke two images, one being putrefaction, the other being male genital function. The early work on polyamines left many biochemists with the impressions that these cations were the end product of a degradative pathway and that the instances of polyamine occurrence in mammalian systems were keyed to bacterial decay or to excretion into seminal fluid. The early efforts of Celia and Herbert Tabor of the National Institutes of Health in elucidating the biosynthetic pathway in bacteria and the work of Seymour Cohen and his group in linking polyamine biosynthesis to RNA metabolism have provided the backbone for the expansion of polyamine research into the mammalian system (1, 2). This expansion was further catalyzed by an article by Dykstra and Herbst (3) expressing the relationship between spermidine synthesis and RNA synthesis in regenerating rat liver. Somehow the stigma of polyamines being involved in decay began to fall away as Dykstra and Herbst showed that the uptake of putrescine and its conversion into spermidine in partially hepatectomized rats was a major event. The large accumulation of spermine in seminal fluid has never been explained and remains one of the unanswered questions.

My own interest in polyamine research occurred during my collaboration with Dr. Solomon Snyder. We were intrigued by the role of histidine decarboxylase in the rapid growth process as postulated by Kahlson (4). In discussion, we postulated that if histidine decarboxylase, an enzyme which forms histamine, a diamine, is important in rapid growth, this should be greatly enhanced in all rapid-growth systems. However, studies had indicated that this was not true. Could it be that histamine serves a function in certain rapidly growing tissues which could be served in other tissues by polyamines? Therefore, we looked at the first enzyme in the polyamine biosynthetic pathway in regenerating rat liver. In a pilot experiment, we assayed ornithine decarboxylase activity in the liver of sham-hepatectomized rats and in the liver of rats that had undergone a partial hepatectomy 24 hr prior to sacrifice. Results were rewarding. It appeared that ornithine decarboxylase activity was very low in the liver of normal rats: the counts were in the range of 200 to 400 cpm. However, the counts for the first 24-hr regenerating liver sample were around 20,000.

After finding this dramatic increase in ornithine decarboxylase activity in regenerating rat liver (5), which is rather unusual since mammalian enzymes usually fluctuate a few-fold and rarely 25-fold such as found for ornithine decarboxylase, I was astounded to find in the literature how widely polyamines had been implicated in cell regulation. The quotation that comes to mind appears in Seymour Cohen's book *The Introduction to the Polyamines* (2) at the beginning of Chapter 1: "All this has been said before – but since nobody listened, it must be said again" (André Gide). It seemed reasonable to suppose that early increases in ornithine decarboxylase activity which lead to such dramatic increases in the putrescine and spermidine pools in growing tissues had to be of great importance. First, polyamines had been implicated in growth processes. Herbst and his collaborators had found that certain bacterial mutants exhibit absolute requirements for polyamines (6). Further, polyamines were implicated by Seymour Cohen and others in the regulation of RNA synthesis (7, 8).

To summarize, then, we found that increased ornithine decarboxylase activity was one of the earliest, marked events that occurs after partial hepatectomy in the rat. Its increased activity appears to parallel the early increase in RNA synthesis, and precedes by many hours the maximal DNA synthesis that occurs in regenerating rat liver (9). We also found a close relationship between ornithine decarboxylase activity and the initiation of rapid growth in chick embryos and tumors (5).

The ability of ornithine decarboxylase activity to fluctuate rapidly in response to the introduction or withdrawal of stimuli suggests that putrescine synthesis is under strict modulation. The rapid turnover rate of hepatic ornithine decarboxylase is the most striking example of this modulation.

We found that ornithine decarboxylase activity declined rapidly in unoperated rats or in hepatectomized rats after cycloheximide administration. The decline had an estimated half-life of 11 min (10). To my knowledge, this is the most rapidly turning over mammalian enzyme known. Further, estimating the half-life of ornithine decarboxylase after growth hormone stimulation and decline, without any inhibitors, led to a similar estimation of a half-life of less than 20 min (11). The only evidence lacking, of course, was evidence of the turnover rate of the purified enzyme. Since ornithine decarboxylase has not been purified to homogeneity, it was impossible to do studies on the purified enzyme. This rapid turnover rate is of great importance because the synthesis of most mammalian enzymes is a linear function of time, whereas enzyme degradation is an exponential function of time. Therefore, rates of change of enzyme levels from one steady state to another are determined solely by the degradation rate of the enzyme. The very high degradative rate of ornithine decarboxylase suggests that its activity changes rapidly in response to stimuli for new synthesis. Taken together, these data suggest that polyamine synthesis is a finely modulated process. Moreover, they suggest that this kind of sensitive regulation of synthesis would be necessary only to control the level of compounds important in the cell stimulatory system. This is of further importance when you consider that in most mammalian tissues there are not known enzymes that degrade or metabolize the polyamines. Therefore, an overproduction of the polyamines could lead to an elevated growth rate for a particular tissue or organ. This could be catastrophic in an adult mammal, since most of the tissues and organs are in dynamic equilibrium, and are not growing *per se*. In the mammalian organism, exceptions to this generalization are proliferating surfaces such as the gut, secreting organs such as the pancreas, and abnormal growths such as cancers (discussed in detail later).

I. HORMONAL REGULATION OF POLYAMINE BIOSYNTHESIS

If polyamine biosynthesis is necessary for growth processes to occur, it should be expected that this biosynthesis would be affected by hormones that regulate growth. Indeed, this is true. Castration in the rat results in a rapid decrease in both ornithine decarboxylase activity and S-adenosyl-L-methionine decarboxylase activity in the rat ventral prostate. When testosterone is administered to the castrated rat, there is a rapid increase in the activities of both ornithine decarboxylase and S-adenosyl-L-methionine decarboxylase (12). In young rats, ornithine decarboxylase activity exhibits an early dramatic induction after growth hormone administration, followed later by a substantial increase in the level of S-adenosyl-L-methionine decarboxylase activity (11, 13, 14). *De novo* synthesis appears to be involved in the enzyme inductions since the administration of RNA and protein inhibitors suggest that both protein synthesis and DNA-dependent

RNA synthesis are necessary for these elevations to occur. Enhancements in the biosyntheses of putrescine, spermidine, and spermine can be shown also in the castrated rat uterus after estradiol administration (15–17). Further, it has been reported that cortisone has an effect on hepatic ornithine decarboxylase activity (18). The mammary gland, which is under strict hormonal control and which can be cycled through growth, lactation, and involution, exhibits intensive polyamine biosynthesis and accumulation during pregnancy and lactation, with the concentration of spermidine reaching levels above 5 mM during midlactation (19). If the number of suckling young is decreased by removing them from the mother, the amount of spermidine drops precipitously and is concomitant with dramatic drops in the amount of RNA present. To my knowledge, there are no growth processes that occur without prior stimulation of polyamine biosynthesis.

II. EMBRYONIC SYSTEMS

One of the compelling reasons to study polyamine biosynthesis and accumulation in embryonic systems is to assess polyamine metabolism in a maximally responding system. The other compelling reason, however, is to understand the growth process *per se*, which is best exemplified here. Further, the embryonic system, considered a normal growth system, most nearly parallels tumor systems. That is, "Resemblance of hepatoma to fetal tissues indicates some resemblance of all tumors to all fetal tissues, a general tendency that can be called the fetalism of tumors" (20). It appears that tumors and fetal systems resemble each other because both fetal tissues and tumors tend to be undifferentiated and therefore exhibit very similar enzyme patterns. Knox (20) states that "undifferentiated tumors are very similar to each other, as similar as some fetal tissues are to one another." Organ-specific components and great diversity disappear in undifferentiated tumors, and are only present in the highly differentiated tumors which are rather rare. These concepts which stress the similarities of tumors and fetal systems are gaining more acceptance from the scientific community. A recent report indicated that several human tumors contain the fetal form of thymidine kinase in contrast to other human tissues which have only a postnatal thymidine kinase (21). Therefore, the finding that polyamine biosynthesis and accumulation is an early marked event in all types of embryos [chick (22, 23), toad (24, 25), rat (26), and sea urchin (27, 28)] has implications for the understanding and the control of the cancer process. The same rapid synthesis and accumulation of polyamines that is exhibited by embryos is also exhibited by tumors in early growth stages (29). An effective inhibitor of putrescine synthesis or spermidine synthesis would appear to be an ideal cancer chemotherapeutic agent. If the levels of polyamines of a particular system could be lowered, it should decrease the viability of the tumor substantially.

We have screened a large number of analogues of ornithine *in vitro* for their ability to inhibit ornithine decarboxylase activity from 24-hr regenerating rat liver. Most of these analogues contained ring structures attached to the delta amino group. Only two were even moderately good inhibitors *in vitro*, α-methyl ornithine (Table 1) and *n*-methyl ornithine. At 10^{-3} M, *n*-methyl ornithine resulted in a 40% inhibition of ornithine decarboxylase. Neither of these analogues changed the ornithine decarboxylase activity during the course of L1210 leukemia in mice nor resulted in an increased survival rate for those leukemic mice receiving the drug(s).

TABLE 1. *Effect of an ornithine analogue on ornithine decarboxylase activity of 24-hr regenerating rat liver*

Inhibitor	Concentration of inhibitor (M)	% Inhibition
α-Methyl ornithine	0	0
	10^{-7}	5.3
	10^{-6}	13.3
	10^{-4}	40.5
	10^{-3}	64.6

III. PHYSIOLOGICAL SIGNIFICANCE OF POLYAMINES

It has been stated before that the accumulation of polyamines is concomitant with RNA synthesis and, of course, with protein synthesis. The correlation of polyamine synthesis with RNA synthesis is parallel in so many systems that it is hard to believe at this point that there is not a cause and effect relationship. Probably because of the tight relationship between RNA synthesis and DNA synthesis in certain systems, there appears at times to be a relationship between polyamine concentrations and DNA synthesis. However, there are systems in which you can uncouple RNA synthesis and DNA synthesis, such as in the heart undergoing hypertrophy after constriction of either the aortic or the pulmonary artery; in this case there is extensive polyamine accumulation which again correlates with RNA synthesis, but there is no concomitant DNA synthesis (30, 31).

The first clue that has come from our work as to one possible physiological role for the polyamines comes from work on developing *Xenopus laevis*. There is an anucleolate mutant of *X. laevis* which is unable to make ribosomal RNA after gastrulation. Therefore, the organism is only partially viable and dies at a particular time when it is a swimming tadpole. The anucleolate mutant is unable to organize a viable nucleolus and synthesize ribosomal RNA. It is also unable to synthesize spermidine, and has lowered amounts of putrescine synthesis (25). Therefore, the deletion of ribosomal DNA in this organism leads not only to a lack of ribosomal RNA synthesis but also to a lack of spermidine synthesis. Therefore, it is entirely likely that the gene sites for ribosomal RNA synthesis contain not only the genes for ribosomal RNA synthesis but also regulator genes for substances necessary to regulate ribosomal RNA synthesis. The understanding of the complex regulation of RNA synthesis, I feel, is one of the most exciting areas for researchers today. Evidence from many laboratories suggests that polyamines are intimately involved in these complex regulations (2). However, we are just beginning to reach a level above that of the correlative stage of research and to gain definitive facts of exactly which is the cause and which is the effect.

In order to understand better exactly what the molecular roles of the polyamines are, we have been studying isolated nuclei from rat liver. Instead of isolating the nuclei with magnesium as a stabilizing cation, we isolate the nuclei with 5 mM spermidine. Much better yields of nuclei are obtained with spermidine than with magnesium, and there are both structural and biochemical changes in nucleolar structure when isolation is with spermidine (32, 33). This will be discussed in detail elsewhere in this volume by Gfeller, but I will review the general findings.

Biochemically, there is approximately two times as much DNA associated with the nucleolus in spermidine-isolated preparations (32). The level of DNA-dependent RNA polymerase is approximately 10-fold that found in nuclei isolated with magnesium. In both preparations the optimal ionic milieu is present in the assay mixture (34). When other biochemical parameters of the nucleolus such as protein and RNA content are assayed, these appear to be similar in both preparations. At this time we feel that the increased nucleolar-associated RNA polymerase activity must be due to the additional DNA associated with the nucleolus, and that this additional DNA must be a direct result of the spermidine isolation procedure. To determine further if polyamines have a direct effect on RNA polymerase I activity, we have isolated nucleolar fractions, have partially purified RNA polymerase I from the nucleolus on DEAE Sephadex columns, and have thereafter found that spermidine will not increase or decrease the activity of the enzyme at this point. It is assumed that this implies that polyamines do not work by a direct effect on the enzyme or, if they do, that we have purified away a subunit or a fraction necessary for the direct polyamine stimulation of the enzyme. This is, of course, a possibility.

IV. BIOSYNTHESIS

Initially it was observed that the mammalian enzyme S-adenosyl-L-methionine decarboxylase, which is involved in spermidine synthesis, was vastly different in its requirements than the enzyme performing the same task in *Escherichia coli* (1, 35). In the mammalian system, evidence indicated that the decarboxylation of S-adenosyl-L-methionine and the transfer of the propylamine moiety from decarboxylated S-adenosyl-L-methionine to putrescine to form spermidine, or to spermidine to form spermine, must be catalyzed by one enzyme or an enzyme complex. This appeared to be a valid assumption since it was not possible to demonstrate any decarboxylated S-adenosyl-L-methionine as a free intermediate in a crude homogenate, and this reaction required either putrescine or spermidine, or a similar substrate, as a receptor molecule for the propylamine moiety (35, 36). This reaction did not require any metal ions as did the enzyme S-adenosyl-L-methionine decarboxylase in *E. coli* (1, 37, 38). There is some question as to whether pyridoxal phosphate is required by the mammalian enzyme complex, but we have found that after this enzyme is purified about 300-fold from rat liver, the activity is enhanced by the addition of small amounts of pyridoxal phosphate (38). This evidence plus the fact that the decarboxylase is strongly inhibited by known inhibitors of pyridoxal phosphate-requiring enzymes make it appear likely that pyridoxal phosphate will be an ultimate requirement of this enzyme (35, 38).

We have stated so far that spermidine synthesis in mammals appears to be catalyzed, at least in crude enzyme preparations, by one enzyme or one enzyme complex. However, the effects of purification on this enzyme system are unclear at this time. Two groups have reported that they can separate S-adenosyl-L-methionine decarboxylase from spermidine synthase (39, 40), and one group has reported the further separation of a spermine synthase (41). This is puzzling for the following reasons. First, the K_m for decarboxylated S-adenosyl-L-methionine in the amino propyl transfer reaction is 2.5×10^{-5} M. This is very high if this compound is going to serve efficiently as a receptor for a system in which S-adenosyl-L-methionine (K_m 5×10^{-5} M) is decarboxylated in the presence of putrescine and there is the formation of spermidine. This has been discussed in detail by Seymour Cohen (2). Further, when the S-adenosyl-L-methionine decarboxylase activity is separated from the propylamine transfer activity, it is stated that spermidine synthase is present in a 50- to 100-fold greater amount than that for the decarboxylase (40). We have found that S-adenosyl-L-methionine decarboxylase from rat liver, which has been purified about 400-fold, is still capable of decarboxylating S-adenosyl-L-methionine and of transferring propylamine moieties to either putrescine or spermidine, depending on which substrate is provided. This purified enzyme has a molecular weight of about 50,000 as estimated by two different methods: that of Martin and Ames (42) utilizing sucrose density gradients, and that of Andrews (43) which involves the determination of molecular weight by column chromatography with standard proteins. It is difficult to see how an enzyme or enzyme complex with this low a molecular weight can indeed be three separate enzymes. Recently it has been reported that when S-adenosyl-L-methionine decarboxylase and spermidine synthase are placed on Sephadex G-200 columns, they elute at identical spots, indicating they have similar molecular weights (40). Further, unpurified S-adenosyl-L-methionine decarboxylase, which still contains spermidine synthase activity and will catalyze the conversion of putrescine to spermidine, has a similar weight to both the separated S-adenosyl-L-methionine decarboxylase and the separated spermidine synthase (40). The unpurified S-adenosyl-L-methionine decarboxylase activity also must contain spermine synthase activity, which presumably has been removed by the time that purified S-adenosyl-L-methionine decarboxylase was obtained.

It is difficult to understand how a possible enzyme complex that weighs 50,000 can be divided into three enzymes each of which weighs about 50,000. It is possible that fractionation procedures could inactivate sites or subunits of a single enzyme complex capable of coupled decarboxylation and polyamine synthesis. This has been postulated by Cohen (2). Certainly further studies are necessary to clarify this most important question. Obviously, the biosynthetic pathway for polyamines in conjunction with the total picture of the physiological roles of polyamines presents a rather complex situation, and the biosynthesis of polyamines and the control of the relative amount of each amine present must be subject to more intricate regulation than we have yet been able to comprehend or question. Again it must be stressed that in the physiological situation, i.e., the crude system in rats, there is no demonstrated uncoupling of decarboxylation and propylamine transfer functions in the synthesis of either spermidine or spermine.

There are several other lines of evidence which indicate that the enzymes responsible for spermidine and spermine synthesis may indeed be one coupled system. We reported (15) that spermidine synthesis (putrescine-stimulated S-adenosyl-L-methionine decarboxylase activity) was stimulated in the castrated rat uterus after intraperitoneal injection of estradiol-17β. We later found (17) that spermine synthesis (spermidine-stimulated S-adenosyl-L-methionine decarboxylase) was also subject to activation or enhancement by the hormone estradiol-17β, and that both putrescine-stimulated and spermidine-stimulated S-adenosyl-L-methionine decarboxylase respond to inhibitors of protein and RNA synthesis in precisely the same manner. Further, the half-life of the spermine-synthesizing enzyme measured after cycloheximide administration is similar to that for the spermidine-synthesizing system, i.e., 60 min (15, 17). Therefore, there is evidence in both rat liver and rat uterus to support the concept of one complex which functions as both a decarboxylase and a propylamine transferase for the synthesis of both spermidine and spermine. We recently performed experiments on the effects of methyl glyoxal-*bis*(guanylhydrazone), a known inhibitor of S-adenosyl-L-methionine decarboxylase, on both putrescine-stimulated and spermidine-stimulated S-adenosyl-L-methionine decarboxylases. This substance has a structure similar to that of spermidine. To my knowledge, this inhibitor has only been tested as an inhibitor of the putrescine-stimulated S-adenosyl-L-methionine decarboxylase of rat ventral prostate and liver (36). This inhibitor affects spermidine-stimulated S-adenosyl-L-methionine decarboxylase in a manner similar to that of putrescine-stimulated S-adenosyl-L-methionine decarboxylase (Heby and Russell, *in preparation*).

The rate-limiting factor in either case will continue to be the activity of S-adenosyl-L-methionine decarboxylase. Therefore, the best enzymic estimates of spermidine and spermine syntheses can be obtained from the measurements of putrescine-stimulated and spermidine-stimulated S-adenosyl-L-methionine decarboxylase, respectively.

V. ELEVATED URINARY POLYAMINES IN HUMAN CANCER PATIENTS

As previously stated, studies of rapid-growth systems, both normal and neoplastic, indicate that polyamine synthesis and accumulation is markedly elevated early after tissue stimulation. We have previously discussed evidence that these compounds may play roles in the control of ribosomal RNA metabolism. Since in the mammalian system there are no known pathways for polyamine degradation, it seemed possible that polyamines could be excreted in the urine and that elevated levels could be present in pathological situations. A pilot study of the urine of a patient with a large metastatic ovarian teratoma indicated astonishingly high levels of putrescine, spermidine, and spermine. This was even more convincing when the surgical removal of the major portion of this solid tumor mass led to a marked depression in urinary levels of polyamines. This initial finding led to a systematic screening of a number of patients with various types of diagnosed cancer to determine if polyamines were excreted by other cancer patients. Indeed, approximately 200 cancer patients and close to 100 normals have been screened, and this generalization still appears to be true (44, 45).

Polyamine concentrations were determined by extraction of the amines into l-butanol and separation by high-voltage electrophoresis (46, 47). To obtain polyamines in a nonconjugated form it is necessary to subject the urine to acid hydrolysis prior to butanol extraction. We have found that the lower limit of detection with this method is approximately 2.5 mg of amine per a 24-hr urine. Many of the normals fell below this range and did not have detectable amounts of putrescine, spermidine, and spermine, as measured by this method. However, separation of polyamines with an amino acid analyzer (48, 49) and by a new method using gas-liquid chromatography (Gehrke, *personal communication*) appears to allow the detection of even lower amounts of polyamines in normal volunteers, and indeed there is no doubt that small amounts of putrescine, spermidine and spermine are characteristic components of human urine (44, 45, 49). In general then, putrescine excretion is somewhat elevated in cancer patients. Characteristically, spermidine levels are elevated in the urine of cancer patients, and recently we have been able to corroborate this with elevations in the serum of human cancer patients (Marton, Vaughn, Hawk, Levy, and Russell, *this volume*). We recently found that the levels of spermine excretion reported for cancer patients utilizing high-voltage electrophoresis might indeed be overestimated. This is probably due to excretion of an unknown which has an identical migratory pattern with spermine in the system, and which is ninhydrin-positive in precisely the same manner as spermine. However, whatever this unknown compound might be, it is not excreted in detectable amounts by normals, and it appears at this time to be characteristic of neoplasia. Further, evidence that this unknown is involved in the cancer process comes from a study of mouse urinary excretion of polyamines after inoculation with L1210 leukemia cells. Both normal and tumor mice excrete rather substantial amounts of putrescine and spermidine (Table 2). They excrete considerable amounts of the unknown that we described above but no detectable spermine. The amount of the unknown excreted appears to increase by day 3 after tumor inoculation and by day 6 has reached a level six times that found in normal nonleukemic mice.

The conclusions at this time are that, certainly in most advanced human tumor patients whose urines we have screened, there is elevated excretion of polyamines, particularly putrescine and spermidine, with the quantification of spermine as yet uncertain. Coupled with elevated polyamine excretion there is elevated excretion of an unknown substance which is ninhydrin-positive and which migrates with spermine, so that the value which we have previously reported is a combined spermine–unknown estimation. This unknown appears to be the major elevated excretory product in the urine of mice inoculated with L1210 leukemia cells. We are currently

TABLE 2. *Effect of L1210 leukemia in mice upon urinary polyamine excretion*

	nmoles/24 hr		
	Putrescine	Spermidine	Unknown (conc. estimated)
Controls	432 ± 75	156 ± 21	234
Day 1	487 ± 223	150 ± 33	152
2	500 ± 236	150 ± 53	210
3	710 ± 49	118 ± 12	261
4	374 ± 112	176 ± 40	312
5	380 ± 83	121 ± 27	331
6	458 ± 179	155 ± 25	1194

10^6 L1210 leukemia cells inoculated on Day 0 to BDF_1 mice. Each value represents the mean ± SEM of 6 determinations of separate samples.

investigating the identity of this compound by collaborating with Gehrke and Waalkes of the University of Missouri and NCI, respectively, in order to elucidate its structure by gas-liquid chromatography-mass spectroscopy.

In summary, there is a potential for two marker systems which might be of value in cancer detection and chemotherapy procedures. Because the control of cancer would be of such great value in the alleviation of human misery, the need for further work in both of these areas is immediately apparent, and to the extent possible we are emphasizing this area at the Baltimore Cancer Research Center. The utilization of the amino acid analyzer technique for polyamine determination in our laboratory at this time, by which we can detect low concentrations of polyamines, with complete separation of the unknown from spermine, is of great value for further studies. The clinical application of routine urinary and blood polyamine assays would be possible only after a rapid simple automated technique for polyamines can be worked out; this area is also being explored.

VI. POLYAMINE PROFILES OF HEPATOMAS AS A FUNCTION OF THEIR GROWTH RATE

We have examined three rat hepatomas with different growth rates. We have found that a fast-growing hepatoma with a doubling time of 2 days has a high spermidine/spermine ratio, whereas a slow-growing hepatoma with a doubling time of 24 days has a low spermidine/spermine ratio (Table 3). It is interesting that the sum of the spermidine and spermine concentrations is essentially similar in all three hepatomas. However, the most rapidly growing hepatoma has the most spermidine. Putrescine is present only in trace amounts in the slower-growing hepatomas, but is present in consider-

TABLE 3. *Polyamine profiles of hepatomas as a function of their growth rates*

Hepatoma	Putrescine	Spermidine	Spermine	Spermidine /Spermine Ratio	Volume Doubling Time
	nmoles/mg protein				
					(days)
16	trace	4.60 ± 0.22	7.86 ± 0.40	0.58	24.46
7800	trace	5.06 ± 0.57	6.32 ± 0.27	0.80	6.07
3924A	1.52 ± 0.19	6.63 ± 0.29	4.10 ± 0.37	1.62	4.35
Liver control	trace	3.10 ± 0.15	4.65 ± 0.20	0.67	------

Each value represents the mean ± S.E.M. of 9 separate determinations.

able amounts in the fast-growing hepatoma. This work was done in collaboration with Dr. William B. Looney (University of Virginia Medical School, Charlottesville). Doubling times for the hepatomas are those measured by his group. Similar results were recently reported by Williams-Ashman et al. (50). An understanding of tumor cell kinetics should ultimately lead to more effective cancer chemotherapy. It has been established recently (29, 51, 52) that relatively effective chemotherapeutic agents decrease polyamine levels in tumor systems (see chapter by Heby, *this volume*).

In summary, there is little doubt that polyamines are intimately involved with the regulation of growth. Further, an understanding of the complex interactions which they influence should have both esthetic value and practical applications.

REFERENCES

1. Tabor, H., and Tabor, C. W., *Pharmacol. Rev.*, **16,** 245 (1964).
2. Cohen, S. S., *Introduction to the polyamines,* Prentice-Hall, New Jersey (1971).
3. Dykstra, W. J., Jr., and Herbst, E. J., *Science,* **149,** 428 (1965).
4. Kahlson, G., *Perspect. Biol. Med.,* **5,** 179 (1962).
5. Russell, D., and Snyder, S. H., *Proc. Nat. Acad. Sci. U.S.A.,* **60,** 1420 (1968).
6. Herbst, E. J., and Snell, E. E., *J. Biol. Chem.,* **176,** 989 (1948).
7. Raina, A., and Cohen, S. S., *Proc. Nat. Acad. Sci. U.S.A.,* **55,** 1587 (1966).
8. Cohen, S. S., and Raina, A., in H. J. Vogel, J. O. Lampen, and V. Bryson (Editors), *Organizational biosynthesis,* Academic Press, New York (1967).
9. Bucher, N., *Intern. Rev. Cytol.,* **15,** 245 (1963).
10. Russell, D. H., and Snyder, S. H., *Mol. Pharmacol.,* **5,** 253 (1969).
11. Russell, D. H., Snyder, S. H., and Medina, V. J., *Endocrinology,* **86,** 1414 (1970).
12. Pegg, A. E., and Williams-Ashman, H. G., *Biochem. J.,* **109,** 32P (1968).
13. Jänne, J., and Raina, A., *Biochim. Biophys. Acta,* **174,** 769 (1969).
14. Russell, D. H., and Lombardini, J. B., *Biochim. Biophys. Acta,* **240,** 273 (1971).
15. Russell, D. H., and Taylor, R. L., *Endocrinology,* **88,** 1397 (1971).
16. Cohen, S., O'Malley, B. W., and Stastny, M., *Science,* **170,** 336 (1970).

17. Russell, D. H., and Potyraj, J. J., *Biochem. J.*, **128,** 1109 (1972).
18. Richman, R. A., Underwood, L. E., Van Wyk, J. J., and Voina, S. J., *Proc. Soc. Exp. Biol. Med.*, **138,** 880 (1971).
19. Russell, D. H., and McVicker, T. A., *Biochem. J.*, in press.
20. Knox, W. E., *Amer. Scientist*, **60,** 480 (1972).
21. Stafford, M. A., and Jones, O. W., *Biochim. Biophys. Acta*, **277,** 439 (1972).
22. Caldarera, C. M., Barbiroli, B., and Moruzzi, G., *Biochem. J.*, **97,** 84 (1965).
23. Snyder, S. H., and Russell, D. H., *Fed. Proc.*, **29,** 1575 (1970).
24. Russell, D. H., Snyder, S. H., and Medina, V. J., *Life Sci.*, **8,** 1247 (1969).
25. Russell, D. H., *Proc. Nat. Acad. Sci. U.S.A.*, **68,** 523 (1971).
26. Russell, D. H., and McVicker, T. A., *Biochim. Biophys. Acta*, **259,** 247 (1972).
27. Manen, C. A., and Russell, D. H., *Amer. Zool.*, **12,** xxxix (1972).
28. Manen, C. A., and Russell, D. H., *J. Embryol. Exp. Morphol.*, in press.
29. Russell, D. H., and Levy, C. C., *Cancer Res.*, **31,** 248 (1971).
30. Russell, D. H., Shiverick, K. T., Hamrell, B. B., and Alpert, N. R., *Amer. J. Physiol.*, **221,** 1287 (1971).
31. Feldman, M. J., and Russell, D. H., *Amer. J. Physiol.*, **222,** 1199 (1972).
32. Russell, D. H., Levy, C. C., and Taylor, R. L., *Biochem. Biophys. Res. Comm.*, **47,** 212 (1972).
33. Gfeller, E., Stern, D. N., Russell, D. H., Levy, C. C., and Taylor, R. L., *Z. Zellforsch.*, **129,** 447 (1972).
34. Roeder, R. G., and Rutter, W. J., *Nature*, **224,** 234 (1969).
35. Pegg, A. E., and Williams-Ashman, H. G., *J. Biol. Chem.*, **244,** 682 (1969).
36. Williams-Ashman, H. G., and Schenone, A., *Biochem. Biophys. Res. Comm.*, **46,** 288 (1972).
37. Feldman, M. J., Levy, C. C., and Russell, D. H., *Biochem. Biophys. Res. Comm.*, **44,** 675 (1971).
38. Feldman, M. J., Levy, C. C., and Russell, D. H., *Biochemistry*, **11,** 671 (1972).
39. Jänne, J., and Williams-Ashman, H. G., *Biochem. Biophys. Res. Comm.*, **42,** 222 (1971).
40. Hannonen, P., Jänne, J., and Raina, A., *Biochem. Biophys. Res. Comm.*, **46,** 341 (1972).
41. Raina, A., and Hannonen, P., *FEBS Letters*, **16,** 1 (1971).
42. Martin, R. G., and Ames, B. N., *J. Biol. Chem.*, **236,** 1372 (1961).
43. Andrews, P., *Biochem. J.*, **91,** 22 (1964).
44. Russell, D. H., *Nature*, **233,** 144 (1971).
45. Russell, D. H., Levy, C. C., Schimpff, S. C., and Hawk, I. A., *Cancer Res.*, **31,** 1555 (1971).
46. Raina, A., *Acta Physiol. Scand.*, **60,** Suppl., 218 (1963).
47. Russell, D. H., Medina, V. J., and Snyder, S. H., *J. Biol. Chem.*, **245,** 6732 (1970).
48. Morris, D. R., Koffron, K. L., and Okstein, C. J., *Anal. Biochem.*, **30,** 449 (1969).
49. Bremer, H. J., and Kohne, E., *Clin. Chim. Acta*, **32,** 407 (1971).
50. Williams-Ashman, H. G., Coppoc, G. L., and Weber, G., *Cancer Res.*, **32,** 1924 (1972).
51. Russell, D. H., *Cancer Res.*, **32,** 2459 (1972).
52. Heby, O., and Russell, D. H., *Cancer Res.*, **33,** 159 (1973).

Polyamines in Normal and Neoplastic Growth. Edited by D. H. Russell. Raven Press, New York © 1973.

Tumor Cells, Polyamines, and Polyamine Derivatives

Uriel Bachrach and Miriam Ben-Joseph

Institute of Microbiology, Hebrew University-Hadassah Medical School, Jerusalem, Israel

I. POLYAMINES AND MALIGNANCY

It has been well established that in rapidly growing tissues the accumulation of polyamines, or their derivatives, parallels the cellular proliferation rate. It is therefore not surprising that a number of investigators tried to establish a similar correlation for tumor cells, which are known to proliferate rapidly.

The occurrence of spermine, or its derivative, in the blood of cancer-bearing individuals had been reported independently by two groups in Japan. Kosaki and associates studied the distribution of a spermine-containing phospholipid in malignant tissues; whereas Tokuoka devised a colorimetric assay to detect spermine in the blood of cancer-bearing individuals.

A. Malignolipin

As part of a series of studies on porphyrins, Kosaki and Saka (1) reported that cancerous tissues but not normal tissues could be stained with protoporphyrin III. This work formed the basis for the theory that cancer cells contain a unique phospholipid called malignolipin, which was believed to be responsible for the differences in staining. This phospholipid, which was detected in various malignant tissues (seminoma, stomach, colon, uterine, and breast tumors and in Hodgkin's malignant granuloma), was basic and upon hydrolysis yielded spermine, choline, and an unidentified fatty acid (2). Malignolipin was further characterized by its picrate (m.p. 123°C) (3) paper chromatography (4) and by infrared spectroscopy. Kosaki et al. (4) have also isolated malignolipin from blood and ascites and suggested that the malignolipin test may serve for the early diagnosis of cancer both in human and in experimental animals (5). After examining more than 1,400 cases, they claimed that clinical findings correlated well with the quantities of malignolipin in the blood and that following surgery or irradiation the quantities of malignolipin decreased significantly. However, in 14% of the cases, in which malignancy was confirmed histologically, the malignolipin test was negative (4).

In attempting to repeat Kosaki's histological work, Hughes (6) found only a transient, differential staining effect with protoporphyrin III, which he attributed to nonspecific biochemical differences between the various tissues. Several workers (7–10) have found the blood test unacceptable, and others, including Kögl et al. (11), Kamat (12), and Sax et al. (13), who used the original procedure, failed to isolate malignolipin. Our attempts to isolate a polyamine-containing lipid from malignant tissues were unsuccessful. Attempts to modify the isolation procedure (7, 8) or the identification techniques (14) were not satisfactory. Even though the occurrence of malignolipin in malignant tissues was questioned by some investigators, others were able to confirm Kosaki's findings (15–23). Kallistratos et al. (24) separated malignolipin by paper chromatography and showed that the test was positive in 82% of the patients. Kallistratos et al. (24, 25) also modified the original isolation procedure and advocated the use of an amino acid analyzer to purify malignolipin.

Little is known about the biological role of malignolipin. According to Kosaki and Nakagawa (26), it promotes the intraperitoneal growth of Ehrlich ascites cells. At present the significance of the malignolipin test remains doubtful, and the structure and function of malignolipin in tumor development or metabolism have to await more rigorous isolation procedures.

B. Tokuoka reaction

In 1950 Tokuoka described a colorimetric test for the detection of spermine in tissues and sera of cancer patients (27). The test is performed by heating serum diluted in distilled water with copper carbonate; the formation of a "reddish-purple-blue" color indicates the presence of spermine. Tokuoka (28) tested the blood of more than 1,000 patients and obtained a positive reaction with the sera of 85.6% of the cases. A false positive reaction was also obtained in cases of pregnancy, epilepsy, and encephaloma.

The value of the $CuCO_3$ test for the diagnosis of malignancy has been disputed. Some authors obtained a high percentage of positive results when testing the sera of cancer-bearing individuals (29–31), although others failed to reproduce these findings (32–34). The percentage of false positive results also remains controversial (35–37).

In our hands a positive reaction was obtained with 65% of the patients, whereas 11% of the normal sera were also positive (38). At present the $CuCO_3$ test is not considered reliable because it lacks specificity and it is difficult to distinguish the "reddish-purple-blue" color.

II. POLYAMINE ANALOGUES AS ANTITUMOR AGENTS

The above-mentioned experiments and reports from the laboratories of Rosenthal (39), Russell (40–42), and others (43–51) strongly suggest that tumor cells are rich in polyamines and that these compounds are synthesized rapidly by malignant cells. If polyamines play an important role in tumor development, then their growth should be inhibited by polyamine analogues. Similarly, antibodies against polyamines or polyamine conjugates should inhibit tumor proliferation. Indeed, Kosaki et al. (52) found that antimalignolipin serum inhibited the development of Yoshida sarcoma cells. Similar results were obtained by Quash (53) who showed that the growth of sarcoma T.R.V.L. 53886 in mice was less rapid when mice were preinoculated with polyamines coupled to protein carriers. Apparently, in both cases the antibodies reacted with polyamines and thus reduced their effective concentrations. A different approach to reduce the intracellular concentration of polyamines was based on the inhibition of their biosynthesis. Such an approach was used by Williams-Ashman and Schenone (54) who used methylglyoxal-bis(guanylhydrazone) to inhibit the synthesis of spermidine and ethionine which was used by Raina et al. (55) to inhibit the biosynthesis of polyamines.

Oxidized polyamines comprise another group of polyamine analogues which are potential antitumor agents. It was shown by Halevy et al. (56) and Alarcon et al. (57) that spermine, after oxidation by serum amine oxidase, was toxic for tumor cells, including L1210 ascites lymphatic leukemia cells. The toxicity of oxidized polyamine for various tumors was confirmed by others (58–61).

During recent years, we have studied the biological activity of oxidized spermine and showed that the primary oxidation product is a dialdehyde.

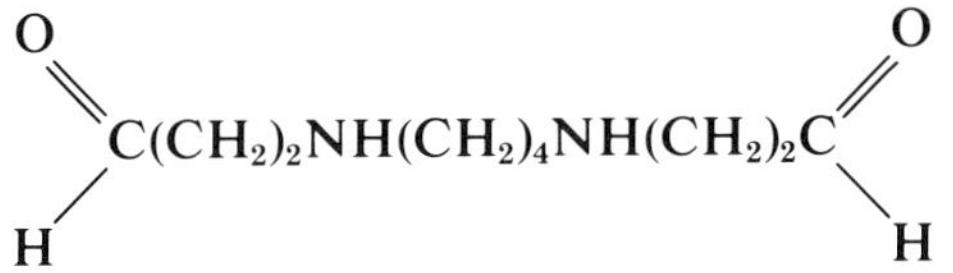

$$OHC(CH_2)_2NH(CH_2)_4NH(CH_2)_2CHO$$

We were able to confirm the toxicity of purified oxidized spermine for Ehrlich ascites cells (50). Figure 1 shows that the multiplication of Ehrlich ascites cells in mice was completely inhibited after exposing the tumor cells to oxidized spermine (5×10^{-3} M) for 60 min.

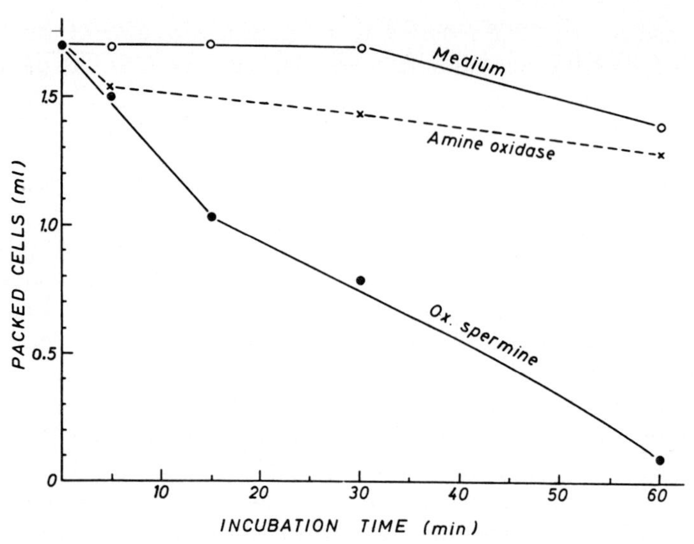

FIG. 1. Cytotoxic effect of oxidized spermine. Oxidized spermine 5×10^{-3}M, was incubated at 37°C with 3.8×10^7 Ehrlich ascites cells. At various times, aliquots of treated cells (0.3 ml) were injected into each of five mice and their packed volume determined after 7 days.

A similar inactivation was also observed when Ehrlich ascites cells were incubated with other polyamines in the presence of purified amine oxidase (Table 1). The antitumor effect of oxidized spermine was not limited to *in vitro* systems. In a typical *in vivo* experiment, Ehrlich ascites cells were inoculated into mice. After 24 hr, mice were injected with 1 μmole oxidized polyamines daily for 5 days. This treatment resulted in a 64% reduction of the volume of the packed tumor cells obtained from the sacrificed mice.

III. POLYAMINES AND TUMOR-BEARING INDIVIDUALS

A. Polyamines in blood

The aforementioned malignolipin and Tokuoka reactions are not specific for polyamines. Indeed, Gropper et al. (34) suggested that the $CuCO_3$ reaction was due to other compounds such as creatine, urea, or amino acids. We tried to clarify this point and determine the polyamine content of the sera which gave a positive $CuCO_3$ reaction. We used both paper and ion-exchange chromatographic procedures and failed to detect polyamines at elevated concentrations (38).

On the other hand, Quash and Wilson (62), who used a serological method, demonstrated the presence of spermine in the sera of patients with cancer

TABLE 1. *Toxic effect of oxidized polyamines on Ehrlich ascites cells*

System tested	μmole/ml	Inhibitory effect: Packed volume (ml)	Inhibitory effect: (%)
Cells + enzyme + diaminodipropylamine	0.77	0.02	99.3
	0.38	0.05	98.0
	0.15	1.86	28.2
Cells + enzyme		2.02	20.0
Cells + diaminodipropylamine	0.77	2.56	–
Cells + enzyme + spermine	0.77	0.44	51
	0.38	0.67	27
	0.15	0.72	22
Cells + spermine	0.77	0.92	–
Cells + enzyme + spermidine	3.1	0.02	98
	1.55	0.22	78
	0.77	0.27	73
Cells + enzyme		0.44	55
Cells + spermidine	3.1	0.98	–

To 1.0-ml portions of Ehrlich ascites cells (8×10^7 to 1.0×10^8 cells) in Krebs–Ringer solution were added 200 units of purified amine oxidase (82 units/mg), 150 units of catalase, and 0.2 ml of polyamine solution in a final volume of 1.3 ml. After incubation at 37°C for 3 hr, 0.2-ml aliquots of treated cells were injected into mice. After 7 days, tumor cells were collected from the peritoneal cavities and their packed volumes were determined by centrifugation. Cells, Ehrlich ascites cells; enzyme, serum amine oxidase. The packed-volume figures represent the average of five experiments.

and chronic infections. Immunoelectrophoretic analyses suggested that spermine did not occur in its free form but was associated with lipids (62).

The discrepancy between our findings and those reported by Quash and Wilson (62) may be explained by differences in the analytical methods. In our experiments, the serum was analyzed for free polyamines, whereas the serological assays indicated the occurrence of a polyamine-complex. According to Russell (63, 64), polyamines are excreted in the urine of cancer-bearing individuals as complexes which yield free polyamines only after hydrolysis. To ascertain if polyamine complexes occur in the blood of patients, the following experiment was carried out: serum was hydrolyzed with HCl and polyamines extracted with n-butanol after adding 3.0 g of salt mixture (Na_2SO_4:$Na_3PO_4 \cdot 12H_2O$; 9:62.5) and 3.7 ml of 20 N NaOH per 10 ml of hydrolysate. The organic phase was then evaporated to dryness after adding a few drops of 5N HCl and analyzed according to the following methods.

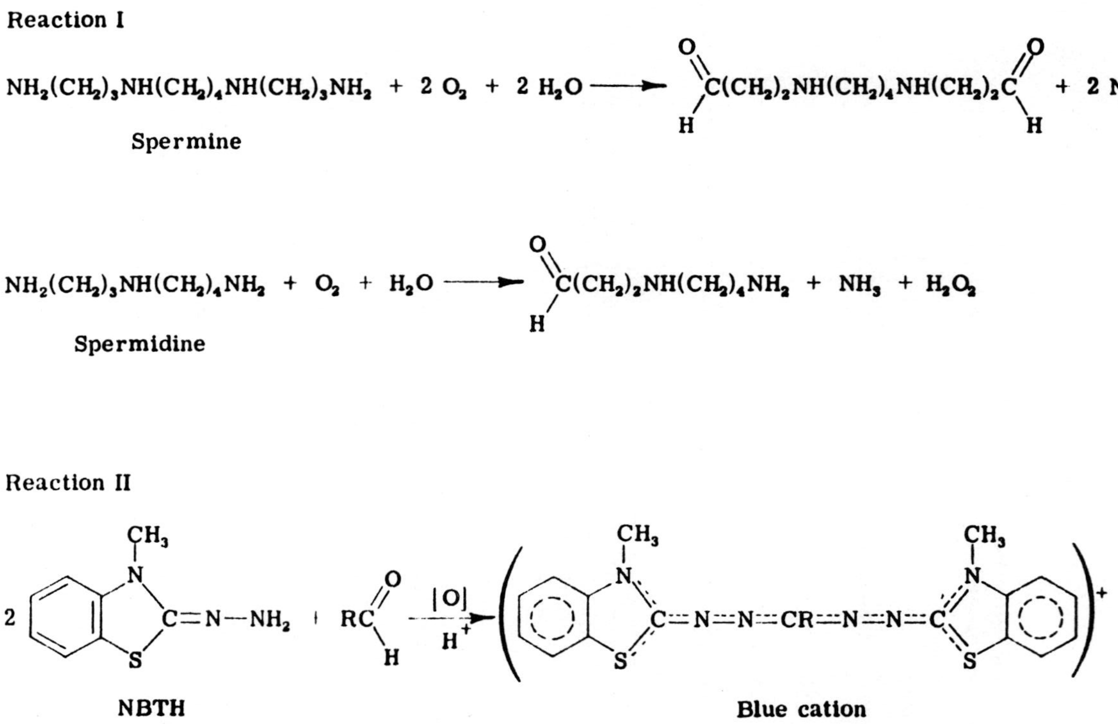

Reaction I

$$NH_2(CH_2)_3NH(CH_2)_4NH(CH_2)_3NH_2 + 2\,O_2 + 2\,H_2O \longrightarrow OHC(CH_2)_2NH(CH_2)_4NH(CH_2)_2CHO + 2\,NH_3 + 2\,H_2O_2$$

Spermine

$$NH_2(CH_2)_3NH(CH_2)_4NH_2 + O_2 + H_2O \longrightarrow OHC(CH_2)_2NH(CH_2)_4NH_2 + NH_3 + H_2O_2$$

Spermidine

Reaction II

$$2\ \text{NBTH} + RCHO \xrightarrow[H^+]{[O]} (\text{Blue cation})^+$$

NBTH

Blue cation

FIG. 2. Enzymatic assay for polyamines (65).

(1) *Enzymatic assay.* This was carried out as described previously (65). An aliquot of the butanol extract of the hydrolysate (0.02 ml out of 1.0 ml) was added to 0.1 ml 0.2M Tris-HCl, pH 7.0, 0.02 ml serum amine oxidase (6 units), and 0.85% NaCl in a final volume of 0.2 ml. After shaking at 37°C for 3 hr, 0.5 ml of NBTH reagent (0.4% N-methyl-2-benzothiazolone hydrazone hydrochloride) was added, and incubation continued for another 30 min. Finally, 2.5 ml of 0.2% $FeCl_3$ was added, and absorbency determined at 660 nm after staying at room temperature for another 15 min (Fig. 2).

(2) *Dansylation.* This was a modification of published procedures (66, 67): an aliquot of the butanol extract of the hydrolysate (0.02 to 0.06 ml) was added to 0.1 ml of 0.1 N $NaHCO_3$ and 0.1 ml of dansyl reagent (1-dimethylamino-naphthalene-5-sulfonyl chloride, 15 mg/ml acetone). After remaining in the dark for 12 to 16 hr, acetone (0.8 ml) was added, and the suspension was clarified by centrifugation. The dansyl derivatives were then separated by thin-layer chromatography on silica plates with ethylace-

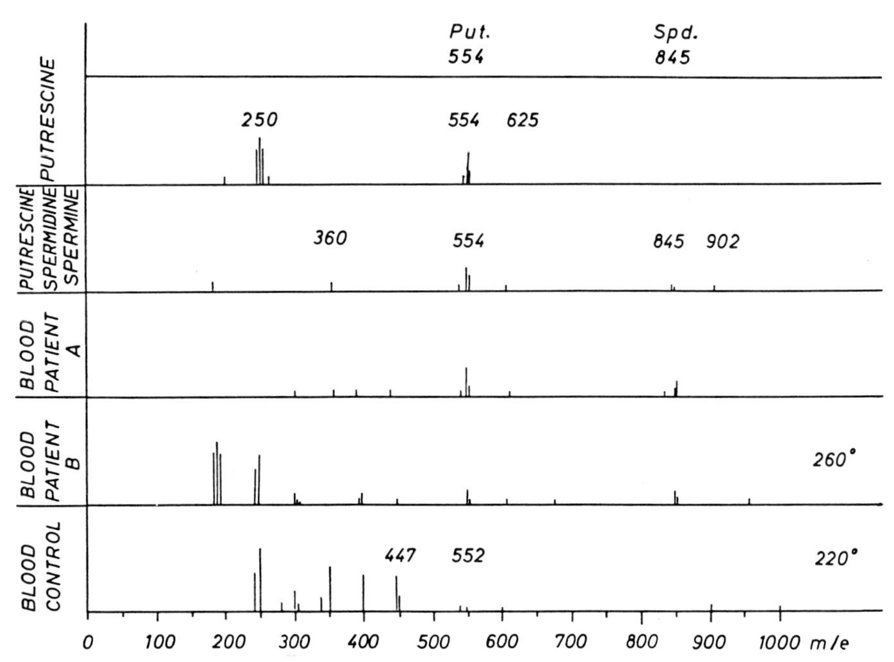

FIG. 3. Mass spectra of dansyl derivatives. Sera from patients and healthy controls were hydrolyzed with HCl and polyamines extracted with n-butanol. Dansyl derivatives were then prepared and analyzed in a CH5 Massenspekrometer Varian Mat (Bremen, Germany).

tate-cyclohexene (2:3) as solvent. The polyamines were finally assayed by scanning the plates in a Turner Model 111 Fluorometer (λ exc. —365 nm, λ emission-505 nm).

The enzymatic assay showed that the sera of healthy individuals did not contain polyamines in significant amounts. On the other hand, some of the cancerous sera showed the presence of polyamines. Thin-layer chromatography of the dansyl derivatives gave similar results: normal sera did not contain polyamines, whereas cancerous sera contained a compound which gave a dansyl derivative resembling that of spermidine.

The presence of polyamines in cancerous sera was also confirmed by mass spectrometry of the dansyl derivatives (68). Figure 3 shows two major peaks, the first at 845 m/e and the second at 554 m/e. These peaks are identical with those reported for the dansyl derivatives of spermidine and putrescine, respectively (68). The serum of a healthy control did not contain spermidine (Fig. 3). Normal and cancerous sera may contain histamine which resembles spermine in its dansyl derivative. The separation between the two derivatives may be accomplished by repeated thin-layer chromatograms.

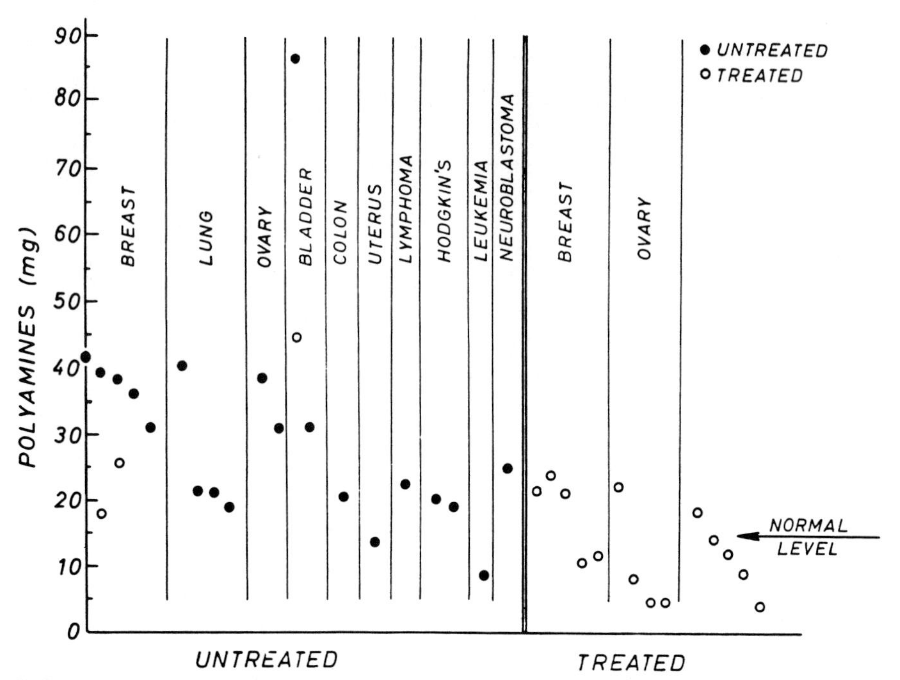

FIG. 4. Polyamines in the urine of cancer patients before and after treatment. Urine was hydrolyzed with HCl and polyamines extracted with n-butanol. The concentration of the polyamines in the extracts was determined by a colorimetric method using plasma amine oxidase.

B. Polyamines in urine

Russell (63, 64) has recently shown that polyamines are excreted in the urine of cancer-bearing individuals. The analytical procedure employed by Russell (63) included hydrolysis of the urine and detection of polyamines by paper electrophoresis. The latter analytical method is not ideal since it lacks specificity and cannot be performed automatically. The above-described enzymatic assay, on the other hand, is specific for polyamines and has the advantage of being a colorimetric assay.

Urine (5 ml) was hydrolyzed with HCl, and the polyamines were extracted with butanol and oxidized by serum amine oxidase. This assay showed that normal individuals excreted less than 15 mg polyamines per day, whereas urine of cancer-bearing individuals contained between 20 and 85 mg polyamines (Fig. 4). It is also evident that urinary polyamine levels declined after surgery or irradiation.

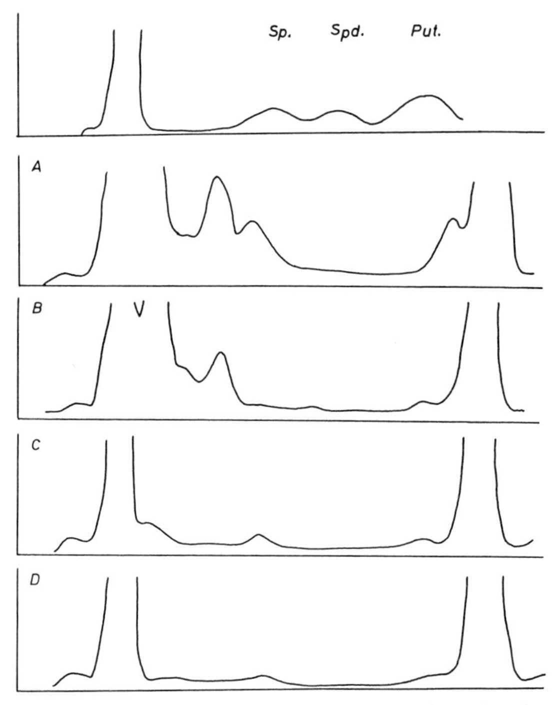

FIG. 5. Polyamines in the urine of cancer patients. Urine was hydrolyzed and polyamines extracted with n-butanol. Dansyl derivatives were then prepared and separated by thin-layer chromatography. The plates were scanned with a Turner Model 111 Fluorometer.

The excretion of polyamines by cancer patients and not by healthy individuals was also confirmed by thin-layer chromatography of the dansyl derivatives (Fig. 5). Thin-layer chromatography also showed that serum amine oxidase oxidized 90 to 95% of the urinary polyamines under conditions used for the enzymatic assay.

It should be remembered that the samples were hydrolyzed prior to their analysis. Table 2 shows that polyamines were not detected in urine samples which were not hydrolyzed properly. It is also evident that alkaline hydrolysis was as effective as hydrolysis with HCl.

TABLE 2. *Effect of hydrolysis on urinary polyamine levels*

		Acid hydrolysis[a]		Alkaline hydrolysis[b]
Sample no.	No. hydrolysis	4 hr	12 hr	4 hr
1	0[c]		135[c]	
2	−20		40	
3		40	220	240
4		30	190	160
5			250	260
6			300	280
7			280	300
8			60	60

[a] Urine samples were hydrolyzed with 6 N HCl for the times indicated. Polyamines were extracted with butanol and assayed enzymatically.

[b] Urine samples were hydrolyzed with 30% KOH.

[c] Absorbency at 660 nm.

It may be inferred from these experiments that polyamines are excreted in the urine of cancer patients in a conjugated form. It is not unlikely that this conjugate is a polyamine-containing phospholipid, as suggested earlier by Kosaki.

Because of the limited number of cases examined by our enzymatic assay, it is difficult to evaluate the significance of our findings. It is striking, however, that more than 90% of the cancer patients excrete polyamines at elevated amounts and that the levels decrease after treatment. It is obvious that a more detailed study is required before any final conclusions can be drawn.

IV. CONCLUSIONS

Our experimental data appear to support the following conclusions. (1) The $CuCO_3$ test for the detection of polyamines in cancerous sera is not specific; neither is the malignolipin reaction. (2) Polyamines are present in the hydrolysates of cancerous sera. This was demonstrated by (a) an enzymatic assay which involved the use of bovine plasma amine oxidase, (b) thin-layer chromatography of dansyl derivatives, and (c) mass spectrometry of the dansyl derivatives. (3) We have confirmed Russell's findings that cancer-bearing individuals excrete conjugated polyamines in their urine. This has been demonstrated by the enzymatic procedure and by thin-layer chromatography of the dansyl derivatives.

REFERENCES

1. Kosaki, T., and Saka, T., *Proc. Japan Acad.*, **34,** 295 (1958).
2. Kosaki, T., Ikoda, T., Kotani, Y., Nakagawa, S., and Saka, T., *Science,* **127,** 1176 (1958).
3. Kosaki, T., Hasegawa, S., Kotani, Y., Nakagawa, S., Uezumi, N., Fujimura, M., Saka, T., and Muraki, K., *Proc. Japan Acad.*, **36,** 516 (1960).
4. Kosaki, T., Hasegawa, S., Nakagawa, S., Uezumi, N., Fujimura, M., Saka, T., Kotani, Y., and Muraki, K., *Mie Med. J.*, **10,** 387 (1960).
5. Kosaki, T., and Kotani, Y., *Mie Med. J.*, **10,** 411 (1960).
6. Hughes, P. E., *Stain Tech.*, **35,** 41 (1960).
7. Gray, G. M., *Biochem. J.*, **81,** 30 p (1961).
8. Hill, J. H., *Cancer,* **16,** 542 (1963).
9. Petering, H. G., Van Giessen, G. J., Buskirk, H. H., Crim, J. A., Evans, J. S., and Musser, E. A., *Cancer Res.*, **27,** 7 (1967).
10. Yamamura, Y., Fujii, S., and Okamura, Y., *Jap. J. Med. Progr.*, **47,** 355 (1960).
11. Kögl, F., Smak, C., Veerkamp, J. H., and van Deenen, L. L. M., *Z. Krebsforsch.*, **63,** 558 (1960).
12. Kamat, V. B., *Nature,* **196,** 1206 (1962).
13. Sax, S. M., Harrison, P. L., Sax, M., and Baughman, R. H., *J. Biol. Chem.*, **238,** 3817 (1963).
14. Miyaki, K., Fuse, Y., and Uzuki, F., *J. Pharm. Soc. Japan,* **82,** 1530 (1962).
15. Hara, Y., and Hasegawa, K., *Gann* (suppl.) **50,** 267 (1959).
16. Tago, H., Takemasa, Y., Koyama, Y., and Ishii, T., *Gann* (suppl.), **50,** 266 (1959).
17. Albertini, E., Della Volpe, R., and Merli, G., *Minerva Med.*, **53,** 1466 (1962).
18. Mirokuchi, M., and Kobayashi, H., *Gann,* **51,** 184 (1961).
19. Nagao, H., *Hokkaido J. Med. Sci.*, **36,** 551 (1961).
20. Okamura, N., and Sato, K., *Sci. Rep. Res. Inst. Tohoku Univ.*, **10,** 274 (1961).
21. Kobayashi, H., *Fukouka Acta Med.*, **54,** 423 (1963).
22. Tomoda, M., and Ikejiri, T., *Chirurg.*, **33,** 495 (1962).
23. Nuvoli, U., and Cassarino, U., *Rif. Med.*, **78,** 1249 (1964).
24. Kallistratos, G., Pfau, A., and Timmermann, A., *Z. Krebsforsch.*, **73,** 387 (1970).
25. Kallistratos, G., and Timmermann, A., (1968) *XII Congress National de Chirurgie.*
26. Kosaki, T., and Nakagawa, S., *Proc. Japan Acad.*, **34,** 293 (1958).
27. Tokuoka, S., *Acta Sch. Med. Univ. Kioto,* **27,** 241 (1950).
28. Tokuoka, S., *Gann,* **47,** 292 (1956).
29. Maggi, A. L. C., Meeroff, M., and Iovine, E., *Sem. Med.*, **106,** 387 (1955).
30. Nakagawa, S., *Gann,* **47,** 294 (1956).
31. Tokuoka, S., *Acta Sch. Med. Univ. Kioto,* **29,** 15 (1951).
32. Ardizzone, G., and Ricci, P., *Quad. Clin. Ostet. Ginec.*, **10,** 717 (1955).
33. Colacurci, A., *Arch. Ostet. Ginec.*, **58,** 334 (1953).
34. Gropper, H., Wittig, R., and Grimalt, F., *Arch. Klin. Exp. Derm.*, **206,** 623 (1957).
35. Fontanili, E., *Riv. Pat. Clin. Sper.*, **13,** 407 (1958).
36. Moggian, G., *Riv. Ital. Ginec.*, **36,** 409 (1953).
37. Perolo, F., *Riv. Ostet. Ginec.*, **7,** 274 (1952).

38. Bachrach, U., and Robinson, E., *Israel J. Med. Sci.*, **1,** 247 (1965).
39. Rosenthal, S. M., and Tabor, C. W., *J. Pharmacol. Exp. Ther.*, **116,** 131 (1956).
40. Russell, D. H., *Fed. Proc.*, **29,** 669 (1970).
41. Russell, D. H., and Levy, C. C., *Cancer Res.*, **31,** 248 (1971).
42. Russell, D., and Snyder, S. H., *Proc. Nat. Acad. Sci. U.S.A.*, **60,** 1420 (1968).
43. Siimes, M., and Jänne, J., *Acta Chem. Scand.*, **21,** 815 (1967).
44. Neish, W. J. P., and Key, L., *Comp. Biochem. Physiol.*, **27,** 709 (1968).
45. Neish, W. J. P., and Key, L., *Int. J. Cancer,* **2,** 69 (1967).
46. Neish, W. J. P., *Biochem. Pharmacol.*, **16,** 163 (1967).
47. Kallistratos, G., Pfau, A., and Timmermann, A., *Proc. 9th Int. Cancer Congress, Tokyo, Japan, 1966,* p. 166.
48. Kremzner, L. T., Duffy, P. E., Defendini, R. F., and Terrano, M. J., *Excerpta Med.*, **193,** 242 (1969).
49. Andersson, G., and Heby, O., *J. Nat. Cancer Inst.*, **48,** 165 (1972).
50. Bachrach, U., Abzug, S., and Bekierkunst, A., *Biochim. Biophys. Acta,* **134,** 174 (1967).
51. Bachrach, U., Bekierkunst, A., and Abzug, S., *Israel J. Med. Sci.*, **3,** 474 (1967).
52. Kosaki, T., Kotani, Y., Saka, T., and Nakagawa, S., *Proc. Japan Acad.*, **36,** 527 (1960).
53. Quash, G. A., *West Ind. Med. J.*, **18,** 1 (1969).
54. Williams-Ashman, H. G., and Schenone, A., *Biochem. Biophys. Res. Comm.*, **46,** 288 (1972).
55. Raina, A., Jänne, J., and Siimes, M., *Acta Chem. Scand.*, **18,** 1804 (1964).
56. Halevy, S., Fuchs, Z., and Mager, J., *Bull. Res. Coun. Israel,* **11A,** 52 (1962).
57. Alarcon, R. A., Foley, G. E., and Modest, E. J., *Arch. Biochem.*, **94,** 540 (1961).
58. Israel, M., and Modest, E. J., *J. Med. Chem.*, **14,** 1042 (1971).
59. Israel, M., Rosenfield, J. S., and Modest, E. J., *J. Med. Chem.*, **7,** 710 (1964).
60. Higgins, M. L., Tillman, M. C., Rupp, J. P., and Leach, F. R., *J. Cell. Physiol.*, **74,** 149 (1969).
61. Miyaki, K., Hayashi, M., Chiba, T., and Nasu, K., *Chem. Pharm. Bull. Tokyo,* **8,** 933 (1960).
62. Quash, G. A., and Wilson, M. B., *West Ind. Med. J.*, **16,** 81 (1967).
63. Russell, D. H., *Nature New Biol.*, **233,** 144 (1971).
64. Russell, D. H., Levy, C. C., Schimpff, S. C., and Hawk, I. A., *Cancer Res.*, **31,** 1555 (1971).
65. Bachrach, U., and Reches, B., *Anal. Biochem.*, **17,** 38 (1966).
66. Cohen, S. S., Morgan, S., and Streibel, E., *Proc. Nat. Acad. Sci. U.S.A.*, **64,** 669 (1969).
67. Dion, A. S., and Herbst, E. J., *Ann. N.Y. Acad. Sci.*, **171,** 723 (1970).
68. Seiler, N., Schneider, H., and Sonnenberg, K. D., *Z. Anal. Chem.*, **252,** 127 (1970).

Polyamines in Normal and Neoplastic Growth. Edited by D. H. Russell. Raven Press, New York © 1973.

Polyamine Metabolism in Normal and Neoplastic Neural Tissue

Leon T. Kremzner

Department of Neurology, College of Physicians and Surgeons, Columbia University, New York, New York 10032

I. INTRODUCTION

It has been amply demonstrated that the polyamines and the enzymes involved in their synthesis are present in neural tissue (1–5). Studies have demonstrated, in a variety of species, the developmental and age dependency of both the amines and the biosynthetic enzymes (6–8), a feature common to the other major organs (6, 8). In addition, it is known that the polyamine content in neural tissue is subject to alteration following nerve injury or section (9, 10), during viral infection (11), and in tumor growth (9). However, it should be noted that the observed alterations in polyamine levels might be referred to as changes in cell morphology. It has not been established that polyamine metabolism is altered in association with normal neurological activity and function. A functional relationship is suggested, however, by the report that norepinephrine stimulates the biosynthesis of spermidine and spermine in the chick embryo (12).

Until recently, virtually nothing was known of the fate of the polyamines in neural tissue. In 1971, Seiler and co-workers (13) demonstrated *in vivo* the incorporation of putrescine carbon into γ-aminobutyric acid (GABA). An additional development was the discovery of N-(4-aminobutyl)-3-aminopropionic acid (putreanine) in the mammalian nervous system (14); this amino acid, a possible catabolite of spermidine, is present mainly in the brain (15).

The studies reported here are directed to the further understanding of polyamine metabolism in normal and neoplastic neural tissue.

II. ANALYTICAL PROCEDURES

In our previous studies (5, 9) the polyamines were separated by column chromatography using phosphoric acid cellulose followed by analysis with

2,4-dinitrofluorobenzene and a fluorometric procedure using orthophthalaldehyde. In these studies an automatic amino acid analyzer procedure was adopted because of its greater resolving abilities, sensitivity, and applicability to the isolation of polyamine metabolites.

The Beckman Model 120C equipped with a scale expansion circuit (4 to 5 mV range) was used. The chromatographic separation was carried out at 39.5°C on a column of Beckman type PA35 cation-exchange resin, measuring 0.9 × 6.3 cm. The initial buffer (0 to 55 min) was 0.35 N sodium citrate, pH 5.25, containing 2% 1-propanol. The succeeding buffer was identical except that it was 2.35 M with respect to sodium chloride. The buffer was pumped at 70 ml/hr; the ninhydrin solution (prepared as recommended in the Beckman Instruction Manual) was pumped at the rate of 35 ml/hr. This procedure is based on a methodology developed by Shull and co-workers (16) for the determination of S-adenosylmethionine. More recently, automated procedures have been described for the analysis of polyamines in *Escherichia coli* (17) and human urine (18).

Following the extraction of the polyamines from biological material with 0.4 M perchloric acid, the supernatant solution, up to 2.0 ml, was applied directly to the column. Standard solutions of amines and amino acids (obtained from Calbiochem, Los Angeles, California) were similarly prepared in 0.4 M perchloric acid. The elution profile from a typical standard run is shown in Fig. 1. The method is quantitated by integration of the area under the curve using an Infotronics Corp. Model CRS-100A digital integrator. When the area under the recorder tracing is too small to be integrated properly, an estimate can be made of the quantity of amine present by matching curves obtained with standards. The duplicability of the method is such that the variation from the mean is generally no more than ±5%. When the concentration of the amine is less than 1 to 2 nmoles, the variation is generally ±10%; below 1 nmole, the error is progressively greater. Recovery of purified ^{14}C-labeled polyamines in the peak area under the tracing was greater than 99%.

In the described procedure, histamine and agmatine are eluted immediately before spermidine and spermine, respectively; a more complete separation of these amines is obtained by programming a temperature change from 39.5 to 45°C at 55 min and increasing the concentration of 1-propanol to 5% in the second buffer. This modification results in a more compact elution profile and is consequently less desirable in metabolic studies when the samples are recovered by splitting the stream from the chromatographic column. The effect of 1-propanol addition is not solely a pH effect. Its use is predicated on the improved resolution of the amines. This effect has also

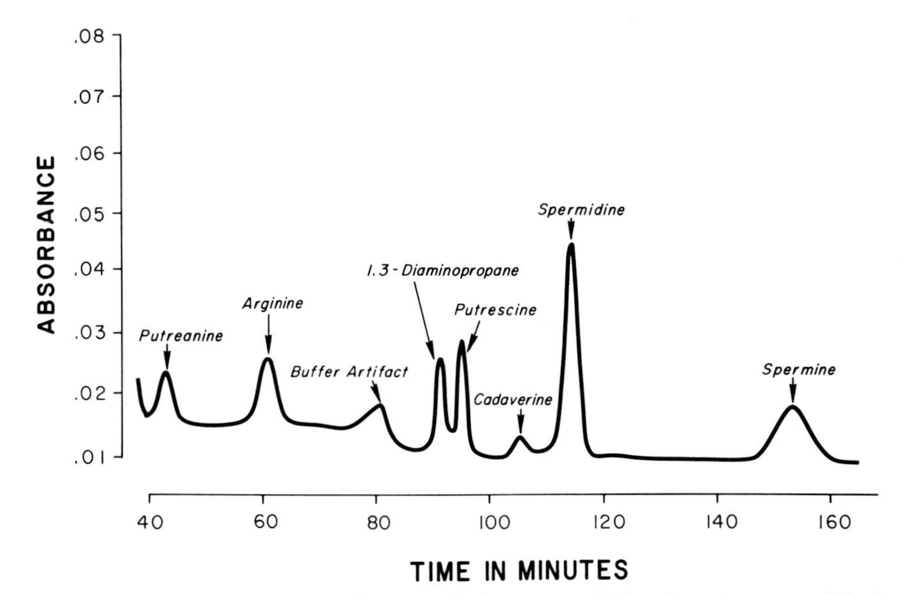

FIG. 1. Elution pattern of polyamines. A solution was applied to the column containing in nmoles: putrescine (4.4), arginine (7.4), 1,3-diaminopropane (3.0), putrescine (3.4), cadaverine (1.4), spermidine (10.8), and spermine (5.4).

been observed for certain amino acids (19). The propanol also retards microbial growth in the buffer solutions.

The total time required for a complete analysis is 160 min. However, if an analysis of only putrescine and the subsequent amines is required, five determinations can be made in an 8-hr period by overlapping in time successive determinations on two columns.

III. ANALYSIS OF HUMAN SPECIMENS

Figure 2 shows the application of the described method to the analysis of a human brain specimen. A trace of putreanine is evident at 42 min, and the following quantities of polyamines in nmoles: putrescine (1.4), spermidine (12.8), and spermine (8.0). The peak immediately following the artifact due to the buffer change (84 min) corresponds in time to the elution position of S-adenosylmethionine. The peak at 92 min corresponds in time to 1,3-diaminopropane, an established constituent of human urine (18). However, there is some evidence, presented later, which indicates that this peak may also correspond to some other polyamine metabolite. At 110 min there

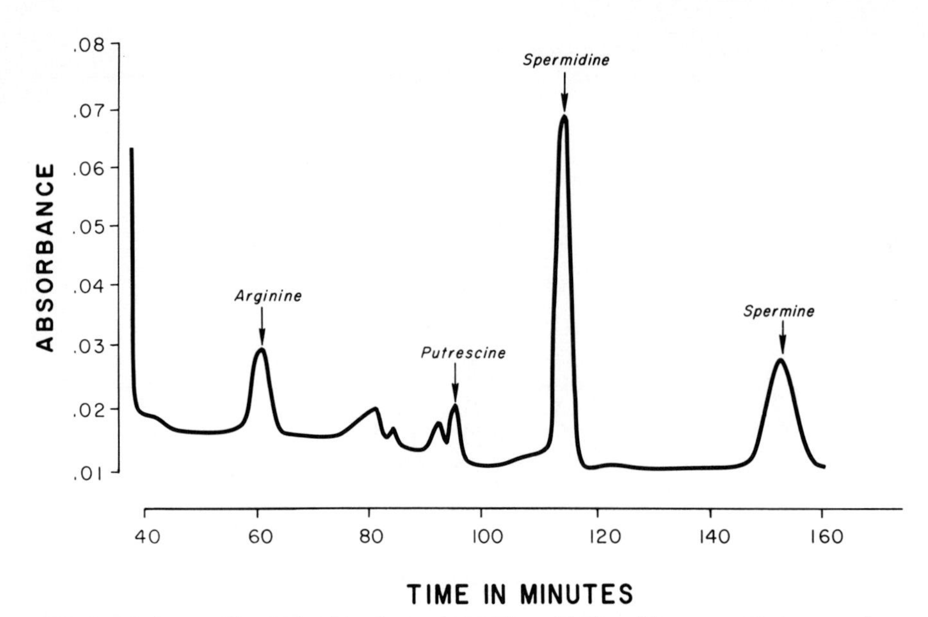

FIG. 2. Elution profile obtained in the analysis of perchloric acid extract of human caudate. The extract applied was equivalent to 62 mg of tissue, wet weight.

is a trace of histamine which is present in the CNS at the level of 0.5 nmole/g tissue (20).

The results of the analysis of discrete areas of human brain material obtained at autopsy are listed in Table 1. The specimens from five cadavers, none of which was known to have a malignancy, were obtained 3 to 8 hr postmortem.

As observed in our earlier studies of sheep brain (9), cortical white matter has a two- to threefold greater content of spermidine than gray matter. Consistent with this observation is the relatively high spermidine content of the pons and medulla, areas rich in white matter. Similarly, the caudate and putamen, predominantly gray areas, have a lower spermidine content. The cerebellum, which is characterized by a high cell density and consequently very rich in DNA, was unique of the areas studied in its relatively high spermine content.

A comparison of the spermidine and spermine content and distribution in human brain with the data previously published for sheep brain (9) or rabbit brain (21) does not indicate well-defined species differences, nor does it suggest a unique function. The consistently high spermidine content of white matter may be related to the metabolic activity associated with myelin formation.

TABLE 1. *Polyamine concentration (nmoles/g) in human brain*

Area	Putreanine[a]	Putrescine	Spermidine	Spermine
Frontal lobe (white)	—	20	610	70
	20	10	630	95
	30	5	545	115
Frontal lobe (gray)	—	25	155	180
	15	35	305	190
	10	30	150	170
Frontal lobe (mixed)	—	20	245	115
	—	10	210	90
Parietal lobe (white)	—	3	855	110
	20	5	750	75
Parietal lobe (gray)	—	12	240	160
	—	30	390	75
Occipital lobe (mixed)	15	40	350	140
	5	30	350	120
	—	15	310	115
Caudate	—	20	170	110
	—	35	280	140
	—	4	155	85
Putamen	—	3	160	110
	—	24	200	120
Thalamus	—	12	645	145
	—	25	435	145
Pons	10	18	570	110
Medulla	—	15	745	125
Cerebellum (mixed)	12	18	360	290
	—	14	350	300
	20	9	240	245

[a] Dash indicates no determination was made.

The putreanine values obtained with this procedure are in agreement with those recently published for human brain (22).

A considerable variation is evident in the putrescine content of the tissues analyzed; the reason for this is not evident. The putrescine values obtained with the above-described procedure are considerably lower than in our previous limited determinations (9), presumably the result of the improved resolutions possible with the amino acid analyzer. The lower putrescine values obtained here cast some doubt on the previously reported unusually high putrescine content observed in some brain tumors (5, 9). This aspect was consequently reinvestigated, as shown in Table 2. The table includes the results of the analysis of normal cortical tissue obtained as the result of surgery not connected with a malignancy rather than autopsy specimens. Although only three such specimens have been obtained, it appears that the polyamine content of tissue removed at surgery and autopsy material is quite similar. The putrescine content of three of the four tumors was found to be markedly elevated over the putrescine values obtained for nonmalignant neural tissue. These data tend to substantiate our earlier conclusions that the putrescine content of many CNS tumors is elevated. The observed high putrescine content of malignant tissue is consistent with the observations of Russell and Snyder (23) that ornithine decarboxylase activity is elevated in certain sarcomas.

TABLE 2. *Polyamine levels (nmoles/g) in normal and tumor cells of the human brain*

Tissue	Putreanine	Putrescine	Spermidine	Spermine
Normal cortex, frontal	6	12	210	155
Normal cortex, frontal (white)	–	18	600	90
Normal cortex, frontal (gray)	–	25	325	190
Glioma	–	3	180	70
Glioblastoma	7	185	350	180
Meningioma	–	66	480	90
Craniopharyngioma	–	88	125	360
Cystic fluid (glioblastoma)	–	0	0	0
Lymphoma (CNS involvement)	–	17	380	990
Epidural carcinoma (metastatic-prostate)	–	15	130	1,010

Included in Table 2 are the results of the analysis of the cystic fluid surrounding glioblastoma. No polyamines were detected. This finding would suggest that the polyamines in the glioblastoma cell remain associated with the cell and are not excreted into the fluid accumulated in the cyst. The two metastatic tumors not of CNS origin are characterized by putrescine values within the normal range for neural tissue; however, the spermine content is unusually high.

IV. POLYAMINES IN CEREBROSPINAL FLUID

Considering the high putrescine content of many types of brain tumors, the possibility existed that the detection of elevated polyamines in the CSF would be of diagnostic value. In collaboration with Dr. Arthur Hays (Department of Pathology, Columbia University), CSF was subject to analysis directly on the amino acid analyzer following centrifugation to remove any cells present and the precipitation of proteins with perchloric acid. The analysis indicated that putrescine, spermidine, and spermine were variably present at barely detectable levels, 0.1 to 0.2 nmole/ml, in some but not all cerebrospinal fluids.

Following the reports of Russell and co-workers (24, 25) of increased polyamine concentration in the acid hydrolysates of urine of cancer patients, this technique was applied to CSF. The CSF, generally 4.0 ml, was hydrolyzed in 6 N HCl and extracted into *n*-butanol; the butanol was evaporated to dryness and the residue resuspended in perchloric acid. The analysis of 15 patients whose CSF was acid hydrolyzed indicates a more consistent presence of putrescine and spermidine at the level of 0.1 to 0.4 nmole/ml of CSF; spermine was less consistently present. When spermine was present, it was present at the level of 0.1 to 0.4 nmole/ml CSF. It was not possible to discern differences between CNS-tumor-bearing and non-tumor-bearing patients. As the analysis is at the limit of sensitivity of the present methodology, these data must be viewed as preliminary in nature.

V. CATABOLISM OF POLYAMINES

Studies of the metabolism of ^{14}C-1,4-putrescine in sheep brain homogenates indicated an extremely low level of catabolism. The occipital cortex, the cortical CNS area with the greatest acitivity, deaminated putrescine at the rate of 0.15 nmole/g tissue wet weight/hr at 37°C in an air atmosphere at a pH of 7.4. These studies demonstrated that the enzyme was particulate in nature, 90% inhibited by pargyline (8×10^{-5}M) and not inhibited by aminoguanidine (10^{-5}M). The activity was greater in gray than in white matter. These observations suggest that the enzymatic activity is due to a monoamine oxidase type of enzyme, and they are consistent with the conclusion that diamine oxidase activity is extremely low or absent in sheep brain. A similar conclusion based on a study of rodent brain tissue was reached by Burkard et al. (26). Fractionation of an aliquot of the putrescine catabolites on the amino acid analyzer, using the standard basic amino acid analysis procedure, showed that the counts were present in several amino acids including GABA and that no single catabolite predominated.

Because of the low catabolic activity in sheep brain, it was clear that another source of tissue would have to be used initially to delineate the pathways.

VI. MOUSE NEUROBLASTOMA CELL STUDIES[1]

This spontaneous neural tumor has retained a high degree of differentiated neuronal properties. *In vitro* this cell line possesses electrically excitable membranes, extends axon-like processes, and synthesizes high levels of the enzymes involved in the biosynthesis of catecholamines and acetylcholine (27).

A brief description of the tissue culture technique is essential to the understanding of these experiments. The details have been described (28). Mouse neuroblastoma (C-1300) cells were grown routinely in Eagle's Minimal Essential Medium supplemented with 10% fetal calf serum, glutamine (2 mM), 100 units of penicillin, and 100 μg of streptomycin per ml of media. The cultures were grown in 250-ml Falcon plastic flasks containing 15 ml of media. The seeding density equalled approximately 100,000 cells per flask. Subculture techniques involved washing the monolayer with a buffered NaCl–KCl solution and incubation in 1.0 ml of a sodium bicarbonate-buffered versene–trypsin solution. The cells were harvested with a rubber policeman after the media in the flask had been decanted. Protein determinations were made using the method of Lowry et al. (29). The harvested cells and filtered culture media were treated with perchloric acid, final concentration 0.4 N. The precipitates were removed by centrifugation and the supernatants frozen for polyamine determination and/or column fractionation.

Under these conditions the doubling time for the cells in these experiments is approximately 24 hr after the cells leave the lag phase. The cells are in the log growth phase during days 2 and 3 and then enter the stationary phase when the media is not changed.

As shown in Fig. 3, the polyamine composition changes to the greatest extent during days 2 and 3, the log growth phase. The data shown in Fig. 3 are the result of a typical single experimental series. A total of 16 separate analyses were performed. The polyamine content was found to vary two-fold between the highest and lowest values for any one day. However, within an experimental cell series it was observed that (1) the putrescine content decreased with increased age, (2) the spermidine content was always highest at day 3 or 4 and then decreased, and (3) the spermidine/spermine ratio decreased from day 2 to day 6 from a high of 1.4 to a low of 0.2. Observations similar to these have been reported in the intact rat (6). In the rat, it was observed that every tissue in the newborn contained more

---

[1] These studies were performed in collaboration with Drs. Eric J. Simon and Jacob M. Hiller of New York University, College of Medicine.

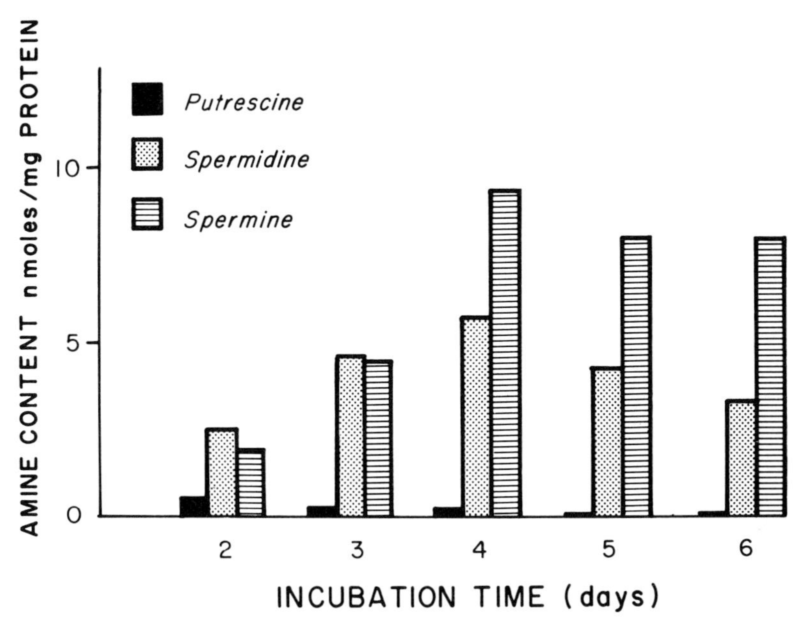

FIG. 3. The polyamine content in nmoles/mg cell protein of mouse neuroblastoma cells as a function of the number of days of incubation following inoculation.

spermidine than spermine and that the spermidine content decreased without exception with increasing age. In general, a high spermidine/spermine ratio is seen in rapidly growing tissue and then decreases to less than 1.0 (6). The results obtained here with neuroblastoma cells are similar to those reported with cultured KB cells (human epidermoid carcinoma) (30). In the KB cell study it was found that the levels of polyamines depend on the state of the medium and that rapidly growing cells tend to contain more spermidine than spermine.

Analysis of the cell culture media at days 2, 4, and 6 indicated that only trace amounts of the polyamines were present, usually only at day 2. The metabolism of putrescine in mouse neuroblastoma cells was investigated by replacing the media in a flask containing 4-day-old cells with media containing ^{14}C-1,4-putrescine. The putrescine (obtained from Amersham/Searle, Arlington Heights, Illinois), specific activity 21.8 mC/mmole, was purified before use by column chromatography with phosphoric acid cellulose (5). The concentration of putrescine in the flask was 5.2×10^{-7}M (approximately 2.5×10^{5}cpm). After the incubation period, the cells were harvested as quickly as possible and treated with perchloric acid. Table 3 shows the results of the fractionation of an aliquot of the cell extract using the amino acid analyzer. The incubation period was 4 hr in this study. The rate of catabolism of ^{14}C-putrescine in the cells was approximately equal to the rate of incorporation of putrescine into spermidine. Spermine was synthesized at about 1/20th of the rate at which spermidine was formed. Most of the catabolites of putrescine were not retarded by the column used for polyamine separation. Two minor unidentified catabolites were retarded. When an aliquot of the cell extract was fractionated on a standard basic amino acid column (0.9 × 20 cm), more than 95% of the non-polyamine counts were recovered at a position corresponding to GABA. The authenticity of this amino acid is now under study. Incubation of the culture media with ^{14}C-putrescine showed some variable amount of catabolic activity but no formation of GABA, spermidine or spermine.

TABLE 3. *Metabolism of ^{14}C-putrescine in mouse neuroblastoma cells*

Column elution time (min)	Compound	CPM/fraction	CPM/fraction/ mg protein	Rate of synthesis (pmoles/mg protein/hr)
10–20	?	33,660	22,600	–
43–52	?	320	220	–
52–58	?	380	260	–
93–99	Put.	22,300	15,950	–
115–121	Spd.	33,860	22,700	200
154–160	Sp.	1,720	1,150	10

Table 4 shows the results obtained when the incubation time of the cells in culture with ^{14}C-putrescine was varied. In this experiment, the perchloric acid extracts of the cells were fractionated on phosphoric acid cellulose columns (5). Spermidine was not separated from spermine. The rate of formation of catabolites was nearly constant as was the rate of formation of spermidine and spermine. Initially, the uptake of ^{14}C-putrescine was extremely rapid, as indicated by the number of counts present in the cells at the end of 0.1 hr, followed by a progressive decline. The rate of catabolism of putrescine in these experiments is approximately 100-fold greater than that observed in the sheep brain homogenate studies. Similarly, the rate of spermidine synthesis, approximately 160 pmoles/mg cell protein/hr, is 100-fold greater than that reported in a previous study of sheep brain (9).

The metabolism of spermidine in mouse neuroblastoma cells was studied in 5- to 7-day-old cells. At this age the cells are in a stationary growth stage. The spermidine [(aminipropyl)-tetramethylene-1,4-^{14}C-diamine · 3 HCl] (obtained from New England Nuclear Corp., Boston, Mass.) had a specific activity of 4.55 mC/mmole and was further purified by column chroma-

TABLE 4. *Rate of* [14]*C-putrescine metabolism in mouse neuroblastoma cells*

Incubation time (hr)	Catabolites (cpm/mg protein)		Putrescine (cpm/mg protein)		Spermidine-spermine (cpm/mg protein)	
	total	per hr	total	per hr	total	per hr
0.1	370	3,700	5,620	56,200	450	4,500
2.0	8,480	4,240	16,450	8,225	8,350	4,175
4.0	18,850	4,710	14,260	3,565	22,200	5,550
6.0	26,050	4,360	9,180	1,530	28,700	4,780

tography (9). The final concentration of ^{14}C-spermidine in the media was 4.6×10^{-6}M (4.2×10^{5} cpm). Following incubation for 2 or 4 hr, the cells were harvested and extracted with perchloric acid, and aliquots of the extract were applied to the amino acid analyzer. Figure 4 shows the distribution of radioactivity obtained. The isotope was found at positions corresponding in time to putreanine, putrescine, spermidine, and spermine, as well as several unidentified peaks. Essentially similar products were found in both the 5- and 7-day cell culture experiments.

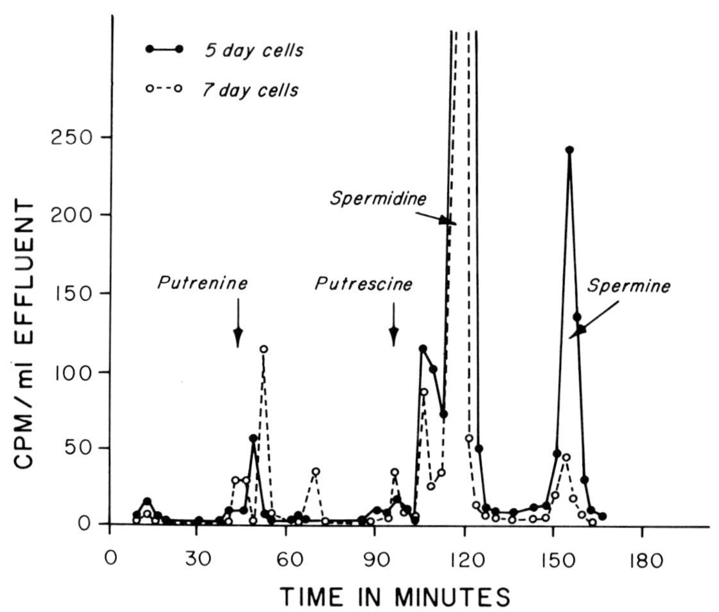

FIG. 4. Amino acid analyzer elution profile obtained in the analysis of a perchloric acid extract of mouse neuroblastoma cells incubated with ^{14}C-spermidine in the media for 2 hr. Fractions were collected at 3-min intervals.

TABLE 5. *Metabolism of ^{14}C-spermidine in mouse neuroblastoma cells*

Cell "age" (days)	Incubation time (hr)	Distribution of counts (%)				^{14}C-uptake (cpm/mg protein)		Spermine synthesis (pmoles/mg protein/hr)
		Unidentified	Put (?)	Spd	Sp	total	per hr	
5	2	3.2	0.3	93.6	2.9	29,200	14,600	70
6	2	2.2	0.2	95	2.6	27,300	13,650	60
6	4	2.1	0.3	92	5.6	34,500	8,650	80
7	2	5	0.7	93	1.3	12,100	6,050	15
7	4	1.9	0.6	94	3.5	24,200	6,050	35
None	2	8	41	51	0	–	–	–

In a similar experiment in which ^{14}C-spermidine was incubated with the culture media without cells, the radioactivity was found predominantly in an area corresponding in time to the elution position of putrescine and spermidine. In addition, two minor peaks were present at 60 and 105 min. It is presumed that the peak corresponding in time to putrescine is due to oxidized spermidine, the product of the amine oxidase present in the fetal calf serum of the media. The results of a series of similar experiments are summarized in Table 5. The data suggest that the rate of uptake of ^{14}C-spermidine decreases with increasing cell age; concomitantly, the rate of spermine synthesis decreased.

VII. CONCLUSIONS

A technique for the analysis of the polyamines and their metabolites is described using the automatic amino acid analyzer. This procedure is sufficiently sensitive and versatile to identify and quantitate the polyamines present in approximately 50 mg of brain tissue and in cells grown in culture. By use of this technique it has been established that the putrescine content of normal human CNS tissue is approximately one-half that previously reported. The putreanine, spermidine, and spermine values are in agreement with those previously published. The analysis of neoplastic tissue verifies that many CNS tumors contain substantially elevated levels of putrescine. Analysis of the CSF of tumor-bearing patients has failed to indicate, possibly owing to insufficient analytical sensitivity, clear differences in the polyamine composition from control patients. The analysis of discrete areas of the human CNS indicates that the polyamines are not uniformly distributed; the significance of these findings is not evident but they do permit some generalizations. It is also apparent that the polyamine content of many discrete areas of the CNS is not markedly different in different animal species.

Normal brain tissue (sheep cortex) appears to have *in vitro* a very limited ability to catabolize putrescine. The limited activity observed suggests that the enzyme is similar to a monoamine oxidase.

Studies, here detailed, with the mouse neuroblastoma cells grown in culture indicate that this system is a good model with which to delineate metabolic pathways in neural tissue. In these cells ^{14}C-putrescine is rapidly catabolized to GABA. The rate of catabolism approximately equals the rate of spermidine synthesis from putrescine. In the mouse neuroblastoma cells, ^{14}C-spermidine is metabolized principally to spermine; little if any putrescine is formed. Studies are in progress to identify the minor catabolites observed. The polyamine content and relative composition were found to be functions of the rate of cell growth and the cell culture age. In general, the spermidine/spermine ratio decreased with less rapid growth. Similar observations have been previously noted in studies of the intact animal.

ACKNOWLEDGMENTS

This work has been supported by grant #NS 05184 from the Clinical Research Center for Parkinson's and Allied Diseases.

REFERENCES

1. Dudley, H. W., Rosenheim, M. C., and Rosenheim, O., *Biochem. J.*, **18**, 1263 (1924).
2. Hamalainen, R., *Acta Soc. Med.*, Duodecim Ser. A, **23**, 97 (1947).
3. Rosenthal, S. M., and Tabor, C. W., *J. Pharmacol.*, **116**, 131 (1956).
4. Kewitz, H., *Naturwissenschaften*, **46**, 495 (1959).
5. Kremzner, L. T., *Fed. Proc.*, **29**, 1583 (1970).
6. Jänne, J., Raina, A., and Siimes, M., *Acta Physiol. Scand.*, **62**, 352 (1964).
7. Shimizu, H., Kakimoto, Y., and Sano, I., *Nature*, **207**, 1196 (1965).
8. Caldarera, C. M., Moruzzi, M. S., Rossoni, C., and Barbiroli, B., *J. Neurochem.*, **16**, 309 (1969).
9. Kremzner, L. T., Barrett, R. E., and Terrano, M. J., *Ann. N. Y. Acad. Sci.*, **171**, 735 (1970).
10. Seiler, M., and Schroder, J. M., *Brain Res.*, **22**, 81 (1970).
11. Giorgi, P. O., Field, E. J., and Joyce, G., *J. Neurochem.*, **19**, 255 (1972).
12. Caldarera, C. M., Giorgi, P. O., and Casti, A., *J. Endocrinol.*, **46**, 115 (1970).
13. Seiler, N., Wiechmann, M., Fischer, H. A., and Werner, G., *Brain Res.*, **28**, 317 (1971).
14. Kakimoto, Y., Nakajima, T., Kumon, A., Matsuoka, Y., Imaoka, N., Sano, I., and Kanazawa, A., *J. Biol. Chem.*, **244**, 6003 (1969).
15. Nakajima, T., and Matsouka, Y., *J. Neurochem.*, **18**, 2547 (1971).
16. Shull, K. H., McConomy, J., Vogt, A., Castillo, A., and Farber, E., *J. Biol. Chem.*, **241**, 5060 (1966).
17. Morris, D. R., Koffron, K. L., and Okstein, C. J., *Anal. Biochem.*, **30**, 449 (1969).
18. Bremer, H. J., Kohne, E., and Endes, W., *Clin. Chim. Acta.*, **32**, 407 (1971).
19. Long, C. L., and Geiger, J. W., *Anal. Biochem.*, **29**, 265 (1969).
20. Green, J. P., in A. Lajtha (Editor), *Handbook of neurochemistry*, Vol. 4, Plenum Press, New York, 1970, p. 221.
21. Shimizu, H., Kakimoto, Y., and Sano, I., *J. Pharmacol. Exp. Ther.*, **143**, 199 (1964).
22. Nakajima, T., Kakimoto, Y., and Sano, I., *J. Neurochem.*, **17**, 1427 (1970).
23. Russell, D. H., and Snyder, S. H., *Proc. Nat. Acad. Sci. U.S.A.*, **60**, 1420 (1968).
24. Russell, D. H., *Nature, New Biology*, **233**, 144 (1971).
25. Russell, D. H., Levy, C. C., Schimpff, S. C., and Hawk, I. A., *Cancer Res.*, **31**, 1555 (1971).
26. Burkard, W. P., Gey, K. F., and Pletscher, A., *J. Neurochem.*, **10**, 183 (1963).
27. Gilman, A. G., in P. Greengard and G. A. Robison (Editors), *Advances in cyclic nucleotide research*, Vol. 1, Raven Press, New York, 1972, p. 389.
28. Augusti-Tocco, G., and Sato, G., *Proc. Nat. Acad. Sci. U.S.A.*, **64**, 311 (1969).
29. Lowry, O. H., Rosenbrough, N. F., Farr, A. L., and Randall, R. J., *J. Biol. Chem.*, **193**, 265 (1951).
30. Pett, D. M., and Ginsberg, H. S., *Fed. Proc.*, **27**, 615 (1968).

Polyamines in Normal and Neoplastic Growth. Edited by D. H. Russell. Raven Press, New York © 1973.

Cations and the Reactivity of Thiouridine in *Escherichia coli* tRNA

Ted T. Sakai and Seymour S. Cohen

© *Department of Microbiology, University of Colorado Medical Center, Denver, Colorado 80220*

I. INTRODUCTION

This laboratory has been studying the reactions of polyamines with tRNA. Such reactions have appeared more amenable to interpretation at the molecular level than many observed effects of the polyamines in the complex systems which polymerize amino acids and nucleotides or in the stabilization of various cellular organelles (1). A reaction of spermidine to activate denatured tRNA was observed quite early (2), and the utility of polyamines in forming well-ordered crystals is well known (3). It had been shown also that spermidine may activate the transfer of an amino acid from an isolated enzyme-amino acid-adenylate complex to acylate tRNA (4). In studies of this reaction it was thought that spermine acted by binding to the tRNA (5). It is known that the fluorescence of the Y base in the anticodon arm of yeast phenylalanyl tRNA and amino acid acceptance by this tRNA can be increased in a parallel fashion by addition of spermidine (6). This result as well is interpreted as the formation of an active conformation of the nucleic acid by addition of the polyamine.

The activation of tRNA methylase by polyamines has been described and significant differences have been found between methylations stimulated by spermidine or Mg^{++} (7). Spermidine can evidently help to organize specific active conformations of tRNA at apparently physiological concentrations of the natural bacterial polyamine (1 to 2 mM).

When tRNA is isolated at low ionic strength from *Escherichia coli* harvested during the synthesis of both tRNA and spermidine, the tRNA is found to contain 2 moles of spermidine per mole of nucleic acid (8). The tRNA of mouse fibroblasts after a similar isolation procedure contains a mixture of spermidine and spermine (9). In recent studies (C. Freda and

S. S. Cohen, *in preparation*), it has been observed that tRNA contains two to three relatively tight binding sites and seven to eight relatively weak binding sites for spermidine (10). Such studies have also led us to explore the binding of the dye ethidium to tRNA. The structure of this dye suggests that it is an analogue of spermidine and, indeed, although some differences in binding have been detected, the stoichiometry of binding of the dye is similar to that of spermidine under comparable conditions (10).

We are attempting to localize the sites of tRNA which bind spermidine tightly. Our working hypothesis is that of the models of Tsuboi and Liquori in which spermidine, in reacting with three phosphates of a double-strand region of nucleic acid, spans a narrow groove and helps to lock the structure (1). We have asked then if the arm of tRNA containing dihydrouracil (DHU), which contains only three or four complementary base pairs (depending on the specific tRNA), uses spermidine to stabilize this limited helical structure. We have studied several activities of 4-thiouridine (Srd) (position 8) which is at the juncture of the DHU and CCA arms of the nucleic acid.

In earlier studies with Pochon (11), we compared the enhancement of fluorescence of Srd in *E. coli* tRNA by Mg^{++} and polyamines. It was found that Mg^{++} gives a threefold enhancement of fluorescence at a ratio of 10 moles per mole. Fluorescence is increased similarly but slightly less by spermine, and very much less by putrescine. Addition of Mg^{++} to the spermidine system increased the fluorescence; this did not occur in the spermine system.

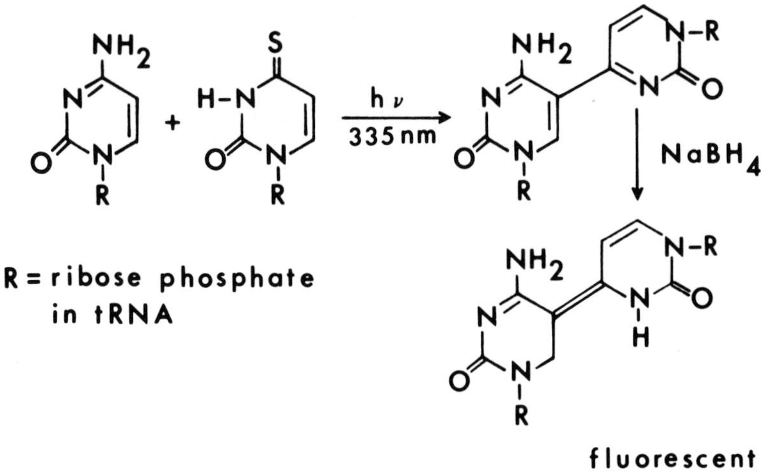

R = ribose phosphate in tRNA

hν 335 nm

$NaBH_4$

fluorescent

FIG. 1. Formation and reduction of the 4-thiouridine-cytidine photoproduct in *E. coli* tRNA.

When small increments of spermidine are added to tRNA titrated with 2, 4, or 6 moles of Mg^{++} per mole, the fluorescence of the system is markedly increased compared with that achieved by Mg^{++} or spermidine alone at the same ratios of cation per mole of tRNA. The two cations act synergistically and a maximum fluorescence appears to be formed cooperatively in the sequence of two atoms of Mg^{++}, eight molecules of spermidine, and two atoms of Mg^{++} per mole of tRNA.

Irradiation of *E. coli* tRNA at 335 nm gives rise to a Srd-Cyt photoproduct, shown in Fig. 1, which spans the postulated helical binding site. This reaction had been known to require Mg^{++} (12, 13) but was shown to occur about twice as rapidly with spermidine (11). The photoproduct is now reducible to a fluorescent derivative by $NaBH_4$. When this photoproduct is generated in the presence of Mg^{++} and reduced with borohydride, the fluorescent tRNA, containing a new covalent cross-link, may be partially denatured to approximately half maximal fluorescence by dialysis against EDTA to remove Mg^{++}. As shown in Fig. 2, the fluorescence of the denatured product is far more rapidly regenerated, although not quite maximally, by addition of spermidine than it is by addition of Mg^{++}.

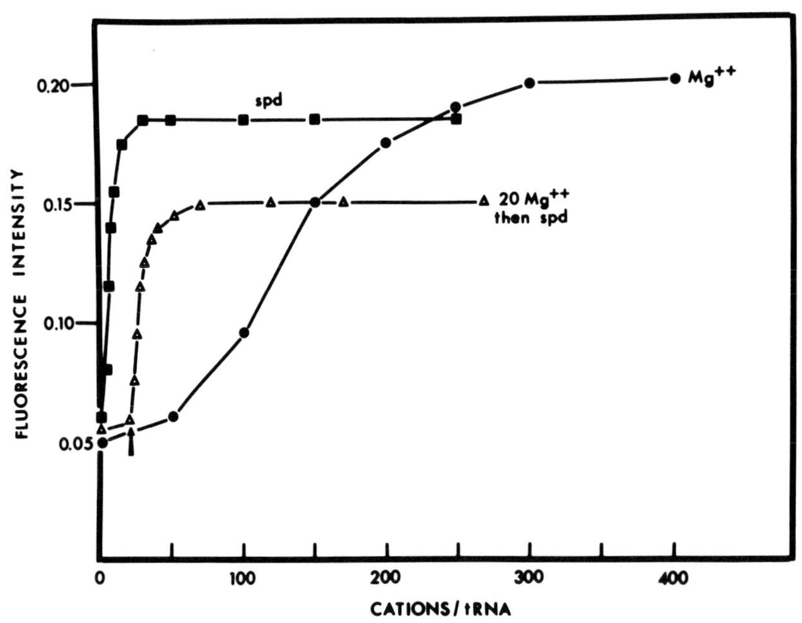

FIG. 2. Effect of cations on fluorescence of reduced Srd-Cyt in *E. coli* tRNA. Samples (3.0 ml) of tRNA (0.0021 μmole) dialyzed against EDTA and 0.01 M K acetate, pH 7 were titrated with Mg^{++} and spermidine (spd). Excitation 400 nm, emission 450 nm.

These studies suggested, therefore, that spermidine did affect the conformation of tRNA, and were consistent with the hypothesis that some molecules of spermidine are involved in the organization of three-dimensional geometry of the limited helical region close to Srd. The shape of the structure into which spermidine fits must be fairly specific since the various properties conferred by this base cannot be exactly matched by Mg^{++} or by the natural bases putrescine and spermine.

In this chapter we shall describe additional studies on the effects of spermidine and Mg^{++} on (a) the ionization of Srd in tRNA, (b) the reduction of the photoproduct with $NaBH_4$, (c) the reactivities of Srd with iodoacetamide and cyanogen bromide, and (d) the rotation of an Srd-substituted spin-label.

II. MATERIALS AND METHODS

Iodoacetamide was obtained from Nutritional Biochemicals and recrystallized twice from water before use. Cyanogen bromide (Pierce) was used without further purification. Spermidine, putrescine, and spermine were obtained as hydrochloride salts from Calbiochem, and selected samples were shown to be at least 98% pure by the dansyl procedure (8). All other chemicals were of reagent grade.

Transfer RNA obtained from General Biochemicals was found to be contaminated with high molecular weight materials. The tRNA was purified by passage through a Sephadex G-100 column (2.5 × 100 cm) eluting with a 0.01 M potassium acetate at pH 7. The large tRNA peak was pooled and lyophilized and then dissolved in the minimal amount of water. The tRNA solution was then dialyzed against 0.1 M NaCl–0.01 M EDTA (pH 7) for 2 days and then against distilled water for 2 days. Isoaccepting $tRNA_f^{Met}$ was the gift of Dr. A. D. Kelmers, Oak Ridge National Laboratories, Oak Ridge, Tennessee.

All solutions were made with glass-distilled water. Buffers used were cacodylate and carbonate as the sodium salts at 0.05 M. Transfer RNA concentrations were determined in 1-cm path length cells using a millimolar extinction coefficient of 570 at 260 nm.

Ultraviolet spectra were determined on a Beckman DU spectrophotometer. Fluorescence measurements were made on a Farrand Mark I or Aminco-Bowman spectrofluorometer. Reactions involving alkylations of Srd in tRNA were monitored on a Beckman DU spectrophotometer in conjunction with a Gilford Model 2000 multiple sample absorbance recorder. Reactions were monitored conveniently by following the loss of absorbance at 335 nm. Reactions were thermostatted at 25°C with the aid of a Haake circulating bath.

Photo-linked tRNA was prepared as described by Favre et al. (14) by irradiation of tRNA in a quartz cuvette at 335 nm for 15 hr. The reaction was carried out in 0.01 M potassium acetate at pH 7 containing 0.001 M $MgCl_2$. The linked tRNA was then dialyzed against 0.1 M NaCl–0.01 M Na_2EDTA (pH 7) and distilled water (2 days each step).

The rate of reduction of the photo-linked tRNA by sodium borohydride was determined fluorometrically at 25°C by following the increase in fluorescence at 435 nm (excitation at 385 nm). Sodium borohydride solutions were prepared freshly for each rate determination using the reaction buffer 0.05 M carbonate at pH 9.4.

Mixed *E. coli* tRNA and isoaccepting *E. coli* $tRNA_f^{Met}$ were labeled with 3-(2-bromoacetamido)-2,2,5,5-tetramethylpyrrolidinyl-l-oxyl (SYVA Research Institute, Palo Alto, Calif.) as described by Hara et al. (15), with the exception that reactions were run at pH 9.5 rather than 8.9. The labeled tRNAs were then dialyzed against 0.1 M NaCl–0.01 M Na_2EDTA and water as described above.

Electron paramagnetic resonance spectra of the spin-labeled tRNAs were recorded in 0.05 M Tris-HCl buffer at pH 7.0 using a Varian E-3 EPR spectrometer operating at 9.1 GHz with 10-mW power and a field setting at 3242 G. Spectra were determined in capillary tubes using 30 μl of the appropriate tRNA solution.

III. RESULTS

A. Cation effects on pK of Srd in tRNA

The effects of Mg^{++} and spermidine on the pK of Srd in tRNA were examined spectrophotometrically by determining the ratio of optical densities at 335 and 310 nm as a function of pH. The data shown in Fig. 3 indicate that in 0.05 M buffer and in the absence of added cations the apparent pK of Srd in tRNA is 9.45. The titration curve is seen to be broad, and the maximum ratio of OD_{310}/OD_{335} is approximately 2.

In the presence of 10 spermidine or 10 Mg^{++} per tRNA (Fig. 3a), the curves were found to sharpen greatly. The apparent pK_a for Srd increases to 9.60 in the presence of 10 spermidine and 9.70 in the presence of 10 Mg^{++}. The OD_{310}/OD_{335} ratio in both cases increases to approximately 2.5 for spermidine-tRNA and 2.3 for Mg^{++}-tRNA.

As the ratio of cation to tRNA is increased to 30 (Fig. 3b), the $OD_{310}/$

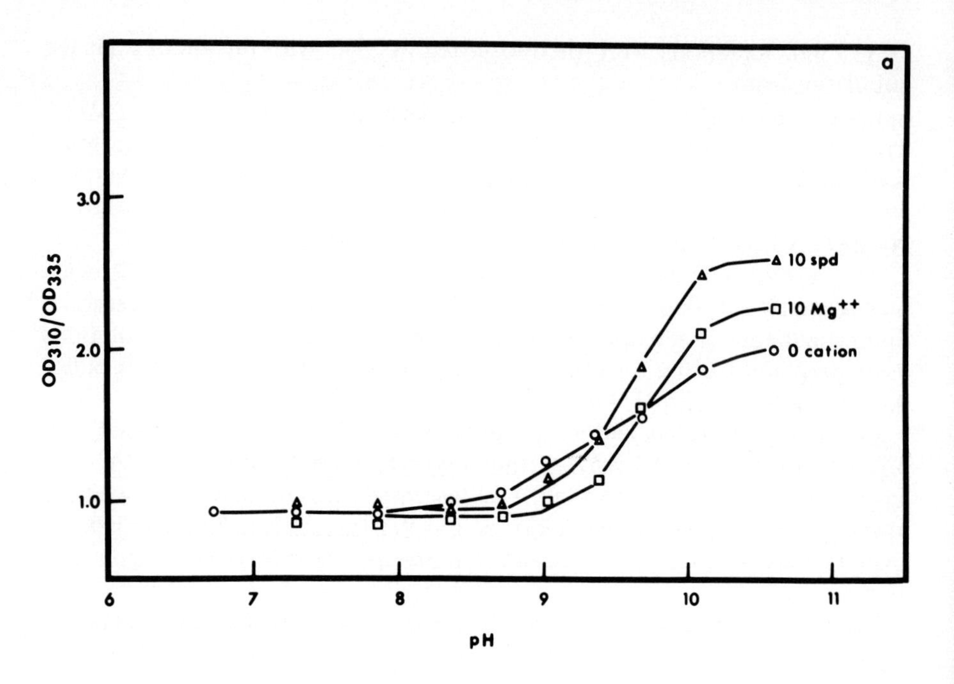

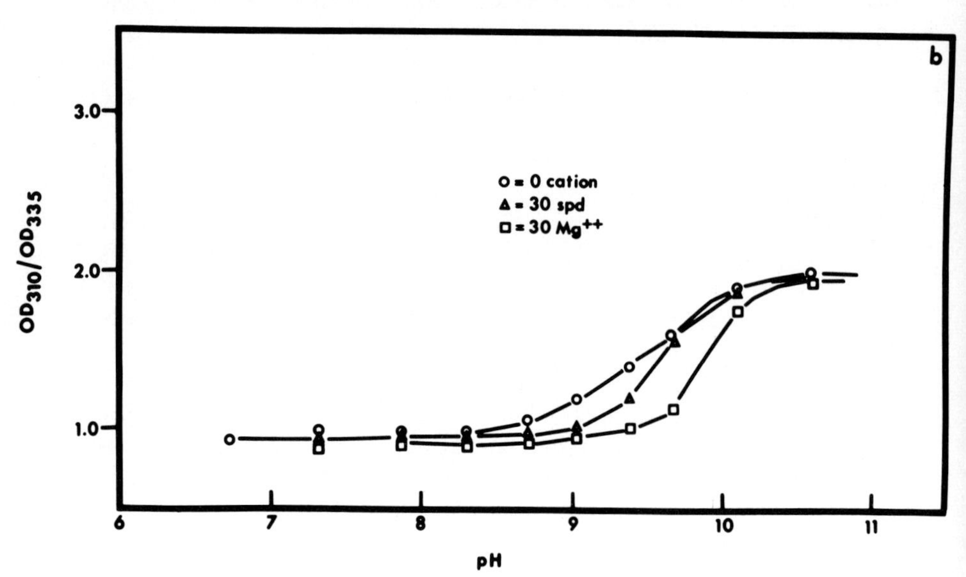

FIG. 3. Effects of Mg^{++} and spermidine (spd) on the apparent pK of Srd in *E. coli* tRNA at (a) 10 moles per mole tRNA and (b) 30 moles per mole tRNA.

OD_{335} ratio decreases to approximately 2 again. The apparent pK of Srd remains relatively unchanged in the presence of spermidine (pK 9.63) and increases significantly in the presence of 30 Mg^{++} to 9.85.

Under the same conditions, the free nucleoside has a pK of 8.30, which is not altered by addition of spermidine.

B. Borohydride reduction of photo-linked tRNA

The effects of the various cations on the reduction of photo-linked tRNA by borohydride are shown in Table 1. Reactions were run at 25°C in 0.05 M carbonate buffer, pH 9.4, to assure reasonable stability of borohydride during the course of the reaction. The results show that for a borohydride concentration of 0.03 M and tRNA concentration of 0.1 μM Mg^{++} affects the reaction to the greatest extent. The effect is a 30% reduction in rate for Mg^{++}/tRNA ratios greater than 10 to 15. Spermidine shows a similar effect although only to the extent of giving a 20% decrease in rate. Spermine shows a slightly smaller effect, and putrescine causes very little decrease in rate.

TABLE 1. *Borohydride reduction of Srd-C linked tRNA*

Cation	Moles of cation/mole of tRNA						
	0	5	10	15	20	25	30
Mg^{++}/tRNA							
$k(min^{-1})$	0.115	0.105	0.091	0.0865	0.0885	0.0885	0.0865
$t_{1/2}$ (min)	6.0	6.5	7.6	8.0	7.8	7.8	8.0
spd/tRNA							
$k(min^{-1})$	0.112	0.102	0.0962	0.0962	0.0935	0.0950	0.0935
$t_{1/2}$ (min)	6.2	6.8	7.2	7.2	7.4	7.3	7.4
pu/tRNA							
$k(min^{-1})$	0.115	0.112	0.110	—	0.110	—	—
$t_{1/2}$ (min)	6.0	6.2	6.3	—	6.3	—	—
spm/tRNA							
$k(min^{-1})$	0.115	0.106	0.106	0.102	0.103	0.102	0.102
$t_{1/2}$ (min)	6.0	6.5	6.5	6.8	6.7	6.8	6.8

C. Iodoacetamide reactions

The reactions of tRNA and iodoacetamide at 25°C as measured by loss of Srd absorbance at 335 nm at pH 9.4 are summarized in Fig. 4. At an iodoacetamide concentration of 4 mM and tRNA concentration of 28 μM, the data show the same effects as observed in the reduction of photo-linked

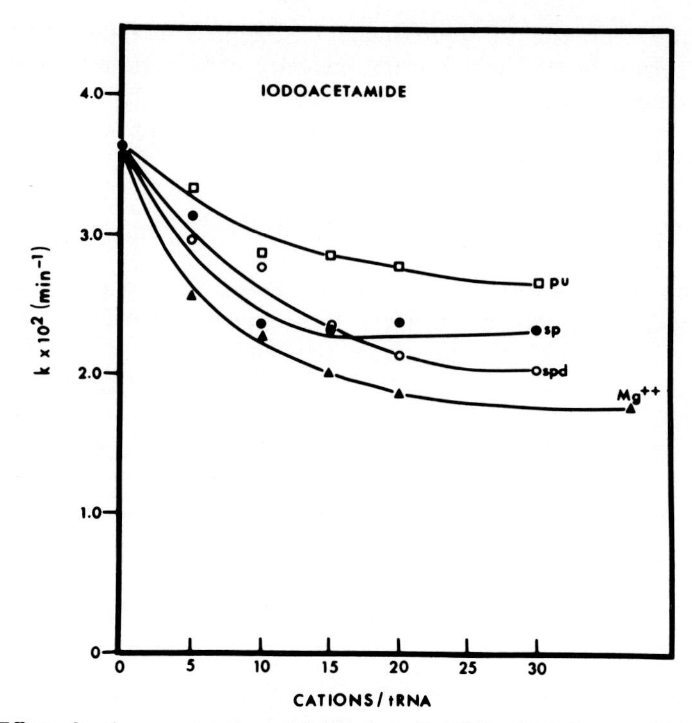

FIG. 4. Effect of cations on reaction of Srd in *E. coli* tRNA with iodoacetamide at pH 9.4, 25°C [iodoacetamide] = 4mM and [tRNA] = 28 μM. spd, spermidine; sp, spermine; pu, putrescine.

tRNA; i.e., Mg^{++} has the greatest effect, followed by spermidine, spermine, and putrescine. Again all effects appear maximal beyond 10 to 15 cations added per tRNA. The rates of alkylation were reduced by Mg^{++}, spermidine, and putrescine by 50%, 45%, and 25% respectively. The greater effect of cations in these cases than in the reduction reactions may reflect the larger size of iodoacetamide relative to borohydride.

D. Cyanogen bromide reactions

Cyanogen bromide is reported to react specifically with 4-thiouridine residues in *E. coli* tRNA (16) to give the 4-thiocyanatouridine derivative, which in turn hydrolyzes readily to uridine in acid or base. The results of the reactions of tRNA (21 μM) and cyanogen bromide (0.25 mM), summarized in Table 2, differ slightly from the iodoacetamide and borohydride reactions. Spermidine and spermine exhibit similar effects, although again both effects are less than shown by Mg^{++}. Putrescine again shows the smallest effect on reaction rate. These results may reflect the fact that these reactions were run at pH 7.4 rather than 9.4 and that the binding constants of the polyamines to tRNA are different at the different pH values. Although consideration of the possible change in binding with pH would be important in comparing data at different pH values, it should be possible to make valid comparisons of the various cations at a given pH.

TABLE 2. *Cyanogen bromide reaction with 4-thiouridine in tRNA*[a]

Cation	Moles of cation/mole of tRNA					
	0	5	10	15	20	25
Mg^{++}/tRNA						
$k(min^{-1})$	0.514	0.406	0.347	0.302	0.248	0.230
$t_{1/2}(min)$	1.35	1.7	2.0	2.3	2.8	3.0
spd/tRNA						
$k(min^{-1})$	0.514	0.396	0.330	0.315	0.315	0.300
$t_{1/2}(min)$	1.35	1.65	2.1	2.2	2.2	2.3
pu/tRNA						
$k(min^{-1})$	0.514	0.496	0.462	—	0.385	—
$t_{1/2}(min)$	1.35	1.4	1.5	—	1.8	—
spm/tRNA						
$k(min^{-1})$	0.514	0.385	0.315	—	0.290	—
$t_{1/2}(min)$	1.35	1.8	2.2	—	2.4	—

[a] 0.05 M cacodylate buffer, pH 7.4

E. Spin-label studies

We wished to explore the openness of the structure in the vicinity of the thiouridine. To that end an appropriate bromoacetamido derivative containing an unpaired electron was added to the thiouridine in tRNA and the restriction of rotation of this derivative (spin-label) was determined.

The data from the spin-label studies with mixed tRNAs and $tRNA_f^{Met}$ are given in Fig. 5. The results are given in terms of the rotational correlation time, τ, a measure of the degree of freedom possessed by the nitroxide radical. The values of τ are calculated from single spectra using the quadratic term as described by Stone et al. (17). The value of τ is inversely related to the rotational freedom of the radical. The calculations show that in mixed tRNAs putrescine confers only a small effect on the rotational freedom of the radical in tRNA, and Mg^{++} has only a slightly greater effect. Spermine gives a greater initial immobilization of spin of the radical than spermidine does, but spermidine shows a larger final effect. It should be

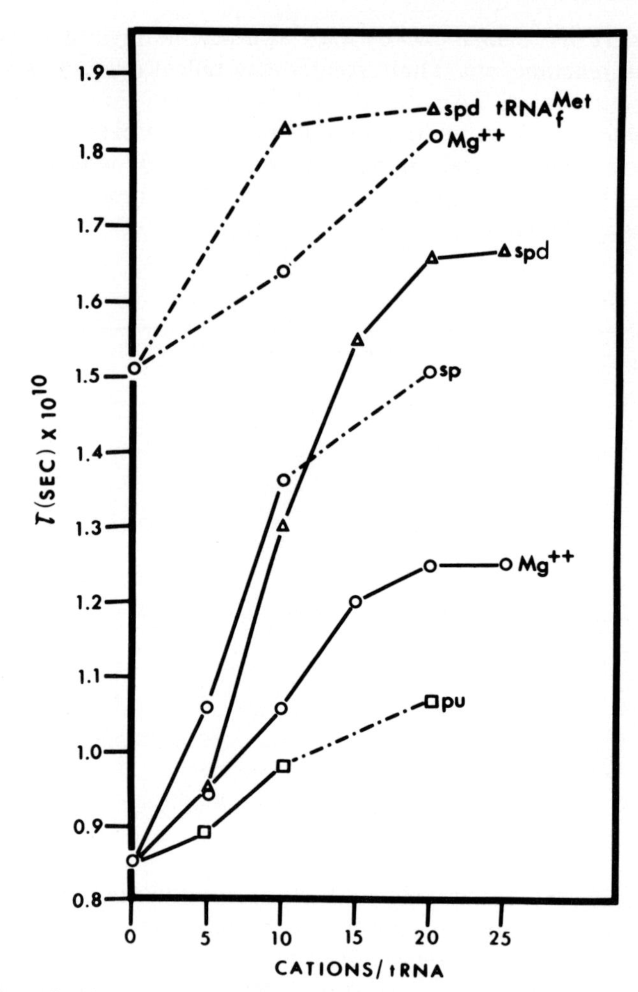

FIG. 5. Effect of cations on the rotational correlation time τ of spin-labeled *E. coli* mixed tRNAs and tRNA$_f^{Met}$. Esr spectra were determined in 0.05 M Tris-HCl buffer, pH 7. Concentration of mixed tRNAs = 30 μM; of tRNA$_f^{Met}$ = μM. spd, spermidine; sp, spermine; pu, putrescine.

noted that the spin-label is immobilized to a great extent by the tRNA structure even in the absence of added cations. A completely free spin-label molecule has a τ value of the order of 10^{-11} seconds, and a spin-label in tRNA without added cations gives a value of approximately 8.5×10^{-11} seconds.

The effect of spermidine at spermidine-to-tRNA ratios greater than about

15 is to restrict the rotation of the spin-label nearly twofold relative to tRNA without spermidine, τ increasing from 8.5×10^{-11} seconds to nearly 1.7×10^{-10} seconds.

The values of τ for spin-labeled $tRNA_f^{Met}$ are relatively high to begin with, indicating that the spin-label is restricted even in the absence of added cations. Mg^{++} and spermidine both restrict rotation further but the increase in τ is not as great as with mixed tRNAs. That the effects on a pure tRNA and mixed tRNAs are different indicates that each species of spin-labeled tRNA may behave differently and that, although all Srd residues occur in the same position in the primary sequence of tRNAs bearing Srd, the tertiary structures may be significantly different.

IV. DISCUSSION

The addition of Mg^{++} and spermidine appears to increase the apparent pK of Srd in tRNA. We can postulate several possible reasons for this observation: (a) the cations bind near Srd and inhibit ionization by a direct physical or electrostatic interaction; (b) the cations alter the composition of the medium such that ionization is suppressed; and (c) the cations may cause a conformational change in the tRNA molecule, altering the local environment near Srd, changing the apparent pK, perhaps by "squeezing out" water from the Srd site.

It is difficult to believe that the observed increase in pK is due to (b), and we suspect that (a) or (c) or both reflect the true explanations. This is primarily based on the fact that the pK of Srd is increased from 8.3 in the free nucleoside to 9.45 in the tRNA molecule. The binding of cations to tRNA further enhances the effect, perhaps by tightening the structure of tRNA and removing water from the interior of the molecule. Since spermidine does not affect the pK of the free nucleoside, the effect of cations on tRNA appears to involve a conformational change in tRNA.

The studies with borohydride and iodoacetamide indicate that polyamines and Mg^{++} affect tRNA such that the Srd region of tRNA is made less reactive. The choice of studying these reactions at alkaline pH seems somewhat questionable in that the polyamines probably are not fully charged at pH 9.4 and that the association constants for tRNA and polyamines are different from the values at physiological pH; nonetheless, polyamines show effects of a magnitude similar to Mg^{++}.

The reactions of cyanogen bromide also show that the cations are altering the reactivity of Srd to this reagent, presumably by reorganization of the tRNA structure.

The differences observed in the magnitudes of the effects of cations on the various alkylating agents may be due in part to the size of the reagents. The effect of cations on iodoacetamide reactivity is greater than on borohydride reactivity because borohydride is smaller than iodoacetamide and, therefore, can reach Srd more readily. With cyanogen bromide which is a very small molecule, the rate is affected significantly, indicating that the tRNA structure has been altered so that the very reactive cyanogen bromide molecule is hindered from reacting. This suggests that tRNA is even more structured at pH 7.4 than at pH 9.4; i.e., the "hole" through which molecules get to Srd is smaller at pH 7.4 than at pH 9.4. This difference probably is a reflection also of the variations in affinity of polyamines for tRNA at different pH values.

Probably the best evidence that polyamines order tRNA structure near Srd comes from the data obtained in the esr studies. The data obtained from the esr spectra are rotational correlation times, τ, and are a measure of the time it takes for a nitroxide radical to turn a certain amount. Thus, an increase in the value of τ indicates a decrease in the rotational freedom of the nitroxide radical or an increase in immobilization.

Polyamines appear to close the structure of the tRNA molecule such that the nitroxide radical is more hindered in its rotation. This interpretation assumes that the tRNA molecule with the nitroxide radical is in the same conformation as a normal tRNA molecule, i.e., that the label has little effect on the secondary and tertiary structure of tRNA. In this regard, it is noted that most of the spin-labeled tRNAs retain their amino acid acceptor activities (15).

The data indicate that spermidine is most effective in causing restriction of the spin-label in mixed tRNAs. Mg^{++} is only half as effective at the same levels per tRNA. Spermine is most effective at lower levels ($<$ 5 to 8 per tRNA) but is not quite as effective as spermidine at higher levels. Putrescine has very little effect. The differences might be attributed to the ability of spermidine and spermine to form a bridge between phosphate groups in separate parts of the tRNA molecule, causing a closing of the structure.

The reactivity of 4-thiouridine in tRNA and rotation of a spin-label molecule attached to Srd in tRNA have been shown to be altered by the presence of polyamines and Mg^{++}. The effects are not tremendously large but they suggest that polyamines bind to tRNA and restructure the tRNA molecule such that one of the effects is to tighten the arms of tRNA adjacent to Srd.

Spermidine in most cases is just as effective as Mg^{++} in restricting the reaction of Srd and, in the spin-label studies, is more effective than Mg^{++} in restructuring tRNA. These data extend observations indicated earlier in which polyamines can replace magnesium ion in various reactions of tRNA and are consistent with the hypothesis that spermidine may play a particular role in organizing the dihydrouridine arm in the vicinity of thiouridine in *E. coli* tRNA.

SUMMARY

Spermidine and Mg^{++} increase the apparent pK of 4-thiouridine (Srd) in *E. coli* mixed tRNAs from the value obtained in the absence of these cations. These substances are also effective in inhibiting the reduction by sodium borohydride of tRNAs in which the Srd residue at position eight and a cytosine residue at position 13 had been linked photochemically. Spermine and putrescine are less effective than spermidine and Mg^{++} in this reduction, as well as in the subsequent reactions. The polyamines and Mg^{++} also inhibit the rate of reaction of Srd in tRNA with iodoacetamide and cyanogen bromide.

The electron paramagnetic resonance spectra of *E. coli* mixed tRNAs and isoaccepting $tRNA_f^{Met}$ labeled at the 4-thiouridine residues with 3-(2-bromoacetamido)-2,2,5,5,-tetramethylpyrrolidinyl-1-oxyl were also studied as a function of added cation. Maximal restriction of the spin-label was observed at spermidine/tRNA ratios greater than about 15. Spermine was nearly as effective as spermidine, Mg^{++} showed only a slight restriction of rotation, and putrescine gave little effect.

The results indicate that the cations restrict the tRNA molecule in such a way that the region around 4-thiouridine is less accessible to external agents. They suggest that polyamines and Mg^{++} may be binding near 4-thiouridine, perhaps in the helical region of the dihydrouridine arm of tRNA, and that the conformation of this arm is particularly sensitive to spermidine.

ACKNOWLEDGMENTS

Dr. Sakai is a fellow of the Damon Runyon Memorial Fund for Cancer Research. This work was supported by a U.S. Public Health Service grant AI 10424–01. We are pleased to acknowledge the help of Dr. Kenneth Rubenstein of SYVA Research Institute, Palo Alto, in determining the electron spin resonance spectra.

REFERENCES

1. Cohen, S. S., *Introduction to the Polyamines,* Prentice-Hall, Inc., Englewood Cliffs, N.J., 1971.
2. Fresco, J. R., Adams, A., Ascione, R., Hemley, D., and Lindahl, T., *Cold Spring Harbor Symp. Quant. Biol.,* **31,** 527 (1966).
3. Hampel, A., and Bock, R. M., *Biochemistry,* **9,** 1873 (1970).

4. Bluestein, H. G., Allende, C. C., Allende, J. E., and Cantoni, G. P., *J. Biol. Chem.*, **243,** 4693 (1968).
5. Igarashi, K., and Takeda, Y., *Biochim. Biophys. Acta,* **213,** 4664 (1970).
6. Robison, B., and Zimmerman, T. P., *J. Biol. Chem.,* **246,** 4664 (1971).
7. Leboy, P. S., *FEBS Letters,* **16,** 117 (1971).
8. Cohen, S. S., Morgan, S., and Streibel, E., *Proc. Nat. Acad. Sci. U.S.A.,* **64,** 669 (1969).
9. Cohen, S. S., *Ann. N.Y. Acad. Sci.,* **171,** 869 (1970).
10. Cohen, S. S., *Adv. Enzyme Regul.,* **10,** 207 (1972).
11. Pochon, F., and Cohen, S. S., *Biochem. Biophys. Res. Comm.,* **47,** 720 (1972).
12. Pochon, F., Balny, C., Scheit, K. H., and Michelson, A. M., *Biochem. Biophys. Acta,* **228,** 49 (1971).
13. Favre, A., Michelson, A. M., and Yaniv, M., *J. Mol. Biol.,* **58,** 367 (1971).
14. Favre, A., Yaniv, M., and Michelson, A. M., *Biochem. Biophys. Res. Comm.,* **37,** 266 (1969).
15. Hara, H., Horiuchi, T., Saneyoshi, M., and Nishimura, S., *Biochem. Biophys. Res. Comm.,* **38,** 305 (1970).
16. Saneyoshi, M., and Nishimura, S., *Biochem. Biophys. Acta,* **204,** 389 (1970).
17. Stone, T., Buchman, T., Nordio, P., and McConnell, H., *Proc. Nat. Acad. Sci. U.S.A.,* **54,** 1010 (1965).

Polyamines in Normal and Neoplastic Growth. Edited by D. H. Russell. Raven Press, New York © 1973.

The Role of Polyamines During tRNA Processing in Leukemic Cells

Bruce Hacker

Division of Oncology, Departments of Biochemistry and Medicine, Albany Medical College, Albany, New York 12208

I. INTRODUCTION

There are now data to indicate that the methyl substituents of tRNA, and hence the corresponding tRNA-methylating enzymes,[1] are vital for acceptance of the proper amino acid during reactions involving aminoacyl-tRNA synthetases (1) as well as for recognition of the correct codon when binding to ribosomes during translation (2–4). The biochemical reactions involved in the processing of tRNA to a functional state are exceedingly complex (Fig. 1). The final stages, starting with "trimmed pre-tRNA" (5), are extremely important, since they are critical for the synthesis of a macromolecule, which must function accurately in its key role during both protein synthesis and transcriptional control. During the past several years, much evidence has accumulated to indicate that there is either increased, new, or aberrant methylase activity during various kinds of differentiation or neoplastic transformation (for reviews, see refs. 6 and 7). Changes have been demonstrated in the quantities and type of tRNA-methylating enzymes in tumor-bearing tissues (8, 9) and cells transformed in culture by the avian leukosis virus of Marek's disease (10). Other studies have shown that tRNA-methylating enzymes are also elevated after hormonal treatment and in tissues that are hormonally dependent or responsive (11–14). The potential for aberrant methylation of tRNA in neoplastic tissue may be mediated in part by the deletion of "inhibitors" (15) or the presence of polyamine "activators" (13). Depending upon the metabolic state and the biological system involved, those biochemical pathways for the biosynthesis of poly-

[1]Suggested trivial nomenclature for that group of transferases (EC 2.1.1.) that methylate unmodified components of pre-tRNA or other similar substrates with S-adenosyl-L-methionine as the donor.

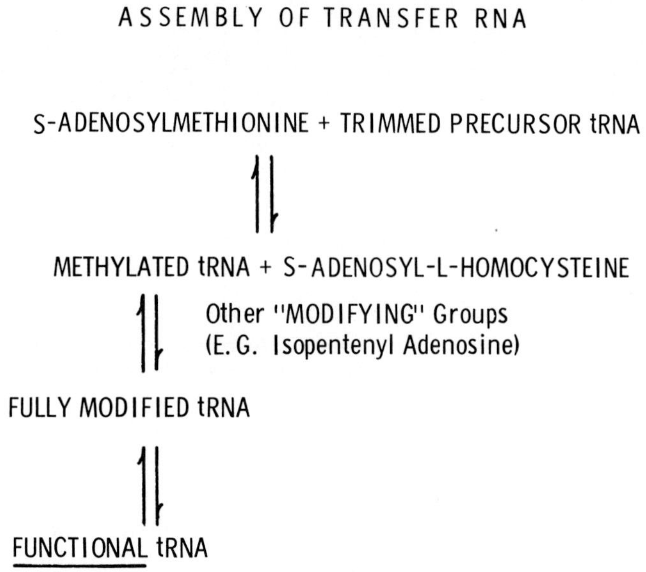

ASSEMBLY OF TRANSFER RNA

S-ADENOSYLMETHIONINE + TRIMMED PRECURSOR tRNA

⇅

METHYLATED tRNA + S-ADENOSYL-L-HOMOCYSTEINE

⇅ Other "MODIFYING" Groups (E. G. Isopentenyl Adenosine)

FULLY MODIFIED tRNA

⇅

FUNCTIONAL tRNA

FIG. 1. Processing of tRNA.

amines will diverge to varying degrees from those presented in Fig. 2. Yet the intricate biochemical relationship between S-adenosylmethionine, the methyl-group donor for tRNA, and the polyamines *per se,* shown herein as "stimulators" of tRNA methylation, indicates the significance of a clearer understanding of the value of polyamines in differentiation and/or neoplasia.

Although it has been reported that changes in the activities of various enzyme systems involved in polyamine synthesis occur in response to treatments with various drugs (16), and that certain drug-resistant sublines of murine leukemic cells do possess altered levels of tRNA methylating enzymes, it would appear that these two phenomena are unrelated, at least in a direct sense. Cells which possess resistance to various types of therapeutic agents are of great concern during the treatment of certain diseases, particularly in cancer. In this connection, resistance to actinomycin D and L-asparaginase in the sublines of L5178Y cells has been found to be coordinated with altered synthesis of membrane glycoproteins (17, 18). As in the case of the tRNA methylases, it has been demonstrated that increases occur in the levels of both polyamines and their synthesizing enzymes under conditions of enhanced growth (19–24). The capacity of the polyamines to stimulate the synthesis of ribosomal RNA and the binding of aminoacyl-tRNA to ribosomes (25–29) suggests a specialized role for the polyamines as regulator molecules during the processing of tRNA. The data presented herein demonstrate that the binding affinity of the polyamines to transfer RNA follows a pattern which is similar to their ability to facilitate the enzymic methylation of tRNA. Evidence is presented which suggests that the changes which occur in the binding of the polyamines to preparations of tRNA from drug-resistant leukemic cells may reflect alterations in the nature of the milieu of isoaccepting species from each source.

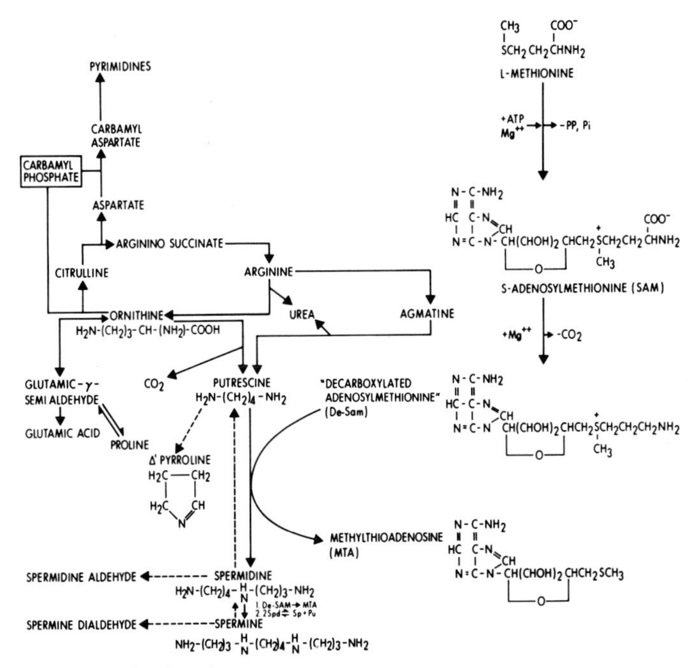

FIG. 2. Biosynthesis of polyamines.

II. EXPERIMENTAL PROCEDURE

A. Transplantation of tumor cells

A series of transplantable mouse leukemias, including those drug-resistant sublines L1210/Cyt and L1210/Ara-C[2] (30, 31), were developed at the Uni-

[2]The abbreviations used for drug-resistant sublines are indicated by a slash followed by the agent, and include: Cyt, Cytoxan® as cyclophosphamide; Ara-C, cytosine arabinoside or 9-β-D-arabinofuranosylcytosine.

versity of Rochester and at A. D. Little Inc. (Cambridge, Mass.), and are being propagated *in vivo* and in culture at the Albany Medical College. All cell lines were transplanted by intraperitoneal inoculation of 2×10^5 cells into recipient male CDF_1 mice (age 5 to 6 weeks; weight about 20 g). Maintenance doses of Ara-C (500 mg/kg) or cyclophosphamide (Cytoxan®; 100 mg/kg) were repeated every 3rd to 4th day thereafter, and cells were transplanted at 8- to 12-day intervals until the fourth generation, at which time the agent was discontinued. Sensitivity and resistance to the various agents were routinely monitored to ensure the maintenance of these properties in the cell lines used for these studies. Animals were maintained on Purina chow for mice and water *ad libitum*.

B. Isolation of tumor cells

Each group of mice was sacrificed by cervical dislocation 5 days after inoculation with tumor cells. Ascites fluid was rapidly removed from each mouse and mixed with about 10 volumes of ice-cold 0.14 M NaCl or Hanks' balanced salt solution (32). The suspended cells were then collected by centrifugation at $1{,}000 \times g$ for 5 min at 2 to 4°C in an International centrifuge (Model PR-2); the resulting supernatant was discarded. Each 2 ml of packed cells was gently resuspended in an equal volume of 0.14 M NaCl solution, followed by the addition of 12 ml of glass-distilled water to lyse contaminating erythrocytes (33). After a 15-sec hypotonic exposure, 4 ml of 0.56 M NaCl was added to restore isotonicity. The tumor cells were finally collected by centrifugation as before, and erythrocyte ghosts were removed.

C. Preparation of tRNA-methylase enzymes

L1210 mouse leukemic tumor cells, isolated during the mid-log phase of growth, were suspended in an equal volume of ice-cold buffer A (0.25 M sucrose, 10 mM sucrose, 10 mM 2-mercaptoethanol, and 20 mM Tris-HCl, pH 7.8). The chilled cell suspension was then sonicated for a period of 10 sec with the aid of a Bronwill Biosonik III sonicator at a setting of 25 using a BP-40T needle probe. After a 1-min cooling period, the sonication-cooling cycle was repeated twice more. The sonicate was initially centrifuged at $10{,}000 \times g$ for 15 min followed by another centrifugation of the initial supernatant fraction at $105{,}000 \times g$ in the L2-65 B Spinco ultracentrifuge for 45 min. The final, clear supernatant fraction (S_{40}), diluted with buffer A when required, constitutes the source of enzyme used throughout this investigation.

D. Protein estimations

Protein measurements were conducted according to the method of Lowry et al. (34) using crystalline serum albumin as the reference standard.

E. Transfer RNA and biochemicals

Unfractionated $tRNA^{Me-}_{E.\ coli\ K_{12}W_6}$[3] preparations possessing $A_{260\ nm}/A_{280\ nm}$ rations of 1.95 were isolated from *E. coli* $K_{12}W_6$ cells which had been grown in a methionine-deficient medium as previously described (8, 35). (Methyl-^{14}C)-S-adenosyl-L-methionine (44.6 mC/mmole) was purchased from the New England Nuclear Corp. The purity of each batch was evaluated prior to use by chromatography on Whatman 3MM paper using a solvent system composed of *l*-butanol:acetic acid:water (60:15:25, V/V/V). This corresponds to a specific radioactivity of 69 cpm/pmole when measured in a scintillation solvent containing Beckman Bio-Solv BBS-3:NEN liquifluor:toluene (100:42:858, V/V/V). ^{14}C-Spermine tetrahydrochloride (9.9 mC/mmole), ^{14}C-spermidine trihydrochloride (10.7 mC/mmole), and ^{14}C-putrescine dihydrochloride (11.27 mC/mmole) were obtained from the New England Nuclear Corp., and were evaluated for homogeneity and purity using paper chromatography. Tris buffer and unlabeled spermidine trihydrochloride were obtained from the Sigma Chemical Co. Putrescine dihydrochloride and spermine tetrahydrochloride were purchased from Mann Research. All other reagents were analytical grade and bought from the Mallinckrodt Chemical Company. Sephadex G-50 (superfine grade) was purchased from Pharmacia Fine Chemicals AB.

F. Measurement of tRNA-methylating enzyme activity

The assay system and procedures for determining the quantities of methylase enzymes present in each preparation have been previously described (8, 11).

G. Large-scale tRNA-methylase assays

For the preparation of $tRNA^{Me+}_{E.\ coli\ K_{12}W_6}$ (L1210) used for those experiments involving the binding of radioactive polyamines, large-scale incubation mixtures containing $tRNA^{Me-}_{E.\ coli\ K_{12}W_6}$ and enzyme isolates from L1210

---

[3] Denotes unfractionated, methyl-deficient tRNA from methionine-starved cultures of *E. coli* $K_{12}W_6$; $tRNA^{Me+}_{E.\ coli\ K_{12}W_6}$ (L1210) refers to the tRNA product isolated after methylation of $tRNA^{Me-}_{E.\ coli\ K_{12}W_6}$ with enzymes prepared from L1210 mouse leukemic cells.

leukemic cells were prepared as previously described (13). Conventional small-scale tRNA methylase assays containing radioactive S-adenosyl-L-methionine as described in the previous section were also conducted in parallel with each large-scale reaction to assess the extent of methylation.

H. Interaction of polyamines with tRNA

For the determination of the quantity of each polyamine bound to tRNA, reaction mixtures contained in 100 μl: 1.0 μmole of Tris-HCl (pH 7.8), 5 μmoles NH_4Cl, 0.6 μmole 2-mercaptoethanol, 40 nmoles ^{14}C-polyamine (0.2 μC), and 4.1 $A_{260\ nm}$ of $tRNA^{Me-}_{E.\ coli\ K_{12}W_6}$ or 4.2 $A_{260\ nm}$ of $tRNA^{Me+}_{E.\ coli\ K_{12}W_6}$ (L1210). Each reaction was incubated for various periods at several temperatures ranging from 0 to 37°C. The extent of binding of each type of polyamine to tRNA was then determined by Sephadex gel column exclusion chromatography for ^{14}C-spermine, and confirmed in experiments using thin-layer gel chromatography, both of which have been previously described. The results of specific experiments to determine the stoichiometry of polyamine binding to various types of tRNA preparations at 37°C are presented in Tables 4 and 5. Binding of the polyamines to tRNA proceeded rapidly, and was complete within 15 min at 37°C. Higher temperatures and longer incubation periods, however, gave rise to instability of the complex and diminished levels of binding.

I. Thin-layer gel chromatography

A suspension of Sephadex G-50 (superfine grade) was prepared by allowing 6.0 g of the dry material to swell in 60 ml of buffer B (50 mM NH_4 Cl, 6 mM 2-mercaptoethanol, and 10 mM Tris-HCl, pH 7.8). The swelling process was facilitated by gently stirring the suspension over a boiling water bath for 90 min. After cooling, the slurry was applied onto glass plates (40 cm) in the cold with a Pharmacia TLG-spreader at a thickness of 0.6 mm using a smooth and rapid motion. The coated plate was immediately transferred to a thin-layer gel filtration apparatus (Pharmacia) inclined at the appropriate angle to be used during the actual chromatography, and equilibrated in the presence of buffer B for a minimum of 16 and no more than 40 hr. Samples were then applied using minimal volumes (5 to 10 μl) with the plates in the horizontal mode, taking special precautions to avoid disruption of the gel surface. Development was initiated by restoring the plates to the same angle used during the equilibration, and conducted at 4 to 6°C. After each run (6 hr), components were located on "paper prints" of each plate prepared according to the method of Radola (36) by visualizing with ultraviolet light and radioactivity determined with liquid scintillation counting in the usual manner. The rates of migration are described elsewhere (13; see also Table 3).

III. RESULTS AND DISCUSSION

The approach used in the present investigation for studying the possible role of the polyamines during the methylation of tRNA represents an indirect one, since it correlates the actual stimulatory effects produced by the polyamines during the enzymic methylation of tRNA with their capacity to bind directly to tRNA from various sources.

Studies by Russell (37) have shown that levels of polyamines and the cognate enzymes ornithine decarboxylase and S-adenosyl-L-methionine decarboxylase are elevated during development of the leukemic state in mouse L1210 cells, and following the administration of various drugs such as phenobarbital, 3-methylcholanthrene, and 3,4-benzpyrene (16). In Fig. 3,

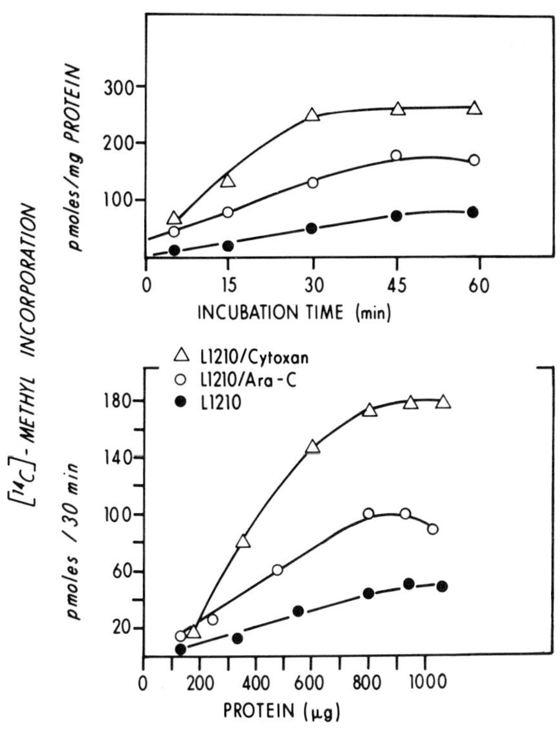

FIG. 3. Kinetics of tRNA-methylase enzymes isolated from sensitive and drug-resistant mouse leukemic cells. Assays conducted as described previously (13).

data are presented to show how drug resistance, in this case to the antitumor agents cyclophosphamide and cytosine arabinoside, is accompanied by a concomitant change in the kinetic properties of the methylating enzymes for tRNA. It has not yet been demonstrated whether these particular observations reflect changes in the requirements for specific isoaccepting species of tRNA utilized during glycoprotein synthesis (see ref. 18) or whether they are more the result of direct intracellular effects of the drug agent. It is clear, however, that any differences in the synthesis of the polyamines are not directly related to the observed changes in tRNA-methylase activities which accompany the development of drug resistance.

A. Effects of polyamines upon tRNA-methylases from leukemic cells

The stimulatory effects of putrescine (1,4-diaminobutane), spermidine, and spermine upon the specific activity of tRNA-methylases from L1210 cells as a function of concentration are depicted in Fig. 4. Concentrations of 60 and 2.0 mM are required for putrescine and spermine, respectively, to attain maximal stimulation. In contrast, activation by spermidine appears to be biphasic, with a maximum of 10 mM. The extent of methylation exhibited by the three polyamines increases in proportion to the number of amino groups present for binding.

The nature of the kinetic curves shown in Fig. 5 reveals that the rate of tRNA-methylation is linear for only about 10 to 15 min in the presence of putrescine at its most stimulatory concentration (60 mM; see Fig. 4). The fivefold increase observed after a 30-min incubation is still on the linear

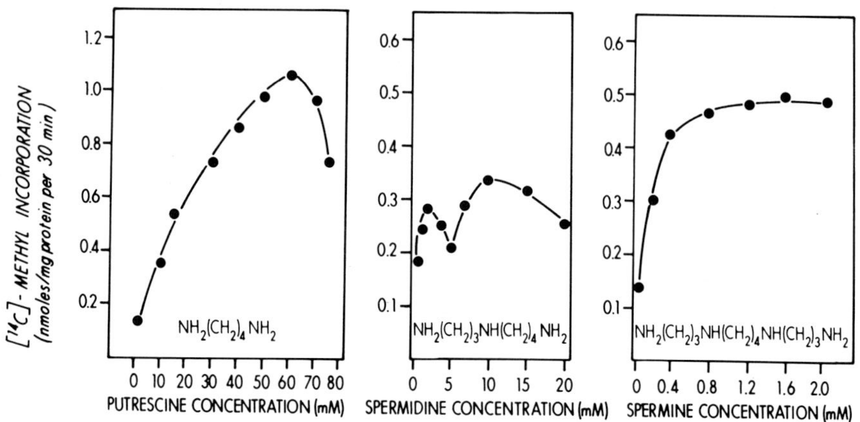

FIG. 4. Augmentation of tRNA methylation by the polyamines. Undialyzed supernatant fractions of L1210 sonicates (200 μg) were incubated at 37°C with each polyamine presented.

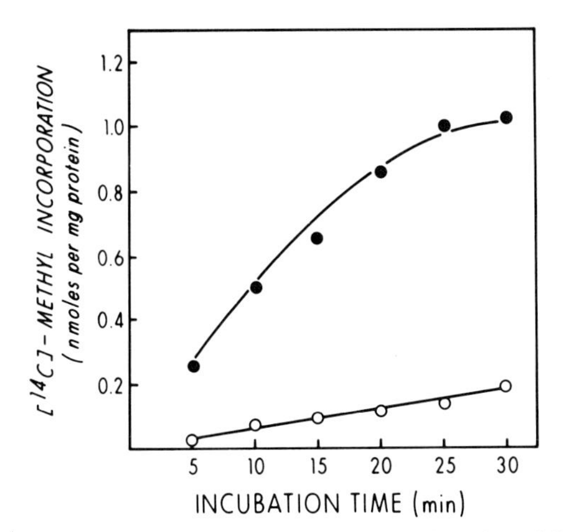

FIG. 5. Extent of methylation by L1210 cell methylases as a function of the incubation period. Standard reaction mixtures containing 40 μg of $tRNA^{Me-}_{E. coli\ K_{12}W_6}$ were incubated at 37°C for 0 to 30 min with 200 μg L1210 enzyme extract. ○, without putrescine; ●, plus putrescine (60 mM). Each value reflects the mean of triplicate determinations, with a mean S.E. ± 0.04.

portion of the rate curve for the unstimulated reaction mixture. The different slopes suggest that in the presence of putrescine the tRNAs may undergo changes in their conformation which make them more amenable to enzymic methylation.

B. Stability of tRNA-methylases

In an attempt to assess the stability of tRNA-methylases from L1210 mouse leukemia cells, dialysis was attempted in the presence or absence of putrescine with various buffers (Table 1). In general, it appears as if the dialyzed tRNA-methylases, or perhaps certain of them, are not very stable or active in the absence of Mg^{++} in the assay *in vitro*. Dialysis in the presence of putrescine or putrescine plus 2-mercaptoethanol appears to restore partially some of the lost capacity for methylation, although complete activity is never fully attained even in the presence of added Mg^{++} in the assay *in vitro*.

C. "Sparing" effect of multiple polyamines upon tRNA-methylases

Previously, we reported that at concentrations of putrescine less than 25 mM, the addition of spermine (0.4 mM) tended to augment the stimulatory effects upon tRNA-methylase activity from L1210 leukemic cells. On the

TABLE 1. *Effect of dialysis upon tRNA-methylating enzymes from L1210 cells*

Dialyzing buffer[a]	Specific radioactivity[b] −Mg^{++}	Specific radioactivity[b] +Mg^{++}
Undialyzed enzyme	295	482
3 mM Putrescine	137	224
0.10 M Sucrose + 3 mM putrescine	211	446
0.10 M Sucrose + 10 mM 2-mercaptoethanol + 20 mM Tris-HCl (pH 7.8)	88	412
0.10 M Sucrose + 10 mM 2-mercaptoethanol + 3 mM putrescine	87	377
3 mM Putrescine + 10 mM 2-mercaptoethanol	177	367
0.10 M Sucrose + 20 mM Tris-HCl (pH 7.8)	95	252
Glass-distilled water	32	237

[a] S_{40} enzyme preparations (0.9 ml) were dialyzed separately over a 15-hr period at 2 to 4°C against two changes of each type of buffer (250 ml).

[b] Specific radioactivity is defined as pmoles of methyl-^{14}C incorporated per mg of protein after incubating the standard assay in the presence of 10 mM $MgCl_2$ for 30 min at 37°C. Each value reflects the average of triplicate determinations for two separate enzyme preparations, each originating from pooled L1210 ascites cells from eight mice 5 days after transplantation with 2×10^5 cells/mouse.

other hand, it was demonstrated that the enhancing effect of spermine was negated at a concentration of putrescine above 25 mM; indeed an inhibition was seen to occur (13). The same sort of "sparing" effect for putrescine and spermidine is presented in Table 2. At a putrescine concentration of

TABLE 2. *Sparing phenomena of multiple polyamines upon tRNA-methylating enzymes from L1210 cells*

Spermidine concentration (mM)	Specific radioactivity[a] 30 mM putrescine	Specific radioactivity[a] 60 mM putrescine
0	0.79	1.38
2	0.78	0.83
4	0.70	0.64
6	0.64	0.55
8	0.58	0.49
10	0.55	0.46
15	0.43	0.36
20	0.38	0.31

[a] Specific radioactivity is defined as the number of nmoles of radioactive methyl groups incorporated per mg protein in 30 min of incubation time. All values reported are for complete assays containing the designated polyamines in place of $MgCl_2$. The complete assay without any polyamines plus 10 mM $MgCl_2$ had a value equal to 0.18 nmole/mg protein/30 min.

60 mM, which is required for maximal stimulation of tRNA methylation by undialyzed L1210 enzyme (Fig. 4), the addition of spermidine results in a continuous drop in enzymic activity. A similar pattern is also seen for reaction mixtures containing 30 mM putrescine. This phenomenon suggests the presence of a regulatory feedback or competition among the polyamines during the processing of transfer RNA.

D. Requirements for stimulation of tRNA-methylases by the polyamines

In a previous publication (13) we showed that the degree of stimulation by putrescine of tRNA-methylases from L1210 cells was approximately the same whether reactions were preincubated or not; i.e., preincubating the tRNA substrate with putrescine did not appreciably enhance the capacity to be methylated. On the other hand, data in Fig. 6 show the results of

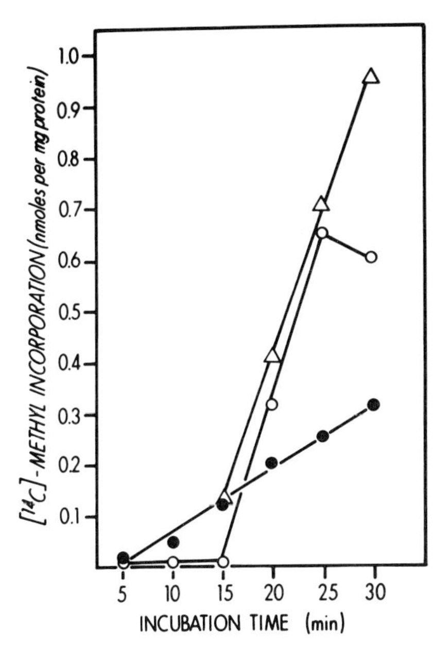

FIG. 6. Kinetics of tRNA-methylases from L1210 cells in the presence or absence of putrescine and/or S-adenosyl-L-methionine. Each assay containing 40 μg of $tRNA^{Me-}_{E.\ coli\ K_{12}W_6}$ was incubated at 37°C in the presence or absence of the following, added at the times indicated: ●, standard assay which is complete at t_0; ○, standard assay with putrescine (60 mM) at t_0 and (methyl-^{14}C)-S-adenosylmethionine (6.3 nmoles) added at t_{15}; △, standard assay at t_0, and putrescine (60 mM) added at t_{15}. Complete reactions containing putrescine incubated for 30 min have values of about 1.14. Each value represents the mean of triplicate determination for each type of experimental assay with a mean S.E. of ± 0.05.

experiments designed to evaluate what requirements and relative timing sequences are necessary for stimulation by putrescine. In the complete reaction mixture without S-adenosylmethionine, until t_{15} there is an apparent loss of overall stimulatory capacity by the polyamine, as shown by the drop in activity at 30 min. In contrast, assays containing radioactive S-adenosylmethionine plus putrescine added at t_{15} have higher final values after the additional 15-min incubation period (i.e., at t_{30}). These data suggest that some sort of conformational changes may occur in the presence of the polyamines in a concerted fashion to expose additional sites for methylation.

E. Binding reactions between polyamines and tRNA

Protocols were next devised to establish the rationale underlying the ability of the polyamines to enhance the methylation of tRNA. Assays were conducted to evaluate the stoichiometry involved during the interaction of a radioactive polyamine and tRNA using column and thin-layer Sephadex gel chromatography to resolve the reaction products (13). As shown in Table 4, polyamine molecules bind in the same relative order as their capacity to stimulate the tRNA-methylase reactions. The values for spermine suggest that on the average there are four to five spermine molecules per molecule of tRNA. This same ratio could also be attained if, for example, 50% of the tRNA molecules had eight to 10 spermines while the remaining 50% had none. This implies the existence of some sort of selectivity for certain iso-accepting tRNA species. This possibility is strengthened by the fact that the affinity ratio value for bound putrescine or spermidine is in the range of 10^{-2} to 10^{-3} (polyamine per 1.0 tRNA) for either type of tRNA. It is very interesting to note that the more highly methylated preparation of tRNA [i.e., $tRNA^{Me+}_{K12W6}$ (L1210)], obtained using enzyme extracts from L1210

TABLE 3. *Thin-layer gel chromatography of tRNA and its reaction products with polyamines*

Inclination angle (degrees)	Migration distance (cm)		
	Polyamines	tRNA	tRNA + polyamine complex
10	5.0	11.2	11.3
15	9.5	19.6	19.7
25[a]	19.0	40	40

Plates (40 cm) coated with Sephadex G-50 (superfine) gel were prepared as previously described (13), and developed for 6 hr at 2 to 4°C.

[a] At an inclination angle of 25°, tRNA and the tRNA + polyamine complex migrate beyond the plate boundary.

TABLE 4. *Binding of polyamines to tRNA products from reaction of transfer RNA from* E. coli *and methylase enzymes of L1210 leukemic cells*

Source of tRNA	Polyamine bound		
	Putrescine	Spermidine	Spermine
$tRNA^{Me-}_{E.\ coli\ K12W6}$	3.7	14.6	405.9
$tRNA^{Me+}_{E.\ coli\ K12W6}$(L1210)	5.4	18.5	484.0

Binding reactions were performed as described in Materials and Methods, using an incubation period of 15 min at 37°C. The reaction products were separated using Sephadex G-50 (superfine) gel column chromatography (13), and confirmed using thin-layer gel chromatography (see Materials and Methods). Values given were determined from two or three separate determinations, and correspond to the mean number of molecules of each polyamine bound per 100 molecules of tRNA.

leukemic cells containing the methylating enzymes, is capable of binding additional molecules of either of the three polyamines (Table 4).

The relative affinities of the three polyamines seem to be the same regardless of the source of tRNA, and considerably tighter for spermine in all instances (Tables 4 and 5). Apparently, there is a somewhat greater binding affinity by putrescine and spermidine for preparations of tRNA from bacteria (*Escherichia coli*) compared to rodent leukemic cells perhaps due to differences in the primary or secondary structural sequences. These differences are not apparent in the case of spermine, however (Table 4).

Since it is apparent that preparations of tRNA from drug-resistant leukemic cell lines bind differently to the polyamines, in a relative sense, the suggestion is that the pattern of tRNAs varies from one subline to another (Table 5). This fact is also borne out by variation in the methylase activities indigenous to each subline (Fig. 3).

It is thus apparent that the polyamines can bind directly to tRNA in the

TABLE 5. *Affinity binding of tRNA from sensitive and drug-resistant lines of leukemic cells with polyamines*

Source of tRNA	Affinity ratio[a]		
	Putrescine	Spermidine	Spermine
$tRNA^{b}_{L1210}$	0.2	7.8	557.0
$tRNA^{b}_{L1210/Cytoxan®}$	2.7	11.3	463.2
$tRNA^{b}_{L1210/Ara-C}$	0.2	9.6	444.0
$tRNA^{Tyr\ I}_{Yeast}$	0.5	1.9	422.2

[a] Number of molecules of polyamine bound per 100 molecules of tRNA
[b] Unfractionated tRNA

order spermine > spermidine > putrescine, and that their ability to stimulate the enzymic methylation of tRNA by extracts of mouse L1210 leukemic cells follows the same order. It is still not yet clear what the relationship is between these observations and the role of the polyamines during tRNA processing in normal, proliferating, or neoplastic cells.

IV. SUMMARY

The present study demonstrated that the tRNA-methylases obtained from extracts of L1210 mouse leukemia cells may be stimulated by various polyamines in the following order: spermine > spermidine > putrescine. In other studies involving nonenzymic binding of each type of polyamine to tRNA, the relative affinity was found to occur in the same order. The extent of binding of each polyamine appeared to be greater if the source of tRNA was more highly methylated. The different binding affinities of the polyamines for preparations of tRNA from drug-resistant lines of mouse leukemia cells suggest that the milieu of isoaccepting species is different among the sublines. These results also imply that the methylation of tRNA in mouse leukemia cells which can be modulated by various factors including the polyamines may be under varying degrees of cellular control.

ACKNOWLEDGMENTS

The excellent technical assistance of Miss Bonnie McDermott is gratefully acknowledged.

This investigation was supported by U.S. Public Health Service grants CA-11198 and CA-14169 from the National Cancer Institute, General Research Support grant RR-05403 from the U.S. Public Health Service, and grant IN-18M from the American Cancer Society.

REFERENCES

1. Shugart, L., Chastain, B. H., Novelli, G. D., and Stulberg, M. P., *Biochem. Biophys. Res. Comm.,* **31,** 404 (1968).
2. Capra, J. D., and Peterkofsky, A., *J. Mol. Biol.,* **33,** 591 (1968).
3. Gefter, M. L., and Russell, R. L., *J. Mol. Biol.,* **39,** 145 (1969).
4. Stern, R., Gonano, F., Fleissner, E., and Littauer, U. Z., *Biochemistry,* **9,** 10 (1970).
5. Altman, S., *Nature, New Biology,* **229,** 19 (1971).
6. Söll, D., *Science,* **23,** 293 (1971).
7. In Symposium on transfer RNA and transfer RNA modification in differentiation and neoplasia, *Cancer Res.,* **31,** 591 (1971).
8. Hacker, B., and Mandel, L. R., *Biochim. Biophys. Acta,* **190,** 38 (1969).
9. Mandel, L. R., Hacker, B., and Maag, T., *Cancer Res.,* **29,** 2229 (1969).
10. Mandel, L. R., Hacker, B., and Maag, T. A., *Cancer Res.,* **31,** 613 (1971).

11. Hacker, B., *Biochim. Biophys. Acta,* **186,** 214 (1969).
12. Hacker, B., Hall, T. C., and Abraham, S., *Clin. Res.,* **19,** 492 (1971).
13. Hacker, B., and McDermott, B. J., *Physiol. Chem. Biophys.,* **4,** 41 (1972).
14. Turkington, R. W., *Tenth International Cancer Congress Abstracts,* University of Texas Press, Austin, 1970, p. 385.
15. Kerr, S. J., *Proc. Nat. Acad. Sci. U.S.A.,* **68,** 400 (1971).
16. Russell, D. H., *Biochem. Pharmacol.,* **20,** 3481 (1971).
17. Kessel, D., and Bosmann, H. B., *Fed. Europ. Biochem. Soc. Lett.,* **10,** 85 (1970).
18. Kessel, D., and Bosmann, H. B., *Cancer Res.,* **30,** 2695 (1970).
19. Dykstra, W. G., and Herbst, E. J., *Science,* **149,** 428 (1965).
20. Jänne, J., and Raina, A., *Acta Chem. Scand.,* **22,** 1349 (1968).
21. Russell, D. H., and Snyder, S. H., *Proc. Nat. Acad. Sci. U.S.A.,* **60,** 1420 (1968).
22. Russell, D. H., Medina, V. J., and Snyder, S. H., *J. Biol. Chem.,* **245,** 6732 (1970).
23. Herbst, R. J., and Bachrach, U., *Ann. N.Y. Acad. Sci.,* **171,** 693 (1970).
24. Russell, D. H., and Levy, C. C., *Cancer Res.,* **31,** 248 (1971).
25. Barros, C., and Guidice, Q., *Exp. Cell Res.,* **50,** 671 (1968).
26. Tanner, M. J. A., *Biochemistry,* **6,** 2686 (1967).
27. Takeda, Y., *Biochim. Biophys. Acta,* **182,** 258 (1969).
28. Igarashi, K., and Takeda, Y., *Biochim. Biophys. Acta,* **213,** 240 (1970).
29. Takeda, Y., and Igarashi, K., *Biochim. Biophys. Acta,* **204,** 406 (1970).
30. Kessel, D., Hall, T. C., and Wodinsky, I., *Science,* **156,** 1240 (1967).
31. Wodinsky, I., and Kensler, C. J., *Cancer Chemother. Rep.,* **43,** 1 (1964).
32. Hacker, B., and Feldbush, T. R., *Biochem. Pharmacol.,* **18,** 847 (1969).
33. Hacker, B., *Biochim. Biophys. Acta,* **224,** 635 (1970).
34. Lowry, O. H., Rosebrough, N. J., Farr, A. L., and Randall, R. J., *J. Biol. Chem.,* **193,** 265 (1951).
35. Fleissner, E., and Borek, E., in G. L. Cantoni and D. R. Davies (editors), *Procedures in nucleic acid research,* Harper & Row, New York, 1966, p. 461.
36. Radola, B. J., *J. Chromatog.,* **38,** 61 (1968).
37. Russell, D. H., *Proc. Amer. Assoc. Cancer Res.,* **11,** 1969 (1970).

Polyamines in Normal and Neoplastic Growth. Edited by
D. H. Russell. Raven Press, New York © 1973.

The Stimulation of RNA Synthesis by Spermidine: Studies with *Drosophila* Larvae and RNA Polymerase

Edward J. Herbst, Craig V. Byus, and Donald L. Nuss

Department of Biochemistry, University of New Hampshire, Durham, New Hampshire 03824

I. INTRODUCTION

The development of *Drosophila melanogaster* is characterized by a rather precise "timetable" of embryonic, larval, pupal, and adult stages (1). We have measured the polyamines in developmental stages after conversion to highly fluorescent dansyl amide derivatives (2). The predominant polyamine in *Drosophila* is spermidine, and lower concentrations of putrescine, but only traces of spermine, occur in all developmental stages. The highest concentrations of spermidine occur during early larval growth; minimum concentrations are found in prepupal stages; there is an increase to intermediate levels during histogenesis of pupae; and low intermediate concentrations are maintained in adult *Drosophila*. These patterns are strikingly similar to the polyamine variations during postembryonic development of the blowfly, *Calliphora erythrocephala,* as recently reported by Heby (3).

Spermidine is taken up by the cells of isolated *Drosophila* salivary glands maintained in liquid culture medium, and the polyamine enters the cell nuclei (4). The incorporation of ^{3}H-uridine into the RNA of salivary gland nuclei is considerably enhanced by the addition of spermidine to the culture medium. The greatest concentration of the ^{3}H-RNA, detected by radioautography, is in the nucleolar region (4).

These observations and the reports from our own and other laboratories (5–9) pertaining to (a) the relationship between polyamine biosynthesis and rapid growth processes and (b) the stimulation of DNA-dependent RNA polymerases by polyamines (10–12) supported the renewed investigations of the effects of polyamines on RNA synthesis by *Drosophila* larvae and a bacterial RNA polymerase which are presented here.

II. MATERIALS AND METHODS

A. Culture and labeling of larvae

Adult flies and all developmental stages were maintained on culture media adapted from Pearl et al. (13) which consisted of sucrose, 46 g; sodium-potassium tartrate, 4.6 g; $(NH_4)_2SO_4$, 1 g; $MgSO_4$, 0.25 g; $CaCl_2$, 0.0125 g; Difco agar, 11.2 g; tartaric acid, 2.5 g; KH_2PO_4, 0.33 g; propionic acid, 2 ml; and distilled water, 1 liter. The medium was heated to a clear solution and dispensed in 500-ml Erlenmeyer flasks for stock cultures and in 13-cm plastic Petri dishes for larval cultures. Microbial contamination of the larval cultures and the liquid medium for labeling studies (same medium as above without the agar) was prevented by the addition of 200 mg penicillin and 50 mg streptomycin per liter of medium.

Developmental stages were maintained on the solid culture media at 25°C. Larvae were washed with sterile liquid culture medium on a 25-mm glass fiber filter and transferred into a 10-ml beaker containing 0.25 ml of the same medium supplemented with ^{3}H-uridine and polyamines as indicated for various experiments. The beakers were covered with Parafilm and gently shaken to prevent "wandering" of the larvae to the walls of the beaker. The radioactive medium was removed from the larvae by washing with distilled water on a glass fiber filter, and the washed animals were rapidly weighed and transferred to a micro Duall (Kontes) homogenizer in an ice bath. Then 0.5 ml modified Greenberg buffer (14) consisting of 0.1 M NaCl, 0.01 M sodium acetate, pH 5.1, 1% sodium dodecyl sulfate (SDS), and 5 μg per ml polyvinyl sulfate (PVS) and 0.5 ml buffer-saturated phenol were added. The mixture was homogenized intermittently for 15 min, and 0.5 ml cold redistilled $CHCl_3$ was added. After it was mixed on a Vortex stirrer, the suspension was centrifuged in the cold at 1,000 rpm and the lower phenol-$CHCl_3$ phase was removed. The aqueous layer and cloudy interphase were rehomogenized in the cold with buffer-saturated phenol + $CHCl_3$ and held on ice for 15 min with intermittent mixing on the Vortex. After centrifugation at 7,000 rpm, the aqueous phase was removed and extracted with two volumes of cold redistilled diethyl ether to remove residual phenol. The aqueous phase was freed of ether *in vacuo,* and the RNA was precipitated by adding two volumes of 80% ethanol containing 0.1 M NaCl and 0.01 M sodium acetate (pH 6) and holding at −20°C for 2 hr. The precipitate was taken up in 0.3 ml of NET (0.1 M NaCl, 1 mM EDTA, 0.01 M Tris, pH 7.5) containing 0.5% SDS and layered onto 17-ml sucrose gradients (5 to 30% w/v in NET–0.5% SDS) and centrifuged for 12 to 13 hr at 27,000 rpm and 22°C in the SW 27.1 rotor. The resolution of molecular species of RNA was monitored with an Isco U.V. Analyzer and external recorder. The radioactivity of the RNA in 1-ml fractions precipitated on glass fiber filters with cold 10% trichloroacetic acid (TCA) after addition of 200 μg of carrier bovine serum albumin (BSA) was determined by liquid scintillation analysis in toluene phosphor counting fluid.

B. Analysis of polyamines

The analysis of polyamine "enrichment" of larvae was performed by a modification of the fluorescence procedure previously described (2). Ten to 25 mg of larvae (wet weight) was homogenized in the micro Duall homogenizer with 0.4 ml of 0.4 N perchloric acid (PCA), and 0.2 ml of the extract was transferred to a small centrifuge tube. Fifty mg of $NaHCO_3$ was added, followed by 0.4 ml of dansyl chloride (10 mg/ml acetone). The tube was capped and shaken gently in the dark for 16 hr at room temperature. Excess reagent was reacted with 0.1 ml proline (100 mg/ml water) for 30 min, and the dansyl amide derivatives of the amines were selectively and quantitatively extracted with 0.5 ml of benzene. Aliquots of the benzene extracts of unknowns and appropriate polyamine standards were applied to silica gel-G thin-layer chromatography (TLC) plates which were developed with ethyl acetate:cyclohexane (2:3) and sprayed with triethanolamine:isopropanol to "stabilize" and intensify the fluorescence. After storage of the plates overnight in a vacuum oven at room temperature, the fluorescent areas were scanned with a Turner model 111 fluorometer equipped with a TLC scanning attachment and external recorder (fluorescence activation at 365 nm and emission measurements above 512 nm). Peak areas of fluorescence traces were measured by planimetry and quantitated from standard curves relating fluorescence to concentration of the dansyl derivatives of the polyamines.

C. RNA polymerase assays

The DNA-dependent RNA polymerase of *Escherichia coli* K-12 was purified by the procedure of Burgess (15). The glycerol gradient centrifugation modification was utilized, and the phosphocellulose chromatography step was not performed. This preparation of the "complete" enzyme (core enzyme plus sigma factor) had a specific activity of 1,050 units/mg protein.

The RNA polymerase assays contained 50 mM Tris buffer, pH 7.9, 5 mM

$MgCl_2$, 1 mM ATP, GTP, and CTP, 0.5 mM ^3H-UTP (2×10^4 dpm per nmole), 0.1 mM dithiothreitol (DTT), 15 μg DNA, and 2.7 μg enzyme protein in a final volume of 0.2 ml. Assays were incubated at 37°C and chilled at the end of the incubation period. Then 10-μl aliquots were spotted on 25-mm Whatman #40 filter paper discs which were air dried before the following precipitation-wash procedure: 2 × cold 10% TCA, 2 × 95% ethanol, 1 × absolute ethanol, and 2 × ether. The air-dried filters were then counted in toluene phosphor by the liquid scintillation procedure. To assay transcription complex-bound RNA, the nitrocellulose filter technique described by Witmer (16) was modified by transferring 10-μl aliquots of the reaction mixture directly to the B6 filter without dilution followed by five washes with 10 ml of 10 mM Tris, pH 7.9, and 10 mM KCl.

Sucrose gradient analyses of RNA polymerase reaction mixtures were performed in 5 to 20% gradients prepared in "gradient buffer" (50 mM Tris, pH 7.9, 40 mM KCl, and 1 mM DTT). An aliquot of reaction mixture was diluted 1:5 with gradient buffer, and 0.2 ml of the diluted solution was layered on the gradients and centrifuged for 105 min at 4°C and 35,000 rpm in the SW-50 rotor. Fractions (0.25 ml) were then collected, and 200 μg BSA was added prior to precipitation with 10% TCA. The precipitates were collected on glass fiber filters and, after three washes with 2.5 ml of 5% TCA, the filters were air dried and counted in toluene phosphor.

D. Templates, isotopes, and other chemicals

T4 DNA was prepared from bacteriophage isolated from infected cultures of *E. coli* B by the procedures described by Thomas and Abelson (17). To prepare T4 ^{32}P-DNA, 2 mC of ^{32}P-phosphate per liter of medium was added at the time of infection with phage. DNA was extracted from phage suspensions at a concentration of 1 to 2×10^{12} phages per ml. To minimize shearing of intact DNA molecules, the solutions were transferred by pouring rather than by pipetting. The T4 DNA showed a single peak at approximately 60S in the 10 to 20% sucrose prepared in gradient buffer.

E. coli DNA was supplied by PL Biochemicals, poly [d(A-T)] by Miles, and the calf thymus DNA was obtained from Worthington and General Biochemicals. Radioisotopes were obtained from New England Nuclear, and dansyl chloride was supplied by Pierce Chemical Co.

III. RESULTS

Larvae were harvested from solid medium 48 to 96 hr after hatching. These animals fed vigorously on the liquid medium and attained a favorable body size (approximately 3 to 4 mm in length and a weight of approximately 1.5 mg) for polyamine analysis and for the isolation of RNA.

A. Rate of feeding on polyamines

The rate of feeding on polyamines was estimated with ^{3}H-putrescine. The 72-hr larvae were incubated in liquid medium containing 1×10^{-3} M putrescine·2HCl and 100 μC of ^{3}H-putrescine. Thirty animals were removed at time intervals up to 12 hr. PCA extracts were prepared, 0.2-ml aliquots of the extracts were dansylated, and the dansyl amides were separated and analyzed by the TLC-fluorescence procedure. The fluorescent spots corresponding to putrescine and spermidine were scraped off the plate, and the silica gel scrapings were extracted with 0.5 ml of dioxane. Aliquots were analyzed in Aquasol counting fluid.

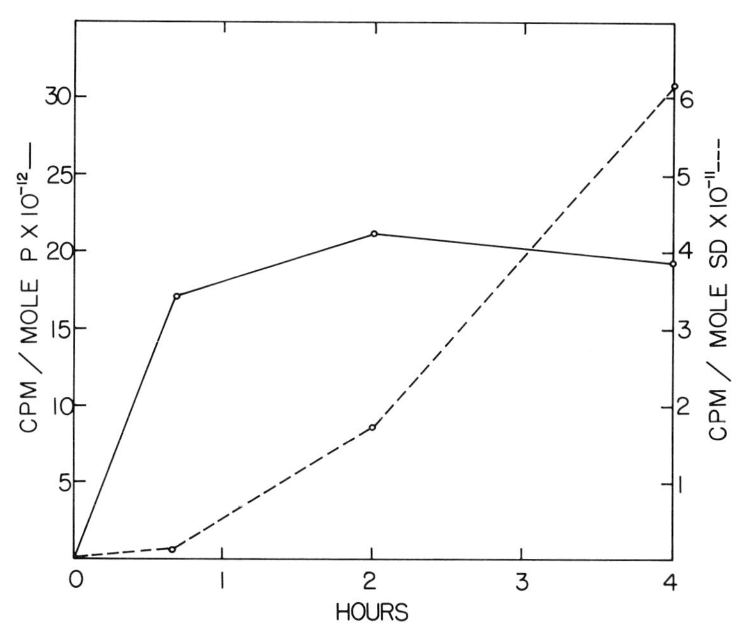

FIG. 1. The rate of uptake of ^{3}H-putrescine and conversion to ^{3}H-spermidine by *Drosophila* larvae. Several hundred 72-hr larvae were incubated in liquid culture medium containing 100μC of ^{3}H-putrescine and 1 mM unlabeled putrescine. Thirty animals were removed at time intervals up to 12 hr; perchloric acid extracts were prepared; and, after conversion of the amines in the extract to dansyl derivatives and separation by TLC chromatography (see Materials and Methods), the fluorescent areas were scraped from the TLC plates, extracted with 0.5 ml dioxane, and radioactivity determined in Aquasol counting fluid. The radioactivity of putrescine obtained at 1 to 4 hr is constant for the 12-hr incubation period, although the radioactivity of spermidine increases at a constant rate during the period 1 to 12 hr.

The labeled diamine was rapidly taken up by the larvae, and saturation of the putrescine pool was attained in approximately 1 hr (Fig. 1). ^{3}H-Spermidine could be detected within 1 to 2 hr, and the labeled intracellular spermidine increased at a constant rate throughout the entire 12-hr period of feeding ^{3}H-putrescine. In view of these uptake data, enrichment of the larvae was determined after 4 hr of preincubation with several levels of polyamines. Control larvae were preincubated for a similar period in the liquid medium without polyamines.

B. Polyamine enrichment of larvae

The enrichment of larvae with polyamines was studied at various time periods during larval development (Table 1). Substantial changes in spermidine could be demonstrated in larvae 72 hr after hatching. The elevation of putrescine after putrescine feeding was demonstrable both in 48-hr larvae and in the older animals. The 72-hr larvae were selected for subsequent experiments involving labeling and analysis of RNA. This choice was predicated by the favorable size of the animals and on the basis of the relatively greater change in spermidine content that could be effected at this age.

TABLE 1. *Enrichment of* Drosophila *larvae with polyamines*

Polyamine in medium	Nanomoles polyamine per animal							
	48-hr larvae		72-hr larvae		95-hr larvae		118-hr larvae	
	SD	P	SD	P	SD	P	SD	P
None	0.64	—	0.34	—	0.32	—	0.33	—
1 mM SD	1.0	—	0.34	—	0.19	—	0.37	—
10 mM SD	0.96	—	0.91	—	0.73	—	0.40	—
1 mM P	0.88	—	0.34	—	0.37	—	0.31	0.18
10 mM P	0.68	0.88	0.26	0.60	—	—	0.33	0.27

Larvae were incubated for 4 hr at 25°C in 0.25 ml of the liquid culture medium in which the concentration of putrescine (P) or spermidine (SD) was varied. The animals were washed, homogenized, and analyzed by the fluorescence procedure described in Materials and Methods.

C. RNA synthesis by intact larvae

The incorporation of ^{3}H-uridine into the RNA of larvae feeding on liquid medium occurred after short "pulses" (15 to 20 min), and the accumulation of the label in all stable molecular species of RNA was maintained for 12 to 14 hr. Sucrose-gradient analysis of the RNA isolated from larvae "pulsed" with ^{3}H-uridine for 20 min in the presence and absence of spermidine is

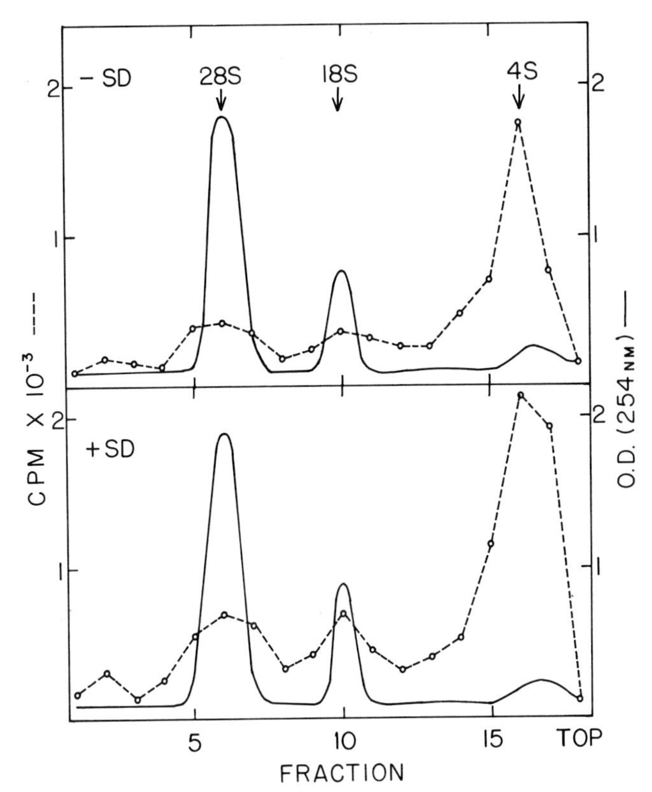

FIG. 2. The effect of spermidine on RNA synthesis by *Drosophila* larvae. RNA was isolated from 72-hr *Drosophila* larvae which had been incubated for 20 min in 0.25 ml of liquid culture medium containing 100 μC of ^{3}H-uridine and either zero (−SD) or 1 mM (+SD) spermidine. (See Materials and Methods for incubation conditions, RNA isolation, and sucrose-gradient analysis.) The peaks identified as 28S, 18S, and 4S RNA in the optical density record represent the rRNA and tRNA components in these preparations of RNA from *Drosophila* larvae. The radioactivity of 1-ml fractions collected from 17-ml 5 to 30% sucrose gradients in NET-0.5% SDS centrifuged for 13 hr at 27,000 rpm and 22°C in the SW 27.1 rotor is superimposed on the O.D. traces of the larval RNA.

shown in Fig. 2. The 260-nm markers at 28S, 18S, and 4S are the "endogenous" *D. melanogaster* RNA components in the labeled preparation isolated from the larvae. The small labeled peak which is larger than 28S is presumed to be the 38S precursor of rRNA identified by Greenberg (14) in *Drosophila virilis* pupae pulse-labeled after injection of ^{3}H-uridine. Rapid labeling of 4S RNA was also observed by Greenberg; however, he reported equivalent incorporation of ^{3}H-uridine into the 38S rRNA precursor during a pulse whereas we found low, barely demonstrable incorporation into an apparent

TABLE 2. *The effect of spermidine on RNA synthesis by* Drosophila *larvae*

Labeling time	Spermidine concentration (mM)	Specific activity			Relative spec. act.		
		4S	18S	28S	4S	18S	28S
20 min	0	5,080	456	452	100	100	100
	1	9,350	612	663	190	134	146
3.5 hr	0	7,200	1,392	1,400	100	100	100
	1	8,560	2,256	2,280	118	162	162
	5	11,430	2,487	2,672	159	178	190
	10	6,680	1,896	2,112	93	136	150
12 hr	0	3,396	3,707	5,504	100	100	100
	1	5,430	8,455	12,000	160	228	215
	5	10,170	9,682	12,242	291	261	220
	120	2,220	653	1,588	65	17	28

RNA was isolated (see Materials and Methods) from 72-hr *Drosophila* larvae which were preincubated for 1 hr in 0.25 ml of liquid "culture medium" containing variable concentrations of spermidine followed by incubation in the presence of 25 μC of ^{3}H-uridine (3.5 or 12 hr) or 100 μC of ^{3}H-uridine (20 min). The RNA was analyzed by sucrose-gradient centrifugation, and the areas of O.D. peaks at 254 nm and of radioactivity profiles on 1-ml gradient fractions were determined by planimetry. The specific activity = cpm per 254 O.D. unit of 4S, 18S, and 28S RNA was calculated, and the "relative specific activity" to zero spermidine controls = 100 was determined.

rRNA precursor. (The labeling of rRNA precursor is not increased by the addition of RNAase inhibitors such as bentonite, heparin, diethyl pyrocarbonate, or combinations thereof to the RNA isolation media which contains SDS and PVS. Turnover of labeled rRNA precursor is, therefore, not believed to be a factor in the maintenance of these low precursor levels which might be related to efficient "processing" of rRNA in the intact larvae.)

The stimulation of ^{3}H-uridine incorporation into all RNA species is apparent in the relative size of radioactivity and absorbance peaks in this pulse experiment. The effect of spermidine has been quantitated by measuring the area of the peaks from the plots of radioactivity and of optical density tracings. The specific activity (cpm per O.D. unit of RNA) of the RNA isolated from 72-hr larvae labeled for different time periods in the presence of different concentrations of spermidine is presented in Table 2. The effect of the polyamine is apparent at all time periods of feeding. An accumulation of label in rRNA, to be expected during extended periods of labeling, is observed, and there is a maintenance of the stimulatory effect of spermidine throughout the longest (12-hr) feeding period. The toxicity associated with high concentrations of spermidine in the medium cannot be properly evaluated. The animals are alive as evidenced by normal movement in media containing 0.12 M spermidine, but the mechanism by which the polyamine depresses ^{3}H-uridine incorporation has not been clarified.

The stimulation of ^{3}H-uridine incorporation into RNA by optimum concentrations of spermidine cannot be ascribed simply to increased feeding rates and enrichment of pyrimidine nucleotide pools. Larvae were labeled with ^{3}H-uridine for 5 hr in the presence and in the absence of 1 mM spermidine. Acid-soluble extracts of the larvae were prepared in 0.5 N PCA, and, after removal of the acid-insoluble residue by centrifugation, the nucleotides in the extract were adsorbed onto acid-washed Norite at 4°C and eluted with ammoniacal ethanol by the method of Tsuboi and Price (18). The eluate was concentrated *in vacuo,* and the nucleotides were separated by high-voltage paper electrophoresis by the procedure of Silver et al. (19). The uridine triphosphate (UTP) spot was well separated from all other nucleotides. After excision the paper was eluted with distilled water, and the absorbance at 260 nm and the radioactivity of the UTP fraction were determined. The specific activity of the ^{3}H-UTP from the spermidine-fed larvae was 55 dpm per O.D. unit; the specific activity of the nucleotide from zero spermidine controls was 48 dpm per O.D. unit. The small difference in the specific activity of the UTP pool cannot be invoked to explain the large stimulation of RNA synthesis by spermidine, and feeding rates are apparently unaffected by the presence of the polyamine in the medium.

D. Specificity of spermidine stimulation of RNA synthesis

The specificity of the effect of spermidine on RNA synthesis was tested with a metabolic precursor (putrescine) and a possible metabolic product (spermine) of the polyamine, as well as with diamines which do not have this precursor-product relationship (cadaverine and ethylenediamine). The results are summarized in Table 3. The specificity experiments, which were carefully controlled by the comparison of 72-hr larvae from the same larval cultures and by labeling the same number of animals under identical experimental conditions, clearly showed a biological preference for spermidine. Spermidine increased incorporation of ^{3}H-uridine into each of the RNA components when "fed" in the range of 10^{-4} to 10^{-2} M. The 0.1 M concentration of spermidine was inhibitory, in confirmation of an earlier result with 0.12 M spermidine in a long-term labeling (12-hr) experiment (see Table 2). Putrescine is quite inhibitory in the range 10^{-3} to 10^{-1} M, as is ethylenediamine. Cadaverine, on the other hand, is completely "inactive." Spermine is very toxic at high concentrations (0.1 M) and appears to stimulate RNA synthesis very slightly at the lowest concentration tested (10^{-4} M).

TABLE 3. *Specificity of the stimulation of RNA synthesis by spermidine*

Amine	Relative spec. act. 4S	18S	28S	Amine	Relative spec. act. 4S	18S	28S
None	100	100	100	None	100	100	100
0.1 M SD	67	61	72	0.1 M CAD	103	89	97
0.01 M SD	151	141	148	0.01 M CAD	92	94	88
0.001 M SD	132	148	155	0.001 M CAD	97	106	99
0.0001 M SD	141	139	135	0.0001 M CAD	99	103	105
None	100	100	100	None	100	100	100
0.1 M P	45	46	55	0.1 M ED	74	41	47
0.01 M P	58	55	68	0.01 M ED	79	87	100
0.001 M P	60	57	57	0.001 M ED	74	75	75
0.0001 M P	75	80	98	0.0001 M ED	93	108	100
None	100	100	100				
0.1 M S	51	13	23				
0.01 M S	92	71	85				
0.001 M S	104	106	110				
0.0001 M S	118	121	127				

Larvae were removed from the solid culture medium 72 hr after hatching and incubated for 3 hr in 0.25 ml of liquid "culture medium" in the presence of 25 μC of ^{3}H-uridine and the appropriate amine at the concentration specified. The conditions of isolation of RNA from washed larvae, sucrose gradient analysis, quantitation of RNA and radioactivity, and calculation of "relative spec. act." (specific activity relative to zero amine controls = 100) are described in Materials and Methods and in the legend of Table 2. (SD = spermidine; P = putrescine; S = spermine; CAD = cadaverine; ED = ethylenediamine).

E. Effect of spermidine on transcription at low ionic strength

The stimulation of transcription by spermidine has been reported with a number of different preparations of *E. coli* RNA polymerase assayed with native DNA at low ionic strength (12, 20, 21). The effect of spermidine on low ionic strength assays of the "complete" enzyme as prepared in our laboratory by the Burgess procedure (15) is shown in Table 4. Increased RNA synthesis is demonstrable in the range of 1 to 8 mM spermidine although the upper limit of stimulation in assays with high molecular weight DNA templates is uncertain because of precipitation induced by the polyamine. The effect of the amine is similar with templates for which sigma-factor dependence has been established, i.e., T4 phage DNA, and with templates on which "initiation" can proceed in the absence of sigma, i.e., calf thymus DNA.

The time course of transcription with supplements of spermidine to low ionic strength T4 DNA assays is illustrated in Fig. 3. The "plateau" kinetics of the polymerase reaction at low ionic strength have been extensively investigated (22), and Fuchs et al. (12) reported the elimination or reversal of

TABLE 4. *Effect of spermidine on transcription of different templates*

Concentration of spermidine (mM)	^{3}H-UMP incorporated (cpm per 20 μl)			
	T4 DNA	Calf thymus DNA	*E. coli* DNA	Poly [d(A-T)]
0	1,284	1,280	437	4,529
1	2,900	2,845	673	5,827
2	4,148	3,238	613	6,903
3	3,460	4,058	838	7,623
4	138[a]	3,386[a]	633	7,446
6	176[a]	639[a]	580[a]	7,962
8	275[a]	264[a]	327[a]	8,414

[a] Precipitate observed in assay mixture

The RNA polymerase assay is described in Materials and Methods. Incorporation of ^{3}H-UMP into RNA was measured after 3 hr of incubation by the filter paper disc assay on 20-μl aliquots from the 0.2 ml reaction mixtures.

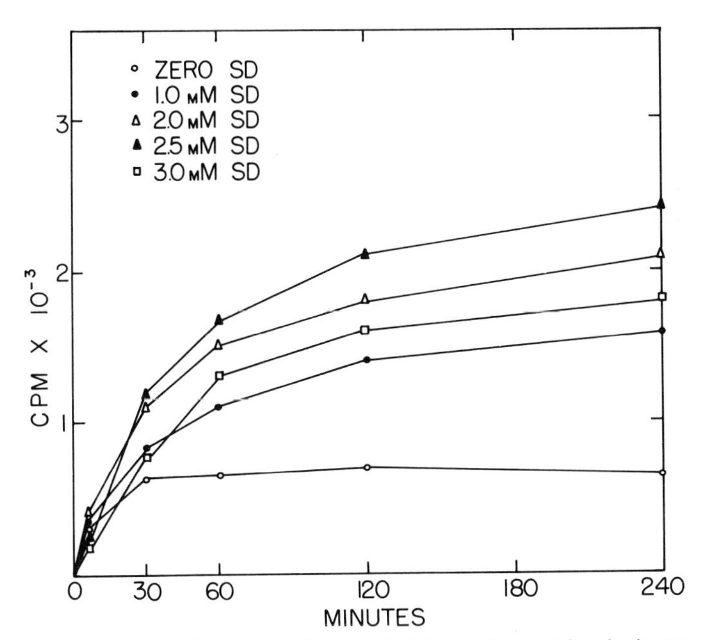

FIG. 3. The effect of spermidine on the kinetics of RNA synthesis at low ionic strength. The RNA polymerase assay medium (see Materials and Methods) was supplemented with variable concentrations of spermidine as indicated, and 10-μl aliquots were removed and analyzed by the filter paper disc method at intervals during the incubation period of 240 min at 37°C.

the plateau by salt supplied as monovalent or divalent ions (0.15 M NH_4Cl or 0.07 M Mg^{++} acetate) or by 8 mM spermidine. Under our assay conditions, nonplateau kinetics were achieved with 2.5 mM spermidine, and higher concentrations were inhibitory apparently because the high molecular weight T4 DNA was precipitated by the polyamine. We also obtained nonplateau kinetics as the ionic strength was increased with KCl or $MgCl_2$. Maximum rates of transcription are achieved and are maintained over at least 4 hr at 37°C with 0.3 M KCl, 80 mM $MgCl_2$, or 2.5 mM spermidine. Higher concentrations of KCl and $MgCl_2$ are inhibitory, but the precipitates seen with excessive concentrations of spermidine are not observed in assays inhibited by high salt.

F. Effect of spermidine on transcription at high ionic strength

The elimination of the plateau in transcription by the low (2.5 mM) concentration of spermidine relative to the effective concentration of salt (0.3 M KCl and 80 mM $MgCl_2$) suggested that the effect of spermidine might be

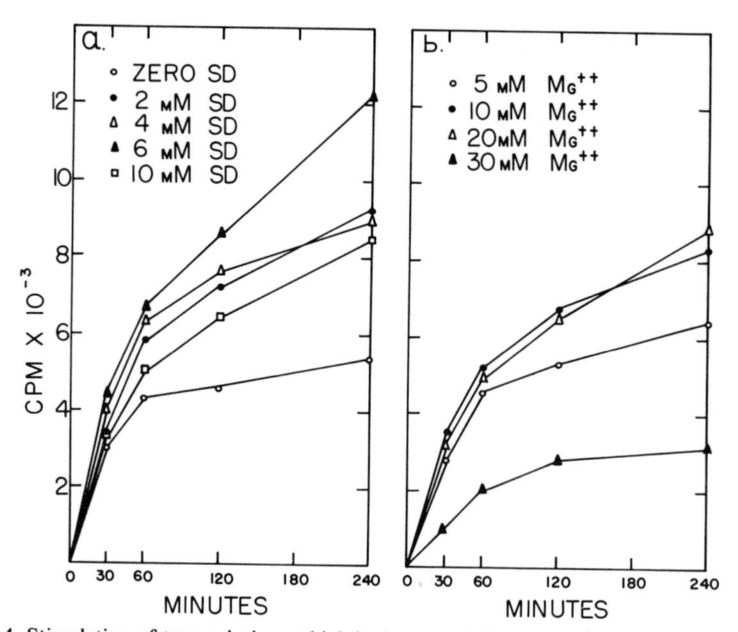

FIG. 4. Stimulation of transcription at high ionic strength by spermidine and magnesium. The RNA polymerase assay medium (see Materials and Methods) was supplemented with 0.3 M KCl and variable concentrations of (a) spermidine (SD) and (b) magnesium as indicated. Ten-μl aliquots were removed and analyzed by the filter paper disc method at intervals during the incubation period of 240 min at 37°C.

somewhat different than the stimulation of the reaction at elevated ionic strength. This interpretation is supported by the time course data of Fig. 4a. A very substantial stimulation of transcription by 2 to 10 mM spermidine occurs at each interval in the period 30 to 240 min of the RNA polymerase reaction at optimum ionic strength (0.3 M KCl). At 2 hr, the stimulation due to spermidine is approximately twofold. Thus the effect of the polyamine is additive and cannot be characterized entirely as contributing to the ionic-strength dependency of the reaction. A similar relationship between ionic strength and Mg^{++} stimulation of transcription is apparent in Fig. 4b. The divalent cation is stimulatory in the range of 5 to 20 mM $MgCl_2$, and there is nearly 50% stimulation of the reaction at the higher level of Mg^{++}. When optimum Mg^{++} (20 mM) and spermidine (2 to 6 mM) are added to assays at optimum ionic strength (0.3 M KCl), the rate of transcription is not increased above the maximum rate achieved with either Mg^{++} or spermidine added separately. It appears, therefore, that the polyamine functions similarly to Mg^{++} in stimulating transcription at optimum ionic strength. The organic cation effect is demonstrable at a lower molar concentration (2 vs. 20 mM), but spermidine does not replace the divalent cation requirement since transcription does not occur in the absence of Mg^{++} or Mn^{++} at any level of the polyamine (12).

G. Effect of congeners of spermidine on transcription

The effect on transcription of congeners of spermidine, including putrescine, the diamine precursor, spermine, a metabolic product, and cadaverine, the next higher diamine homologue of putrescine, was compared with spermidine in low salt and high salt assays (Table 5). Spermine was stimulatory at the lowest concentration, but, as with spermidine, it was impossible to establish the optimum level because of the problem of precipitation of the template by the polyamine in low ionic strength assays.

Putrescine and cadaverine were also active but at much higher concentrations (30 to 120 mM) in the low salt assay. The incorporation of ^3H-uridine monophosphate (UMP) into RNA relative to the low salt control was considerably greater with the diamines than with the polyamines. These incorporation data obtained with high concentrations of the diamines are difficult to relate to the effect of the polyamines at substantially lower concentrations. Diamines do not cause precipitation of the DNA template (as do the polyamines) and consequently can be studied over a wider concentration range.

Each of the compounds increased RNA synthesis in the high salt (0.3 M KCl) assay, and the maximum stimulation was achieved with the two poly-

TABLE 5. *Relative stimulation of transcription by polyamine congeners*

Amine in assay	Relative incorporation of ^{3}H-UMP		Amine in assay	Relative incorporation of ^{3}H-UMP	
	Low salt	High salt		Low salt	High salt
None	100 (0)	100 (0)	None	100 (0)	100 (0)
Spermidine	245 (1)	171 (2)	Putrescine	226 (30)	135 (10)
	382 (2.5)	260 (6)		662 (60)	169 (20)
	102 (3.5)	156 (10)		387 (120)	74 (30)
Spermine	263 (0.1)	137 (0.5)	Cadaverine	176 (30)	156 (10)
	370 (0.25)	220 (2)		317 (60)	121 (20)
	29 (0.5)	248 (6)		710 (100)	86 (30)

Numbers in parentheses indicate mM concentration of amine.

The low salt assay is the standard RNA polymerase assay (see Materials and Methods) with the T4 DNA template. The high salt assay contains 0.3 M KCl which provides the optimum ionic strength. Relative incorporation of ^{3}H-UMP is derived from [cpm (+Amine) ÷ cpm (−Amine)] × 100 at the end of the 4-hr incubation period.

amines. In this assay, incorporation was inhibited with 30 mM putrescine or cadaverine without evidence of precipitation of the template.

H. Mechanism of action of spermidine

In their analysis of T4 DNA transcription *in vitro*, Bremer and Konrad (23) reported that RNA molecules are not released from their complex with enzyme and DNA during the course of the reaction at low ionic strength. Richardson (22) showed that the complex does dissociate to some extent in low-salt reaction mixtures, whereas Millette and Trotter (24) reported that the major portion of the RNA product is released from the complex with equal facility in both low and high ionic strength assays. More recently, Witmer (16) presented data which confirmed the initial interpretation of Bremer and Konrad of product release at high ionic strength and failure to release RNA from the DNA-enzyme–RNA complex at low ionic strength. Each of these studies was restricted to comparisons of low-salt and high-salt assays.

We have reexamined the question of RNA product release throughout the incubation period (0 to 120 min) of transcription of T4 DNA at low ionic strength, and we have tested the effect of an optimum concentration of spermidine (2.5 mM) on this phase of the reaction. Both the nitrocellulose filter technique (25) for the determination of complex-bound RNA employed by Witmer (16) and sucrose-gradient analysis of reaction mixtures utilized by Witmer as well as Millette and Trotter (24) were adapted to

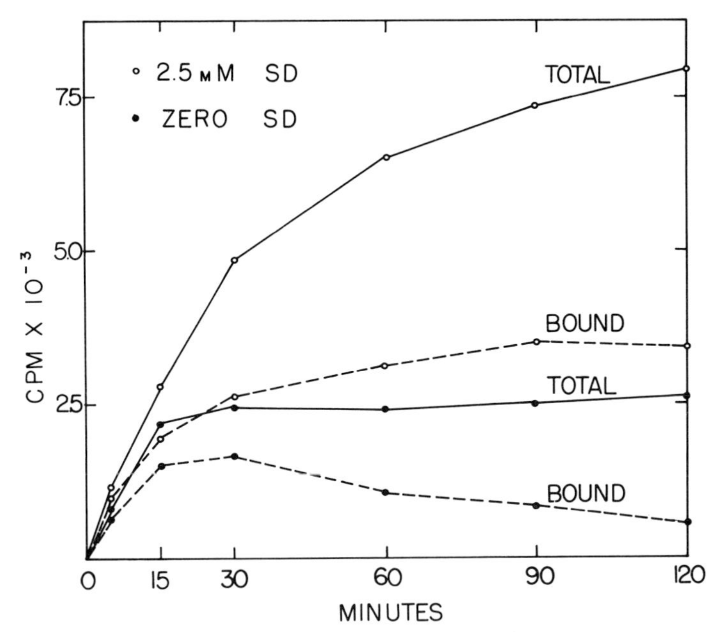

FIG. 5. Maintenance of an active transcription complex by spermidine. RNA polymerase assays (Materials and Methods) unsupplemented (zero SD) or supplemented with spermidine (2.5 mM SD) were analyzed by the removal of 10-μl aliquots at intervals during the incubation period of 120 min at 37°C. The aliquots were subjected to either the filter paper disc method for the determination of total ^{3}H-UMP incorporation into RNA or the nitrocellulose filter technique for the determination of bound ^{3}H-RNA in the transcription complex of RNA, DNA, and RNA polymerase retained by the filter.

studies on the mechanism of the effect of spermidine on transcription at low ionic strength.

Experiments which involved the nitrocellulose filter techniques are illustrated in Fig. 5. The effect of 2.5 mM spermidine on total RNA synthesis at low ionic strength is indicated by the elimination of the "plateau" kinetics and the threefold stimulation of incorporation during the 2-hr reaction period. There is a clear indication that the polyamine maintains an active transcription complex throughout the course of the reaction since the maximum level of filter-bound labeled RNA achieved at approximately 30 min is *maintained* for the entire 2-hr period. In the absence of spermidine, the low ionic strength assay has a lower and a diminishing transcription complex as is evident from (a) the reduced fraction of labeled RNA in a complex with DNA and enzyme at 30 min and (b) the decline throughout the remainder of the 2-hr period of the nascent RNA in this complex.

Further evidence for the involvement of spermidine in the maintenance of

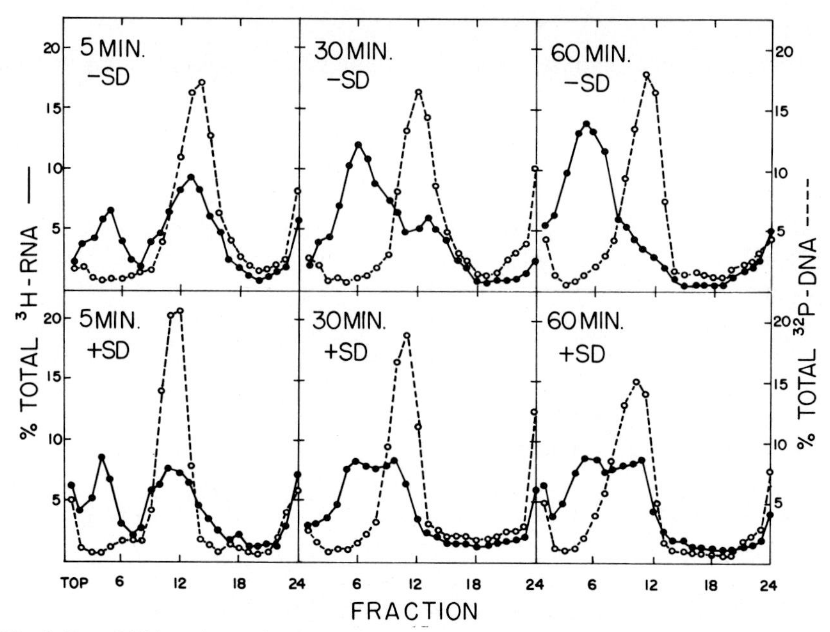

FIG. 6. Free RNA and complex-bound RNA during transcription in the presence or absence of spermidine. RNA polymerase assays (see Materials and Methods) containing T4 ^{32}P-DNA templates and either unsupplemented (−SD) or supplemented with 2.5 mM spermidine (+SD) were analyzed by the removal of aliquots of the assay mixture (after 5, 30, and 60 min at 37°C); these aliquots were diluted 1:5 with gradient buffer and layered on 5 to 20% sucrose gradients prior to centrifugation for 105 min at 4°C and 35,000 rpm in the SW-50 rotor. The gradients were fractionated into 0.25-ml aliquots in which ^{3}H-RNA and ^{32}P-DNA were precipitated in the presence of 200 μg BSA and collected on glass fiber filters. The washed, air-dried filters were analyzed in toluene phosphor counting fluid.

an active transcription complex at low ionic strength was obtained by sucrose-gradient centrifugation of reaction mixtures containing T4 ^{32}P-DNA templates. In Fig. 6 the fraction of free RNA to complex RNA (RNA associated with T4 ^{32}P-DNA template and presumably in a ternary complex with the enzyme) is compared at different time intervals in the absence and presence of spermidine. There is complete confirmation of the participation of spermidine in the maintenance of an active transcription complex as indicated by the constant fraction of nascent RNA in the complex with T4 DNA; in the absence of spermidine, there is rapid loss of nascent RNA from the complex which is essentially complete in 60 min.

The combined data strongly indicate that spermidine is an effective molecule in the prevention or reversal of the disruption of the active transcription complex. Furthermore, the data support the evidence presented by Millette and Trotter (24) that RNA is readily released irrespective of the suboptimum ionic strength which limits the continuation of RNA synthesis.

IV. DISCUSSION

The incorporation of ^{3}H-uridine into RNA by *Drosophila* larvae is increased by supplementing the culture medium with 10 mM spermidine and inhibited by 10 mM putrescine. Larvae feeding on the amine-supplemented medium retain the amines in excess of the concentrations found in control animals. The enrichment level of spermidine is very similar to the highest concentrations of the polyamine reported in an earlier analytical study on *Drosophila* developmental stages (2). The enrichment of the larvae in putrescine, on the other hand, is considerably above the normal concentrations reported in the same study.

The increased incorporation of ^{3}H-uridine by the spermidine-enriched larvae occurs in a cellular environment of spermidine which is, therefore, in the high normal range. The inhibition of RNA synthesis in the putrescine-enriched larvae results from what appears to be an excessive retention of putrescine (Table 1). Perhaps more importantly, the ratio of spermidine to putrescine in these animals decreased from a normal ratio of approximately 5:1 (in 48-hr larvae) or 3:1 (72-hr larvae) to 0.75:1 and 0.5:1, respectively, in the putrescine-supplemented animals.

Putrescine antagonism of spermidine has been cited previously in microbial systems which respond to polyamine regulation or which are nutritionally dependent upon polyamines.

Raina et al. (25) demonstrated putrescine inhibition of the relaxation of stringent control of RNA synthesis by spermidine in *E. coli* 15 TAU deprived of arginine. Inouye and Pardee (26) presented evidence that competition between putrescine and spermidine is an important factor in bacterial cell division. They proposed that a specific ratio of putrescine to spermidine must be attained prior to the initiation of the event of cell division. Another study (27) showed that the induction of morphogenetic events in the production of fruiting bodies and refractile microcysts by *Micrococcus xanthus* is regulated by the spermidine:putrescine ratio in media on which this saprophytic bacterium is cultured. Putrescine (0.05 M) induced the formation of microcysts; spermidine, at a concentration of 0.002 M, caused a 10-fold reduction in putrescine-induced microcysts. With 0.005 M spermidine, an even greater reduction is achieved. Several earlier observations in our laboratory suggest a similar modulation of *Drosophila* growth and function by regulated ratios of spermidine to putrescine.

Dion (28) cited the very excellent agreement between high rates of RNA synthesis and high spermidine-to-putrescine ratios. Low ratios accompanied the low rates of RNA synthesis in 48-hr larvae and in middle to late pupal stages. High ratios, on the other hand, persisted in all other stages of active RNA synthesis. Adult females have high (6:1) ratios of spermidine:putrescine whereas adult males store essentially equivalent molar ratios of the two amines. Investigations by Balazs and Haranghy (29) demonstrated that females undergo an extensive maturation process involving RNA and DNA synthesis which lasts for weeks. No such phenomenon occurs in males.

These examples of experimental observations in which important regulatory functions appear to be associated with precise polyamine ratios serve to indicate the significance of many biological phenomena which are accompanied by fluctuating patterns of polyamine metabolism. The regulation of polyamine synthesis and excretion or modification would seem to offer unique and potentially wide-ranging opportunities for the control or disruption of normal growth processes. The numerous correlations between polyamine patterns and abnormal growth and the possible implications for the regulation of cancer cited in this volume are evidence of the continuing significance of investigations of polyamine function.

The biological effects of polyamines on *in vivo* processes of RNA synthesis in *Drosophila* and in the chick embryo (30, 31) would seem to be related to the *in vitro* stimulation by spermidine of RNA polymerases purified from several eukaryotic tissue or cell sources (32, 33). The precise relationship between our results on spermidine stimulation of RNA synthesis by intact *Drosophila* larvae and by *E. coli* RNA polymerase is more tenuous.

There are multiple forms of eukaryotic RNA polymerases (34) which have characteristic ionic-strength and divalent-cation requirements. The different enzymes have specific nucleolar and nucleoplasmic distributions, thus imposing still another barrier and variable for the possible regulation of RNA synthesis by polyamines in eukaryotes.

In those instances in which spermidine stimulation of purified eukaryotic RNA polymerases is observed, however, the effect of the polyamine in the enzyme assay is remarkably similar to the polyamine activity in assays of bacterial polymerases. For example, spermidine and spermine are stimulatory at lower concentrations than putrescine (32), and the effect of polyamines is demonstrable only in assays primed with native DNA. There is reason to expect, therefore, that the favorable influence of polyamines on transcription of native DNA templates by prokaryotic polymerases might be equally significant in eukaryotic transcription.

Very recently Schekman et al. (35) reported yet another exciting involvement of polyamines in macromolecular synthesis. Spermidine and, to a lesser extent, spermine, and putrescine stimulate the conversion of phage ϕX 174 single-strand DNA to its double-strand replicative form. This study established that DNA synthesis with purified enzyme preparations shows an invariable and almost absolute requirement for spermidine. This polyamine dependence is probably explicable by the requirement for an RNA primer to initiate DNA synthesis.

The effect of polyamines on transcription will possibly have even a greater impact in its indirect role on the initiation of DNA synthesis by RNA priming. The implications for the functional involvement of polyamines for effective cancer regulation are certainly impressive, and continuing investigative efforts directed toward the mechanism of action of polyamines in macromolecular synthesis seem wholly justified.

V. SUMMARY

D. melanogaster larvae were cultured on defined medium containing variable concentrations of spermidine or putrescine. Seventy-two-hour larvae accumulated the amines from media containing 1 to 10 mM spermidine or putrescine and were enriched with respect to control animals incubated in amine-free cultures. The incorporation of ^{3}H-uridine into RNA of 72-hr larvae was increased threefold by 1 to 10 mM spermidine, whereas 1 to 10 mM putrescine inhibited RNA synthesis. The effect of spermidine and putrescine was demonstrable in all labeled molecular species of larval RNA separated by sucrose density-gradient centrifugation. DNA-dependent RNA polymerase of *E. coli* was stimulated in the transcription of a variety of double-strand DNA templates by the addition of polyamines to either low or high ionic strength assays. Putrescine (60 mM) and cadaverine (100 mM) were equivalent to KCl (0.3 M) in the stimulation of transcription at low ionic strength. Spermine (0.25 mM) and spermidine (2.5 mM) increased RNA synthesis three- to fourfold at low ionic strength, and 2 to 6 mM concentrations of the polyamines doubled RNA synthesis at optimum ionic strength. Plateau kinetics of the RNA polymerase reaction were eliminated by either salt ($MgCl_2$ or KCl) or spermidine. RNA was released from the transcription complex of DNA-enzyme-RNA at low ionic strength, and transcription was terminated after 30 min. The addition of spermidine (2.5 mM) to low ionic strength assays maintained an active transcription complex with the continuation of RNA synthesis and release for 4 hr.

ACKNOWLEDGMENT

This work was supported by U.S. Public Health Service grant AI-05397 from the National Institute of Allergy and Infectious Diseases and by Hatch grant H-170. This article is published with the approval of the Director of the New Hampshire Agricultural Experiment Station as Scientific Contribution No. 648.

REFERENCES

1. Bodenstein, D., in M. Demerec (Editor), *Biology of drosophila,* Haftner Publishing Company, Inc., New York, 1965, p. 356.
2. Herbst, E. J., and Dion, A. S., *Fed. Proc.,* **29,** 1563 (1970).
3. Heby, O., *Insect Biochem.,* **2,** 13 (1972).
4. Dion, A. S., and Herbst, E. J., *Proc. Nat. Acad. Sci. U.S.A.,* **58,** 2367 (1967).
5. Raina, A., *Acta Physiol. Scand.,* **60,** Suppl. 218, 1 (1963).
6. Dykstra, W. G., and Herbst, E. J., *Science,* **149,** 428 (1965).
7. Russell, D., and Snyder, S. H., *Proc. Nat. Acad. Sci. U.S.A.,* **60,** 1420 (1968).
8. Jänne, J., Raina, A., and Siimes, M., *Biochim. Biophys. Acta,* **166,** 419 (1968).
9. Russell, D. H., *Proc. Nat. Acad. Sci. U.S.A.,* **68,** 523 (1971).
10. Krakow, J. S., *Biochim. Biophys. Acta,* **72,** 566 (1963).
11. Fox, C. F., and Weiss, S. B., *J. Biol. Chem.,* **239,** 175 (1964).
12. Fuchs, E., Millette, R. L., Zillig, W., and Walter, G., *Europ. J. Biochem.,* **3,** 183 (1967).
13. Pearl, R., Allen, A., and Penniman, W. B. D., *Amer. Naturalist,* **60,** 347 (1926).
14. Greenberg, J. R., *J. Mol. Biol.,* **46,** 85 (1969).
15. Burgess, R. R., *J. Biol. Chem.,* **244,** 6160 (1969).
16. Witmer, H. J., *Biochim. Biophys. Acta,* **246,** 29 (1971).
17. Thomas, C. A., and Abelson, J., in G. L. Cantoni and D. R. Davies (Editors), *Procedures in nucleic acid research,* Harper and Row, New York, 1966, p. 553.
18. Tsuboi, K. K., and Price, T. D., *Arch. Biochem. Biophys.,* **81,** 223 (1959).
19. Silver, M. J., Rodalewicz, I., Douglas, Y., and Park, D., *Anal. Biochem.,* **36,** 525 (1970).
20. Abraham, K. A., *Europ. J. Biochem.,* **5,** 143 (1968).
21. Petersen, E. E., Kröger, H., and Hagen, O., *Biochim. Biophys. Acta,* **161,** 325 (1968).
22. Richardson, J. P., in J. N. Davison and W. E. Cohn (Editors), *Progress in nucleic acid research and molecular biology,* Academic Press, New York, 1969, p. 75.
23. Bremer, H., and Konrad, M. W., *Proc. Nat. Acad. Sci. U.S.A.,* **51,** 801 (1964).
24. Millette, R. L., and Trotter, C. D., *Proc. Nat. Acad. Sci. U.S.A.,* **66,** 701 (1970).
25. Raina, A., Jansen, M., and Cohen, S. S., *J. Bacteriol.,* **94,** 1684 (1967).
26. Inouye, M., and Pardee, A. B., *J. Bacteriol.,* **101,** 770 (1970).
27. Witkin, S. S., and Rosenberg, E., *J. Bacteriol.,* **103,** 641 (1970).
28. Dion, A. S., Ph.D. dissertation, University of New Hampshire, 1968.
29. Balazs, A., and Haranghy, L., *Acta Biol.,* **15,** 343 (1965).
30. Moruzzi, G., Barbiroli, B., and Caldarera, C. M., *Biochem. J.,* **107,** 609 (1968).
31. Caldarera, C. M., Moruzzi, M. S., Rossoni, C., and Barbiroli, B., *J. Neurochem.,* **16,** 309 (1969).
32. Frederick, E. W., Maitra, U., and Hurwitz, J., *J. Biol. Chem.,* **244,** 413 (1969).
33. Ballard, P. L., and Williams-Ashman, H. G., *J. Biol. Chem.,* **241,** 1602 (1966).
34. Roeder, R. G., and Rutter, W. J., *Nature,* **224,** 234 (1969).
35. Schekman, R., Wickner, W., Westergaard, O., Brutlag, D., Geider, K., Bertsch, L., and Kornberg, A., *Proc. Nat. Acad. Sci. U.S.A.,* **69,** 2691 (1972).

Polyamines in Normal and Neoplastic Growth. Edited by
D. H. Russell. Raven Press, New York © 1973.

Influence of Polyamines on the Hydrolysis of Polynucleotides by *Citrobacter* Ribonuclease

Carl C. Levy, William E. Mitch, and Morton Schmukler

Section of Enzymology and Drug Metabolism, Laboratory of Pharmacology, Baltimore Cancer Research Center, National Cancer Institute, Baltimore, Maryland 21211

I. INTRODUCTION

The polyamines have been implicated in a variety of enzyme reactions centering around RNA metabolism (1, 2). Evidence has been presented, in fact, that RNA polymerase I activity is stimulated to a great extent by spermidine (3). Although the concentration of RNA or, for that matter, most constituents of a cell, can be thought of as one of dynamic equilibrium between what is synthesized and what is degraded, little attention has been focused on the effects of these cations on the degradative process, i.e., on RNase activity. The reports that have addressed themselves to this area have dealt either with bovine pancreatic RNase or with an alkaline RNase obtained from liver extracts (4, 5). In general, the polyamines spermine and spermidine were found to inhibit the activities of both enzymes, particularly with respect to the digestion of larger substrates such as Poly A and RNA (4). On the other hand, pancreatic RNase activity toward smaller substrates, CpA and C-cyclic-p, were stimulated by the two cations. To a considerable extent, the activity of the crude liver RNase could be stimulated also, although in this case increased activity was found with cadaverine or putrescine (5).

In an attempt to extend these observations, the effects of the polyamines on a microbial RNase were examined. The enzyme, isolated from the soil organism *Citrobacter sp.*, has been found to exhibit a great deal of selectivity for phosphodiester linkages containing uridylic acid residues (6). Preliminary studies indicated that the specificity exhibited by the enzyme toward uridylic acid bases could be altered in response to spermidine with an accompanying enhancement of enzyme activity. In this chapter we wish to report the effects that these cations have on both the activity and specificity of the *Citrobacter* enzyme.

II. EXPERIMENTAL PROCEDURE

The sources of yeast RNA and ribonucleotide polymers as well as the methods used in their purification have been described previously (7). All polyamines were purchased from K & K Laboratories (Plainview, N.Y.). $^{14}C_2$-Spermidine and ^{14}C-putrescine were purchased from New England Nuclear (Boston, Mass.); Dowex-1 resin AG-1-X8 (200 to 400 mesh in the chloride form) was purchased from Bio-Rad Laboratories (Richmond, Calif.). CNBr-activated Sepharose-4B was obtained from Pharmacia (Piscataway, N.J.).

III. METHODS

A. Isolation and growth of bacteria

The organism *Citrobacter sp.* was isolated from garden soil by enrichment culture in which the sole source of carbon and nitrogen was 0.1% poly-uridylic acid. After isolation the bacteria were transferred from a slant to 100 ml of medium of a composition previously described (6). The culture was allowed to grow at 30°C on a rotary shaker and then transferred as a 10% inoculum to a larger flask of the same medium. Cells were allowed to grow until stationary phase (16 hr), then harvested by centrifugation (Sharples). The yield of cells, approximately 1 g (wet weight) per liter, was kept frozen at −20°C until used.

B. Purification of the enzyme

The methods used for purification of the enzyme from *Citrobacter sp.* have been described elsewhere (6). Unless stated otherwise, all studies were performed with an enzyme preparation (specific activity 4,000 units per mg of protein) obtained in the final step of the purification procedure (6). An enzyme unit is defined as the amount of activity required to increase the absorbance at 260 nm by 0.1 O.D. unit under the conditions of the assay (6).

C. Reaction of *Citrobacter* nuclease with Poly ACGU

The hydrolysis of Poly ACGU by the *Citrobacter* nuclease was carried out by incubating at 37°C a standard reaction mixture containing 4 mg of

Poly ACGU, 100 μmoles of Tris-HCl buffer, pH 7.5, and 10 units of enzyme in a volume of 1 ml. When the effect of spermidine on the reaction was to be studied, 1 μmole of the polyamine was added to a similar reaction mixture. After 4 hr incubation at 37°C, the reaction was stopped by the addition of 1.0 ml of a water-saturated solution of phenol and the mixture was shaken for 10 min (8). The aqueous and phenolic layers were separated by centrifugation, and the digestion products within the aqueous layer were separated and purified by chromatographic systems described in detail elsewhere (7). Similarly, the methods employed in the characterization of these products also have been described (7).

D. Preparation of a Sepharose-Poly C column

The column was prepared as described by Poonian et al. (9) except that 1 g of CNBr-activated Sepharose 4B was suspended in 5 ml of 0.05 M phosphate buffer, pH 8.0. This suspension was added to 40 μmoles of Poly C dissolved in 5 ml of 0.05 M phosphate buffer, pH 8.0, and the mixture stirred for 48 hr at 4°C. The reaction slurry was then poured into a column (0.9 × 2.0 cm) and washed with a sufficient amount of 0.05 M phosphate buffer (50 ml) until noncovalently bound Poly C could no longer be detected in the washings. The difference between the amount of Poly C used in the original slurry and the amount washed from the column indicated the amount of Poly C which had been covalently coupled to Sepharose. This method, essentially, was followed in binding other synthetic polyribonucleotides to Sepharose.

E. Binding of spermidine to a Sepharose-Poly C column

To a slurry of activated Sepharose containing 3.8 μmoles of covalently bound Poly C, 0.5 ml of $^{14}C_2$-spermidine trihydrochloride (specific activity 300,000 cpm/μmole) was added. After being shaken for 24 hr, the slurry was washed with sufficient 0.01 M Tris-HCl buffer, pH 7.5, until radioactivity could no longer be detected in the washings. Based on the counts within the wash fluid, 0.43 μmole of spermidine was bound to Poly C.

IV. RESULTS

In an earlier study (6), the *Citrobacter* nuclease was purified to homogeneity and found to have an absorption spectrum of a typical protein with an absorbance maximum and minimum of 281 nm and 248 nm respectively. A solution of the nuclease containing 1 mg of protein per ml exhibits an ab-

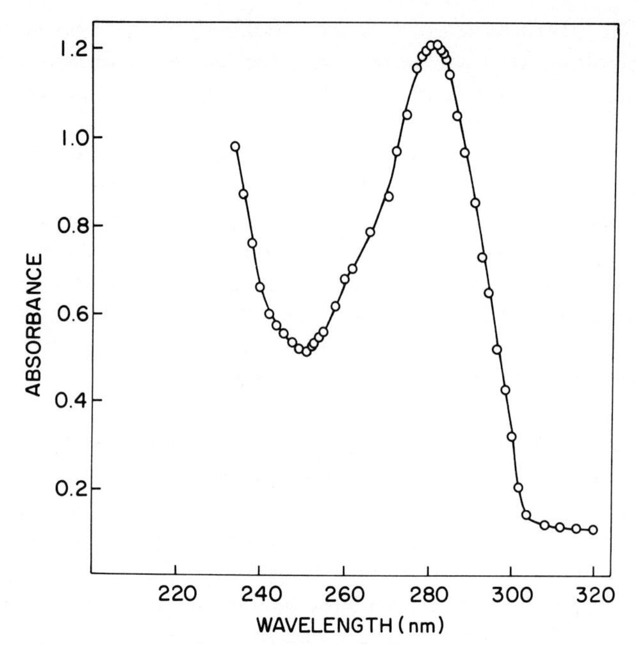

FIG. 1. Absorption spectrum of *Citrobacter* nuclease (1.0 mg of protein per ml) in 0.1 M Tris-HCl buffer, pH 7.5. Several enzyme preparations purified by affinity chromatography on a Sepharose-Poly G column (6) were dialyzed and then reduced in volume by lyophilization until the requisite protein concentration was reached. The spectrum of the enzyme was then recorded.

TABLE 1. *Effect of buffer concentration on hydrolysis of Poly U by* Citrobacter *nuclease*

Buffer concentration (μmoles/ml)	U-cyclic-p released (μmoles)		
	10 min	30 min	60 min
10	0.25	0.51	0.82
100	0.52	0.86	1.80
200	0.58	0.91	1.74

The reaction mixtures contained 2.0 μmoles of Poly U, an amount of Tris-HCl buffer, pH 7.5 as indicated below, and 3 units of enzyme in a volume of 1 ml. Incubation at 37°C was terminated by the addition of 0.05 ml of concentrated NH_4OH at times indicated, and the amount of U-cyclic-p released was estimated by chromatography on Dowex-1 resins (7).

sorbance of 1.21 at 280 nm and has a 280:260 nm absorbance ratio of 1.74 (Fig. 1). The enzyme was shown to have a distinct propensity for internucleotide linkages containing uridylic acid at either end of the phosphodiester bond (6). Moreover, the rate of nuclease activity toward yeast RNA was found to be dependent upon the molarity of the buffer used in the reaction mixture (6). As is apparent from Table 1, the rate of degradation of Poly U, when judged by the release of U-cyclic-p, is similarly dependent upon the molarity of the buffer. When the buffer concentration was raised from 0.01 to 0.1 M, a twofold increase in the rate of release of U-cyclic-p could be recorded (Table 1). Increasing the molarity of the buffer beyond this point, i.e., from 0.1 to 0.2 M, did not appear to alter the reaction rate significantly, since the amount of U-cyclic-p released in each case was the same.

Aside from responding to the ionic strength of the reaction medium, the enzyme activity toward Poly U is affected by the presence of polyamines. The addition, for example, of 0.5 μmole of spermidine to an incubation mixture in which the buffer concentration was kept at 0.01 M caused a release of U-cyclic-p equivalent in amount to that seen when the buffer concentration was some 10-fold higher (i.e., 0.1M) (Table 2). Under similar conditions, the addition of the same amount of putrescine can also increase the rate of enzyme activity, but not to the same extent as spermidine does. If, indeed, the buffer concentration was raised to 0.1 M, the hydrolytic activity in the presence of spermidine was increased by an additional 80% (Table 2). Although the increase in the rate of hydrolysis can be mediated by other polyamines (Table 3), spermidine appears to have the greatest effect. Other cations tested, however, with the exception of the ammonium ion, had no discernible influence on the release of U-cyclic-p.

TABLE 2. *Effect of polyamines on hydrolysis of Poly U in different buffer concentrations*

Buffer concentration (μmoles/ml)	Additions		
	None	Spermidine	Putrescine
10	0.43	0.87	0.68
100	0.91	1.60	1.32

Conditions were the same as in Table 1 except that two of the reaction mixtures at each buffer concentration contained either 0.5 μmole of spermidine or putrescine. After incubation at 37°C for 30 min, the reactions were stopped and U-cyclic-p was estimated as described previously (7).

TABLE 3. *Effect of polyamines and other cations on hydrolysis of Poly U by* Citrobacter *nuclease*

Additions	U-cyclic-p released (μmole)
None	0.82
1,3 Diaminopropane	1.24
1,4 Diaminobutane (putrescine)	1.20
1,5 Diaminopentane (cadaverine)	1.35
Spermidine	1.58
NH_4^+	1.25
Mg^{++}	0.80
Mn^{++}	0.85
Ca^{++}	0.76
Zn^{++}	0.84
Co^{++}	0.88

To the standard reaction mixture (1 ml) consisting of 2.0 μmoles of Poly U, 100 μmoles of Tris-HCl buffer, pH 7.5, and 3 units of *Citrobacter* nuclease, 0.5 μmole of cation in the form of its chloride salt was added. Incubation at 37°C was stopped after 30 min by the addition of 0.05 ml of concentrated NH_4OH and the amount of U-cyclic-p was measured as described elsewhere (7).

A. Alteration in enzyme specificity by spermidine

In addition to an increase in the rate of reaction, the nature of the hydrolytic attack on polynucleotides is affected by spermidine. It was shown earlier (6) that the *Citrobacter* enzyme is quite specific for Poly U and does not attack Poly C, Poly A, or Poly G. In the presence of spermidine, however, Poly C is hydrolyzed to a considerable extent (Table 4). The alteration in enzyme specificity as judged by the hydrolysis of Poly C seems to be unique to spermidine, since other polyamines or other cations do not mediate the hydrolysis of the cytidylic acid polymer. Alterations in enzyme specificity were not confined solely to the hydrolysis of Poly C, however: copolymers in which cytidylic acid was a component part were also degraded in the presence of spermidine (6).

Although neither Poly A nor Poly G was attacked by the *Citrobacter* nuclease under any conditions, spermidine conferred on the enzyme a considerably greater degree of random behavior than was at first appreciated in light of the observations made with the purine polymers. Structural analysis of fragments released over the course of hydrolysis of Poly ACGU demonstrated this clearly. In the absence of spermidine, the products are either U-cyclic-p or terminate in uridylic acid, whereas in the presence of the polyamine the products terminate in any one of the four nucleotides (Table 5).

TABLE 4. *Effect of spermidine on* Citrobacter *nuclease activity toward Poly C*

Conditions	C-cyclic-p released (μmoles)		
	10 min	30 min	60 min
Standard reaction mixture	0.0	0.0	0.0
" " " + spermidine	0.12	0.37	0.82

Standard reaction conditions were as described in Table 3 except that 2.0 μmoles of Poly C were used as substrate. When the effect of spermidine was to be tested, 0.5 μmole of the polyamine was added to the standard reaction mixture. Incubation at 37°C was stopped at the times indicated by the addition of 0.05 ml of concentrated NH_4OH and the amount of C-cyclic-p released was estimated by chromatography on Dowex-1 resins (7).

TABLE 5. *Products of the digestion of Poly ACGU by* Citrobacter *nuclease in the presence and absence of spermidine*

Conditions	Products[a]
Standard reaction mixture	U-cyclic-p, 1.3%; ApUp, 0.5%; GpUp, 0.3%; ApCpUp, 0.6%
Standard reaction mixture plus spermidine	U-cyclic-p, 2.5%; C-cyclic-p, 0.6%; A-cyclic-p, 0.4%; G-cyclic-p, 0.1%; ApUp, 0.6%; ApCp, 0.2%; ApGp, 0.2%; ApCpUp, 0.8%; ApGpAp, 0.3%

Reaction conditions were as described in Experimental Procedure. The methods used in the separation and characterization of the products have been described elsewhere (7).

[a] The amount of material in the product was estimated from the spectral data of Stanley and Bock (11) and of Warshaw and Tenoco (12) and is presented as a percentage of the initial Poly ACGU.

B. Behavior of Citrobacter nuclease on Sepharose-Poly C columns

In an attempt to explain the broadened specificity of the enzyme in the presence of spermidine, the relationship between the binding properties of Poly C, enzyme, and polyamine was examined. It was reasoned that the enzymic hydrolysis of Poly C in the presence of spermidine may be the result of a change in the binding capacity of the enzyme for the polymer. To test this possibility, the method of affinity chromatography was employed. If, for example, Poly C were covalently bound to an activated Sepharose column (see Experimental Procedure), then in the absence of spermidine, the enzyme should pass rapidly through such a column with no evidence of enzyme retention by binding to Poly C. If, on the other hand, spermidine had been bound previously to Poly C, then the enzyme should be retained, or at least delayed in its passage through the column. When 500 units of enzyme were added to a Sepharose-Poly C column, the enzyme passed through quickly and was recovered quantitatively in the first 5 ml of effluent. In contrast, enzyme activity was recovered over some 40 ml of effluent material after passage through a Sepharose-Poly C-spermidine column. Contained within the enzyme eluate from this latter column was a large amount of C-cyclic-p. Although the enzyme was not retained in the strict sense of the word, its passage through the column was certainly delayed. Presumably, as the enzyme passed through the column, Poly C was degraded, releasing both the enzyme and C-cyclic-p. As might be anticipated, the nuclease was

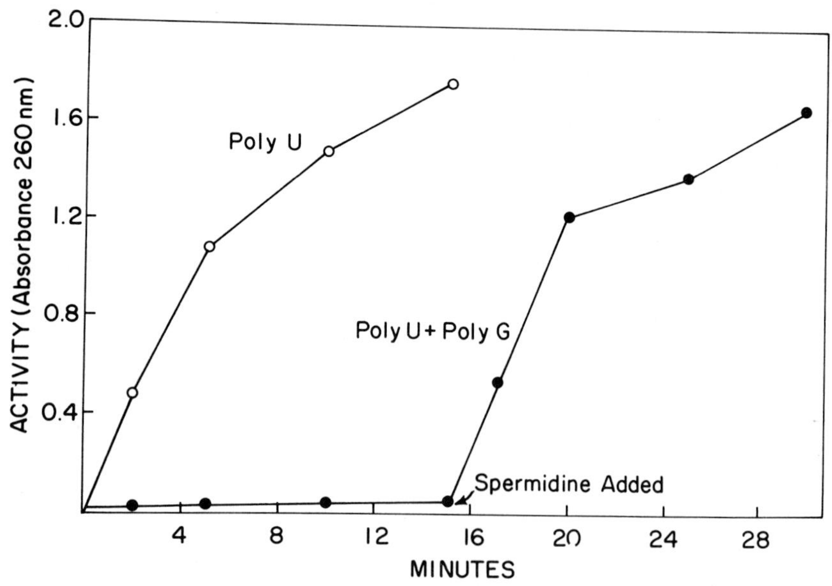

FIG. 2. Reversal by spermidine of the inhibition of *Citrobacter* nuclease activity by Poly G. The hydrolysis of Poly U was followed in a set of reaction mixtures, each of which consisted of 0.6 μmole of Poly U, 100 μmoles of Tris-HCl buffer, pH 7.5, and 3 units of enzyme in a volume of 1 ml. After incubation at 37°C, a reaction was stopped at the time intervals indicated, by the addition of 1 ml of 12% perchloric acid containing 20 mM lantinum nitrate. After cooling in an ice bath for 20 min, the cloudy mixture was clarified by centrifugation and the acid-soluble nucleotides measured at 260 nm (10). A second set of reaction mixtures similar to the first, except that each contained 0.5 μmole of Poly G, was incubated simultaneously, and enzyme activity was measured at appropriate intervals. After 15 min, 0.5 μmole of spermidine was added to each of the remaining second group of reaction mixtures and enzyme activity assayed. ○—○, Poly U; ●—●, Poly U and Poly G.

neither delayed in its passage through the column nor was there any degradation of the polymer when putrescine was substituted for spermidine in a similar type of experiment.

C. Inhibition of Citrobacter nuclease activity by synthetic polyribonucleotides

The *Citrobacter* enzyme has been shown to be inhibited by relatively low concentrations of Poly G, as well as by the complex formed between Poly A and Poly U (6). Binding studies between polymer, enzyme, and spermidine conducted on a Sepharose-Poly G column demonstrated that spermidine could prevent the binding of the enzyme to Poly G (6). Moreover, if in the absence of the polyamine, the enzyme was bound to the polymer, then it could be released quantitatively by the addition of spermidine.

The role that spermidine plays in changing the binding properties between enzyme and polymer also could be demonstrated by following the hydrolysis of Poly U in the presence and absence of inhibitor. As can be seen in Fig. 2, the digestion of Poly U is inhibited completely in the presence of Poly G. After 15 min, addition of spermidine to the inhibited reaction mixture restores activity completely and the hydrolysis of Poly U proceeds in a manner closely paralleling the reaction seen in the absence of Poly G. Experiments of this kind extending for 24 hr gave essentially the same results.

Substitution of putrescine for spermidine under similar experimental conditions restored about 30% of enzyme activity.

V. DISCUSSION

In contrast to the observations made by previous workers with bovine pancreatic ribonuclease (4) and alkaline ribonuclease (5), *Citrobacter* nuclease activity is stimulated to a great extent by polyamines. The stimulation occurs whether the substrate is yeast RNA (6) or Poly U.

Evidence presented elsewhere (6) indicated that the concentration of the polyamine is of critical importance in determining the effect on enzyme activity. Thus, at concentrations approaching 10^{-2} M (a concentration of polyamine which was used with both mammalian systems), *Citrobacter* enzyme activity is severely inhibited (6). The inhibition seen with the mammalian nucleases may be caused by the high concentration of polyamine employed.

Since *Citrobacter* nuclease activity is dependent on the ionic strength of the reaction medium, an argument could be made that the addition of the polyamines simply raises the ionic strength and thus enhances the activity of the enzyme. Several lines of evidence, however, make this argument spurious. First, enzyme activity after the addition of 0.5 μmole of spermidine to an incubation mixture containing 0.01 M Tris-HCl buffer is raised to the same extent as if the buffer concentration had been increased 10-fold. Spermidine at the concentration used could raise the ionic strength of the medium only a fraction of the percent that a tenfold increase in buffer concentration did. Second, at buffer concentrations which in terms of ionic strength are optimal for enzyme activity, spermidine can stimulate hydrolytic rates an additional 80%. Finally, with the exception of the ammonium ion, attempts at stimulating enzyme hydrolysis with other cations (e.g., Mg^{++}, Mn^{++}) at concentrations at which the polyamines were effective proved fruitless.

The effect of spermidine in enhancing enzyme activity, although not understood, is not one of contributing significantly to the ionic strength of the reaction medium. Aside from stimulation of enzyme activity by the polyamines, what might prove to be of considerably greater importance is the effect of spermidine on broadening the specificity of the *Citrobacter* enzyme. The nuclease, having a distinct proclivity for uridylic acid residues at either end of the phosphodiester linkage, could, in the presence of spermidine, hydrolyze Poly C as well. Because the product of this hydrolysis was established as C-cyclic-p, the type of hydrolytic attack on the phosphodiester bond is one presumed to occur between a nucleoside 3′-phosphate and a 5′-hydroxyl group of the adjacent nucleotide (13). An attack of this kind appears in all respects to be similar to the hydrolysis of Poly U so that, although a broadened specificity in enzyme action has occurred, there is no indication of a change in the presumed endonucleolytic nature of the hydrolysis. The change in specificity is not confined to the hydrolysis of the homopolymers but is also found in the attack on Poly ACGU. Analysis of the digestion products of this reaction showed that in the absence of spermidine all products terminated in uridylic acid or U-cyclic-p, whereas in the presence of spermidine, any one of the four nucleotides could be the terminal one. Essentially the same results were obtained after enzymic digestion of yeast RNA (6). Although the mechanism by which this alteration in enzyme specificity occurs is unknown, some suggestive clues come from the binding study between enzyme and Poly C on activated Sepharose columns. The delay in passage of the nuclease through the column together with the enzymic breakdown of the covalently bound Poly C that took place only in the presence of spermidine suggests that the polyamine alters the binding properties between the polymer and the enzyme. Evidence of a similar change also has come from inhibitions studies with Poly G (6). The ability of spermidine to reverse the inhibition of enzyme activity caused by Poly G and the inability of the enzyme to bind to Poly G in the presence of polyamine (6) tend to support the view that spermidine affects the binding relationship between the enzyme and polymer. Although it is clear that spermidine can bind directly to either polynucleotide, and presumably affect its binding capacity, what is not clear is if there is any direct effect of spermidine on the enzyme.

In any case, the profound effects the polyamines have on enzyme activity argue strongly for an important role for these substances in regulating cell metabolism.

VI. SUMMARY

A ribonuclease isolated from *Citrobacter sp.* catalyzes the stoichiometric conversion of Poly U to U-cyclic-p but does not hydrolyze Poly A, Poly C, or Poly G. Evidence obtained from studies with synthetic polynucleotides and yeast RNA (6) indicates that the enzyme has a predilection for internucleotide linkages containing uridylic acid at either terminus of the phosphodiester bond. The polyamine spermidine seems to alter the high degree of specificity exhibited by the enzyme to an activity which is considerably more random in nature. In addition, spermidine and putrescine enhance the rate of hydrolytic activity several-fold. Enzyme activity inhibited by Poly G can be restored completely with spermidine but only partially with putrescine.

REFERENCES

1. Cohen, S. S., *Introduction to the polyamines*, Prentice-Hall, New Jersey, 1971.
2. Russell, D. H., *Biochem. Pharmacol.*, **20,** 3481 (1971).
3. Russell, D. H., Levy, C. C., and Taylor, R. L., *Biochem. Biophys. Res. Comm.*, **47,** 212 (1972).
4. Gabbay, E. J., and Shimshak, R. R., *Biopolymers*, **6,** 255 (1968).
5. Brewer, E. W., *Exp. Cell Res.*, **72,** 586 (1972).
6. Levy, C. C., Mitch, W. E., and Schmukler, M., *J. Biol. Chem.*, in press.
7. Levy, C. C., and Goldman, P., *J. Biol. Chem.*, **245,** 3257 (1970).
8. Rushizky, G. W., Knight, C. A., and Sober, H. A., *J. Biol. Chem.*, **236,** 2732 (1961).
9. Poonian, M. S., Sclabach, A. J., and Weissbach, A., *Biochemistry*, **10,** 424 (1971).
10. Kalnitsky, G., Hummel, J. P., and Dierks, C. J., *J. Biol. Chem.*, **234,** 1512 (1959).
11. Stanley, W. M., Jr., and Bock, R. M., *Anal. Biochem.*, **13,** 43 (1965).
12. Warshaw, M. W., and Tenoco, I., *J. Mol. Biol.*, **20,** 29 (1966).
13. Anfinsen, C. B., and White, F. H., in P. Boyer, H. Lardy, and K. Myrback (Editors), *The enzymes*, Vol. 5, Academic Press, New York, 1961, p. 96.

Polyamines in Normal and Neoplastic Growth. Edited by D. H. Russell. Raven Press, New York © 1973.

In Vitro Studies of RNA Methylation in the Presence of Polyamines

Phoebe S. Leboy and Pamela Piester

School of Dental Medicine, University of Pennsylvania, Philadelphia, Pennsylvania 19104

A volume such as this is apt to be a little disconcerting to those of us engaged in studies involving polyamines. The effect of polyamines on such a myriad of biochemical reactions leads one to suspect either that they are at the heart of much of cell regulation or that our *in vitro* studies may bear little relationship to metabolic events within the cell.

There is even more cause for concern when the *in vitro* reactions under investigation are those involving tRNA methylation. It has been 10 years since evidence was presented that tRNA from tumors contained a higher percentage of methylated bases than tRNA from non-neoplastic tissues (1). Soon afterward, Srinivasan and Borek (2) proposed that tRNA methylation is involved in regulatory mechanisms in the cell. Studies carried out in our laboratory (3, 4) and by others (5–8) have demonstrated that polyamines can have a profound effect on tRNA methylation, on the rate of the reaction as well as what methylated products are formed. These *in vitro* results, combined with the correlation between those cell systems showing changes in polyamine content (9–12) and those reported to contain more extensively methylated RNA (13, 14), have led us to postulate that cellular polyamine content may be a controlling factor in the extent of tRNA methylation in the cell (4).

However, mammalian tRNA methyltransferases appear particularly refractory to isolation and purification. This has led to *in vitro* studies carried out on unpurified enzyme extracts, with resulting concern as to which effects are operating directly on the methyltransferases, and which may be secondary effects of ion interaction with other components in the enzyme preparation. Several of the possible interactions of polyamines with RNA and S-adenosylmethionine, substrates for methylation, are discussed in this volume.

The fact that polyamines influence several reactions which are interrelated with tRNA methylation would not necessarily indicate that their apparent effect on tRNA methylation *in vitro* is an artifact. It does suggest that considerable caution should be exercised in interpreting results of experiments using unpurified tissue extracts to explore possible tissue or tumor-specific differences in RNA methylation. We have been examining *in vitro* methylation reactions in order to determine what conditions produce apparent differences in methyltransferase activity, and to try to ascertain the mode of action of polyamines in stimulating tRNA methylation.

I. THE DIFFERENT EFFECTS OF POLYAMINES AND MAGNESIUM ACETATE ON METHYLATION PATTERNS

The methylation of tRNA involves the modification of the preformed macromolecule by addition of methyl groups from S-adenosylmethionine. The "methylatability" of a given tRNA preparation will depend on its origin and the origin of the methylase enzymes being used. Rat liver tRNA will not serve as substrate for rat liver methyltransferases because the RNA has already been fully methylated by these enzymes during maturation. Since bacteria apparently contain methyltransferases with specificity significantly different from those from mammalian sources, bacterial tRNA is a good substrate for mammalian enzymes. In our laboratory we improved the methyl-acceptor ability of the substrate by using methyl-deficient bacterial tRNA, prepared by starving methionine-requiring *Escherichia coli* for this amino acid. Recently, we have also begun using purified single species of tRNAs as substrates.

The various base-specific tRNA methyltransferases present in an extract can be assayed by methylating tRNA with ^{14}C-S-adenosylmethionine and analyzing the radioactive RNA for the presence of radioactive methylated bases. We hydrolyze the RNA to nucleosides with snake phosphodiesterase and alkaline phosphatase, and separate the nucleosides on a Dowex 50 column (4). Six major radioactive peaks are seen, corresponding to 2-methylguanosine, 2-dimethylguanosine, 7-methylguanosine, 1-methylguanosine, 1-methyladenosine, and 5-methylcytidine. A small amount of radioactivity is also seen in 5-methyluridine (ribothymidine) and 1-methylinosine. From work done on the methylated base content of purified rat liver tRNA (15), we might also expect to have enzymes causing the formation of 3-methylcytidine as well as a variety of ribose-methylated nucleosides; however, we have not observed these compounds in any significant quantity.

Most of the studies designed to explore the relationship between neoplasia and increased tRNA methyltransferase activity have employed reaction mixtures containing 4 to 50 mM magnesium salts. However, in order to demonstrate the presence of all of the methyltransferases in a dialyzed rat liver extract, we find it necessary to carry out methylation in the presence of either polyamines or a high concentration of a monovalent cation, e.g., ammonium acetate (16). If one attempts to methylate tRNA using magnesium salts and a dialyzed enzyme preparation, the only enzymes showing appreciable activity are those methylating guanine residues (4). Several of the factors contributing to the final methylation pattern are explored in Table 1. An ammonium sulfate fraction (30 to 48%) from a high-speed supernatant of rat liver homogenates shows traces of methyltransferase activity after 4-hr dialysis against 0.01 M Tris buffer, pH 8.1. Dialysis of this extract against buffer containing 1 mM EDTA results in a preparation which has no detectable methyltransferase activity in the absence of added ions. When methylation is carried out in the presence of 10 mM magnesium acetate, no significant difference between EDTA-treated and untreated preparations can be seen; this is the expected result if the small amount of methylation seen in non-EDTA-treated samples incubated without added ions is due to traces of residual magnesium after dialysis.

TABLE 1. *Effect of dialysis of enzymes against 1 mM EDTA on patterns of methylation*

Incubation with	EDTA exposure	Concentration (pmoles $^{14}CH_3$/10 μg tRNA)					
		m^2_2Guo	m^2Guo	m^7Guo	m^1Guo	m^1Ado	m^5Cyd
–	–	2	10	0	0	2	1
–	+	0	0	0	0	0	0
10 mM Mg	–	29	28	5	7	1	1
10 mM Mg	+	30	22	7	3	1	3
0.5 mM Spd	–	31	52	19	13	50	7
0.5 mM Spd	+	31	90	17	10	52	36

All enzymes were dialyzed 30 to 48% ammonium sulfate fractions from a high-speed supernatant of rat liver. Four mg enzyme protein was incubated with 10 μg methyl-deficient *E. coli* K12 tRNA in a 1-ml reaction mixture containing 0.03 M triethanolamine buffer (pH 7.9), 4 mM mercaptoethanol, and 21 mM (methyl-^{14}C)-S-adenosylmethionine.

Abbreviations are EDTA, ethylenediamine tetraacetate; guo, guanosine; ado, adenosine; cyd, cytidine; urd, uridine; spd, spermidine.

When methylation is carried out in the presence of 0.5 mM spermidine or 10 mM putrescine, the capacity of an enzyme preparation to methylate bacterial tRNA is increased three- to fourfold over that seen with 10 mM Mg (3). Several lines of evidence suggest that the difference between Mg and polyamine-stimulated methylation is not simply one of efficacy of the stimulating factor. The specific activity of an EDTA-treated preparation measured in the presence of spermidine or putrescine is almost twofold higher than that of the preparation before EDTA exposure (3). Comparison of non-

EDTA- and EDTA-treated preparations incubated in the presence of spermidine shows that formation of two nucleosides, 2-methylguanosine and 5-methylcytidine, is increased after EDTA treatment (Table 1), indicating that traces of Mg depress the activity of the methyltransferases forming these two compounds. Experiments in which spermidine-mediated methylation is carried out with addition of 1 mM Mg result in a significant decrease in 2-methylguanosine and 5-methylcytidine (P. S. Leboy, *unpublished results*). These observations suggest that it would be incorrect to consider that Mg^{++} and polyamines are both operating similarly in the methylation reactions, with polyamines acting more efficiently than the divalent cation. Rather, they indicate an apparent antagonism, in the case of some methyltransferases, with Mg diminishing the polyamine-stimulated formation of at least two nucleosides.

II. USE OF POLYAMINES IN METHYLTRANSFERASE PURIFICATION PROCEDURES

The differences in patterns of methylation seen with Mg vs. polyamines in the reaction mixture have some very practical implications for *in vitro* studies of tRNA methyltransferases. If undialyzed homogenates are used, apparent differences in the methylating enzyme activity of the tissues may be reflections of variations in the polyamine and Mg content. If, on the other hand, procedures are employed which would remove tissue cations, several of the methyltransferases will show little or no activity in the presence of Mg alone, giving the appearance of partial fractionation of methylating enzymes.

The latter situation has arisen in experiments designed to remove natural inhibitors of methylation. Kerr (17) has reported that a pH 5 precipitation will produce methyltransferase activity free of an inhibitor, and furthermore will result in only 2-methylguanine methyltransferase recovered in the precipitate. We have prepared the pH 5 precipitated enzyme according to two procedures suggested by Kerr (17, 18) and have examined the methylating activity in the presence of spermidine and putrescine as well as magnesium acetate. Like the ammonium sulfate fractions routinely used in our laboratory, the pH 5 preparation had specific activities with polyamines fourfold higher than those seen with Mg ion. When the pattern of methylation was examined under conditions of excess enzyme and limiting RNA, the precipitated enzyme proved to have retained all of the methyltransferase activities found before precipitation, and the pattern of methylation was similar to that seen using other procedures for preparation of total methyltransferase activity.

An analysis of the relative specific activities of the various methyltransferases in the presence of 40 mM putrescine is presented in Table 2. Whereas the specific activity of the enzyme forming 2-methylguanosine is approximately threefold higher in the pH 5 precipitate than that seen in an ammonium sulfate fraction, comparable enrichment is also seen for the methyltransferases forming 1-methylguanosine and 5-methylcytidine. Formation of 2-methylguanosine represented 48% of the enzyme activity in the ammonium sulfate fraction, and 53% of the activity after pH 5 precipitation of a high-speed supernatant. Thus, although the precipitation procedure results in a source of enzymes with approximately 2.5 times the specific activity of ammonium sulfate fractions, such a technique will not fractionate any of the component methyltransferases.

TABLE 2. *Comparison of specific activity of pH 5 precipitate with EDTA-treated ammonium sulfate enzyme preparation from rat liver*

	Concentration (pmoles $^{14}CH_3$/200 μg enzyme protein)						
Enzyme	m^2_2Guo	m^2Guo	m^7Guo	m^1Guo	m^1Ado	m^5Cyd	m^5Urd
pH 5 precipitate	20	130	3	33	16	42	2
EDTA-ammonium sulfate	14	45	3	6	11	14	2

Incubation (30 min at 37°C) was carried out using 200 μg enzyme protein and 150 μg methyl-deficient *E. coli* tRNA in a 250-μl reaction mixture containing 40 mM putrescine, 21 mM (methyl-^{14}C)-S-adenosylmethionine (100 cpm/pmole $^{14}CH_3$), 0.03 M triethanolamine (pH 7.9), and 4 mM mercaptoethanol.

The more limited methyltransferase activity seen with Mg ion can be used for studying some guanine-specific enzymes. Kuchino et al. (19) have been studying guanine methyltransferase from rat liver and hepatomas using $tRNA^{fmet}$ and other purified tRNAs as substrates. The methylation of $tRNA^{fmet}$ by rat enzymes in the presence of polyamines results in formation of 1-methylguanine, 2-methylguanine, 1-methyladenine, and 5 methylcytosine; however, in the presence of Mg, only 1-methyl- and 2-methylguanine methyltransferase activity is observed (20, P. S. Leboy, *unpublished results*). The Japanese group, using a total enzyme extract dialyzed against buffer containing EDTA and assaying in the presence of 10 mM $MgCl_2$, obtained only methylated guanine products. Using this system, they concluded that the specificity of these guanine-methylating enzymes is the same in several hepatomas as it is in normal rat liver. The probability that these results reflect only a portion of the guanine methyltransferase activity should, however, be kept in mind.

III. POSSIBLE MECHANISMS OF POLYAMINE STIMULATION

The interaction of polyamines with nucleic acids is well established (for references see ref. 21). A straightforward hypothesis to explain the stimulatory effects of polyamines on methylation reactions would be that altered conformation of tRNA in the presence of polyamines facilitates methylation. Support for this was provided by the observation that the concentration of spermine or spermidine required for maximal stimulation of methylation appeared dependent on the amount of *E. coli* tRNA in the reaction medium (3, 5). However, attempts to confirm that polyamines were functioning solely by complexing with tRNA were unsuccessful (22).

Optimal polyamine concentrations have now been reexamined using several purified tRNAs. When these single species of tRNA are used as substrates for methylation, the optimal spermidine concentration is one-tenth that observed with equivalent amounts on a methyl-deficient tRNA mixture (Fig. 1). The spermidine concentration for maximal activity (0.5 mM) remains constant over a 10-fold range of RNA concentration, using either *E. coli* $tRNA^{fmet}$, yeast $tRNA^{asp}$, or yeast $tRNA^{phe}$. One possible explanation is that the binding affinities of the various tRNA species are different (23). Alternatively, unfractionated tRNA preparations may contain non-tRNA material which binds relatively large amounts of polyamines.

Recent work in our laboratory has suggested yet another possible role for polyamines in methylation reactions. Application of EDTA-treated ammonium sulfate extracts to a DEAE-Sephadex column and elution with Tris-EDTA buffer containing 0.18 to 0.22 M NaCl result in an enzyme preparation with higher specific activity than the original extract and little or no fractionation of the methyltransferases (3). Maximal activity of the DEAE peak requires polyamines (Table 3). However, unlike the enzyme preparation from which it was derived, the DEAE peak from rat liver exhibits significant methyltransferase activity in the absence of added ions (Table 3). The methylating activity seen under these conditions is highly specific, permitting the formation of only 1-methyladenosine and 5-methylcytidine. It therefore appears that at least some methyltransferases lose their absolute requirement for stimulatory ions after partial purification.

These observations indicate that polyamines may be involved in annulling the inhibitory effect of some inhibitor(s) present in unpurified preparations. The nature of the inhibitory substance is presently under investigation. It may be one or several compounds which interact with differing affinities to the various methylating enzymes. One possibility we are exploring is that endogenous RNA associates with methyltransferases and that it must be removed by cations before added RNA can serve as substrate.

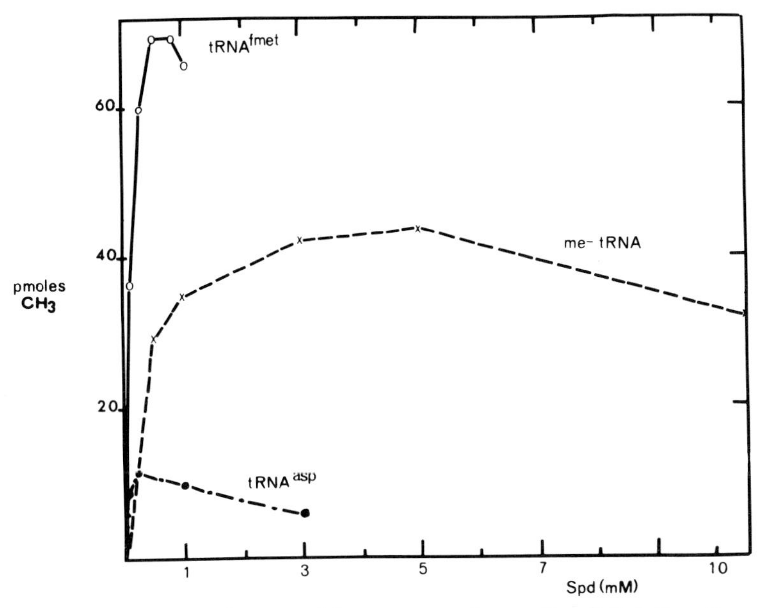

FIG. 1. Effect of increasing spermidine concentration on the methylation of several tRNA preparations. The reaction mixture (0.125 ml) contained 6 μg of tRNA and 60 μg of protein from an ammonium sulfate fraction of rat spleen homogenate. Spd = spermidine.

TABLE 3. *Patterns of methylation after isolation of liver methyl transferases from a DEAE-Sephadex column*

Enzyme	Polyamine added	Concentration (pmoles $^{14}CH_3$/10 μg RNA)		
		m^1Ado	m^5Cyd	mGuo[a]
Ammonium sulfate fraction	—	0	0	0
Ammonium sulfate fraction	0.5 mM Spd	52	36	151
DEAE peak	—	59	30	0
DEAE peak	0.5 mM Spd	100	56	157

Each reaction tube contained 1 to 2 mg enzyme protein and 5 to 10 μg methyl-deficient *E. coli* tRNA in the reaction system described in Table 1.

[a] Includes m^2Guo, m^2_2Guo, m^1Guo, and m^7Guo.

Several factors causing inhibition of *in vitro* tRNA methylation have recently been studied (24–27). These all appear to have a common effect of depleting the amount of S-adenosylmethionine available as methyl donor. In contrast, the inhibitory material seen by studying polyamine requirements seems to have differential effects on the various methyltransferases. To what extent these effects might explain different patterns of methylation, both *in vitro* and *in vivo*, remains to be explored.

ACKNOWLEDGMENTS

We are indebted to Dr. Gordon Tener, University of British Columbia (Vancouver), for a generous gift of yeast $tRNA^{asp}$. This research was supported by a grant from the National Institutes of Health CA–10586. P. S. L. is a recipient of a Public Health Service Career Development Award (Number 09805) from the Institute of General Medical Sciences.

REFERENCES

1. Bergquist, P. L., and Matthews, R. E. F., *Biochem. J.*, **85,** 305 (1962).
2. Srinivasan, P. R., and Borek, E., *Proc. Nat. Acad. Sci. U.S.A.*, **49,** 529 (1963).
3. Leboy, P. S., *Biochemistry*, **9,** 1577 (1970).
4. Leboy, P. S., *Fed. Eur. Biochem. Soc. Lett.*, **16,** 117 (1971).
5. Pegg, A. E., *Biochim. Biophys. Acta*, **232,** 630 (1971).
6. Pegg, A. E. (1972) *Fed. Eur. Biochem. Soc. Lett.*, **22,** 339 (1972).
7. Young, D. V., and Srinivasan, P. R., *Biochim. Biophys. Acta*, **238,** 447 (1971).
8. Hacker, B., and McDermott, B. J., *Physiol. Chem. Physics*, **4,** 41 (1972).
9. Jänne, J., Raina, A., and Siimes, M., *Acta Physiol. Scand.*, **62,** 352 (1964).
10. Neish, W. J. P., and Key, L., *Internat. J. Cancer*, **2,** 69 (1967).
11. Russell, D. H., and Levy, C. C., *Cancer Res.*, **31,** 248 (1971).
12. Russell, D. H., and McVickers, T. A., *Biochim. Biophys. Acta*, **259,** 247 (1972).
13. Borek, E., in *Exploitable molecular mechanisms and neoplasia*, Williams & Wilkins, Baltimore, Maryland, 1970, p. 163.
14. Kerr, S. J., and Borek, E., *Adv. Enzymol.*, **36,** 1 (1972).
15. Rogg, H., and Stalhelin, M., *Eur. J. Biochem.*, **21,** 235 (1971).
16. Kaye, A. M., and Leboy, P. S., *Biochim. Biophys. Acta*, **157,** 289 (1968).
17. Kerr, S. J., *Biochemistry*, **9,** 690 (1970).
18. Kerr, S. J., *Proc. Nat. Acad. Sci. U.S.A.*, **68,** 406 (1971).
19. Kuchino, Y., Endo, H., and Nishimura, S., *Cancer Res.*, **32,** 1243 (1972).
20. Pegg, A. E., *Biochim. Biophys. Acta*, **262,** 283 (1972).
21. Cohen, S. S., in *Introduction to the polyamines*, Prentice-Hall, Englewood Cliffs, New Jersey, 1972, p. 109.
22. Leboy, P. S., *Ann. N.Y. Acad. Sci.*, **171,** 895 (1971).
23. Hacker, B., this volume.
24. Murai, J. T., Jenkinson, P., Halpern, R. M., and Smith, R. A., *Biochem. Biophys. Res. Comm.*, **46,** 999 (1972).
25. Gross, H. J., and Wildenauer, D., *Biochem. Biophys. Res. Comm.*, **48,** 58 (1972).
26. Kuchino, Y., and Endo, H., *Biochem. J.*, **71,** 719 (1972).
27. Kerr, S. J., *J. Biol. Chem.*, **247,** 4248 (1972).

Polyamines in Normal and Neoplastic Growth. Edited by
D. H. Russell. Raven Press, New York © 1973.

RNA and Protein Synthesis in a Polyamine-Requiring Mutant of *Escherichia coli*

David R. Morris

Department of Biochemistry, University of Washington, Seattle, Washington 98195

In vitro studies of polyamines have implicated them in a plethora of processes, ranging from nearly all reactions involving nucleic acids to the stabilization of membranous structures (reviewed in ref. 1). Many of these polyamine effects can be mimicked by magnesium ion and high concentrations of potassium ion, indicating a nonspecific salt effect. In other reactions, the specificity of the polyamine requirement has been inadequately studied. Clearly, in order to relate these *in vitro* interactions of polyamines to their *in vivo* function, an additional approach is needed. One would like to be able to make cells polyamine deficient and study the influence of this deficiency on cellular processes. Perhaps the most selective way of bringing about starvation for polyamines is through mutation. We designed a screening procedure for the isolation of mutants blocked in one of the pathways of putrescine synthesis (2). Using this procedure, we isolated a mutant of *Escherichia coli* which showed a strong requirement for putrescine or spermidine (3). The specificity of this requirement suggested that spermidine, but not putrescine, was required for optimal growth of this organism. In the studies described below, this mutant has been used to gain insight into the *in vivo* role of polyamines in *E. coli.*

I. RATES OF RNA AND PROTEIN ACCUMULATION AND SYNTHESIS

The mutant which we originally isolated (strain CJ352) had an incomplete block in the biosynthetic arginine decarboxylase and showed approximately a threefold reduction in growth rate on starvation for polyamines (3). There is a distinct advantage to working with partially blocked mutants, in that a situation of steady state polyamine-limited growth can be obtained. The mutation in strain CJ352 was transferred by phage-mediated transduction into a strain of *E. coli* B which was highly permeable to a variety of antibiotics including streptolydigen and rifampicin which were used in these experiments. The resulting strain, DK6, showed approximately a twofold dependence of growth rate on added putrescine (Table 1). After overnight starvation for polyamines, these cells grew exponentially over the interval required for a 20-fold increase in mass and appeared to be in a steady state since the macromolecular composition was unchanged over the culture period. As expected, the starved cells had undetectable levels of putrescine and somewhat less than 10% the wild type spermidine level (Table 1). The cellular content of RNA and protein was unchanged on polyamine starvation (Table 1), and the stable RNA species appeared normal on the basis of size distribution and degree of methylation (3). When we calculated the net rates of accumulation of RNA and protein from growth rate and cellular composition (Table 1), both were decreased in proportion to the growth rate.

TABLE 1. *Composition of strain DK6*

	+Putrescine	−Putrescine
Growth rate (μ)[a]	2.2	1.3
Putrescine[b]	9.7	<0.1
Spermidine[b]	3.8	0.27
RNA[c]		
Content	61	58
Accumulation (hr^{-1})	134	75
Protein[c]		
Content	116	99
Accumulation (hr^{-1})	255	129

Cultures were starved for polyamines and grown as previously described (3). The culture medium contained 0.1 M HEPES, pH 7.3, 1×10^{-3} M K_2HPO_4, 0.08 M NaCl, 0.02 M NH_4Cl, 3×10^{-3} M Na_2SO_4, 1×10^{-3} M $MgCl_2$, and $FeSO_4$ (0.1 μg/ml). The medium was supplemented with all 20 amino acids (3) and 0.2% glycerol. Where indicated, putrescine supplementation was at a level of 100 μg/ml. The cellular content of putrescine, spermidine, RNA, and protein was measured as previously described (3).

[a] Doublings/hour

[b] nmoles/ml/A_{540} of culture

[c] μg/ml/A_{540} of culture

Unfortunately, the net rate of accumulation tells us very little, since it is a function of the rates of both breakdown and synthesis. Therefore, it was essential to measure either the actual rate of synthesis, or the degree of turnover. For protein, the latter approach was technically much easier. Cells

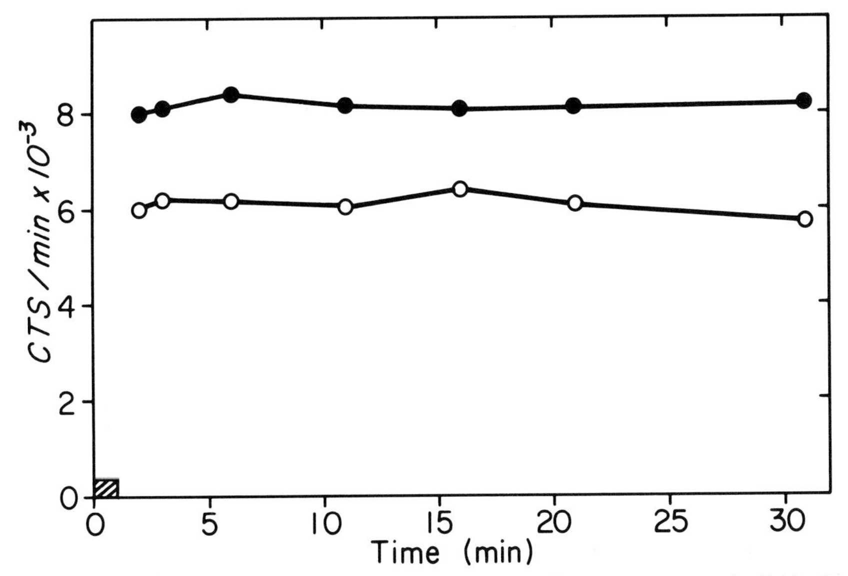

FIG. 1. Protein turnover during polyamine starvation. Cells were grown as in Table 1 to $A_{540} = 0.2$ with (●) or without (○) putrescine. At zero time, ^{3}H-proline (10 μC/μg) was added at a concentration of 0.05 μg/ml. After 1 min, unlabeled proline was added at a final concentration of 1.4 mg/ml, and 1-ml samples were taken at the indicated times into 3 ml 5% trichloroacetic acid. These samples were filtered on cellulose acetate filters and counted in a liquid scintillation counter.

were pulsed with a low concentration (0.05 μg/ml) of ^{3}H-proline and then chased with a high concentration of cold proline (1.4 mg/ml). This experiment is illustrated in Fig. 1. Throughout the period of the experiment no protein turnover was detectable by this technique. Therefore, one can conclude that the actual rate of protein synthesis was equal to its accumulation and that in polyamine-starved cells the rate of protein synthesis was decreased proportionally to the growth rate (Table 1).

The question of the rate of RNA synthesis during polyamine starvation was approached in a different fashion. The incorporation of ^{3}H-uridine into RNA in a 15-sec labeling period was measured. Over the same labeling period, the average relative specific activity of the UTP pool was measured by averaging samples taken at 5, 10, and 15 sec. As can be seen in Table 2, this measurement of precursor pool specific activity was imperative since the average specific activity was considerably higher in the starved cells. This was due, at least in part, to the fact that the UTP pool was lowered approximately 25% by polyamine starvation. However, it should be pointed out that supplementation of putrescine-starved cells with uracil at 20 μg/ml, a concentration which swelled the UTP pool, produced no stimulation of

TABLE 2. *Rate of RNA synthesis*

Addition	μ^a	^{3}H-Uridine incorporation (CPM/15 sec)	Relative UTP spec. act. $^3H/^{32}P$	Relative rate
+Putrescine	2.2	8400	2.1	4,000
−Putrescine	1.3	7160	3.4	2,100

Cultures were grown as in Table 1 except with 2×10^{-4} M K_2HPO_4 in the medium. For measurement of the incorporation rate, 1-ml samples of the culture were pulsed with ^{3}H-uridine (2 μC/μg) at a concentration of 1 μg/ml. The pulses were stopped after 15 sec by the addition of 3 ml 5% trichloroacetic acid containing 50 μg unlabeled uridine per ml. The samples were filtered on cellulose acetate filters and counted in a liquid scintillation counter. For measurement of UTP pool specific activities, cells were prelabeled with ^{32}P (100 μC/μmole) for several generations. ^{3}H-Uridine (40 μC/μg) was then added at a final concentration of 1 μg/ml, and 25-μl samples of the culture were taken into 25 μl ice-cold 2 M formic acid at 5, 10, and 15 sec. The nucleotide triphosphate pools were then separated by thin-layer chromatography (4), and the UTP spot was extracted and counted in a liquid scintillation counter. The UTP specific activity is expressed as the ratio of ^{3}H to the constant ^{32}P label.

[a] Doublings/hour

growth. It can be seen from Table 2 that when the incorporation into RNA was corrected for the relative pool specific activities, the relative rate of RNA synthesis was decreased by 47% during polyamine starvation. This inhibition of the rate of RNA synthesis correlates well with the 44% decrease in the rate of RNA accumulation during putrescine starvation (Table 1). The identity of these decreases clearly shows that there was no massive change in the rate of RNA turnover in the polyamine-starved cells.

Although it is clear that the rates of RNA and protein synthesis were reduced during starvation for polyamines, in a manner proportional to the reduction in growth rate, this observation is not particularly informative in considering the site of polyamine action. There are two possible mechanisms for these reductions in rate: there could be a 50% reduction in the rates of RNA and polypeptide chain elongation, or, alternatively, during polyamine starvation there could be a 50% decrease in the number of active RNA polymerases and ribosomes, with unchanged elongation rates. Both mechanisms would yield an overall 50% reduction in rate. The experiments outlined below were designed to test these alternatives.

II. RATE OF POLYPEPTIDE CHAIN ELONGATION

Now that we know the actual rate of protein synthesis, one way to assess the polypeptide chain elongation rate is to measure the level of active ribosomes and then calculate the rate of amino acid polymerization per active

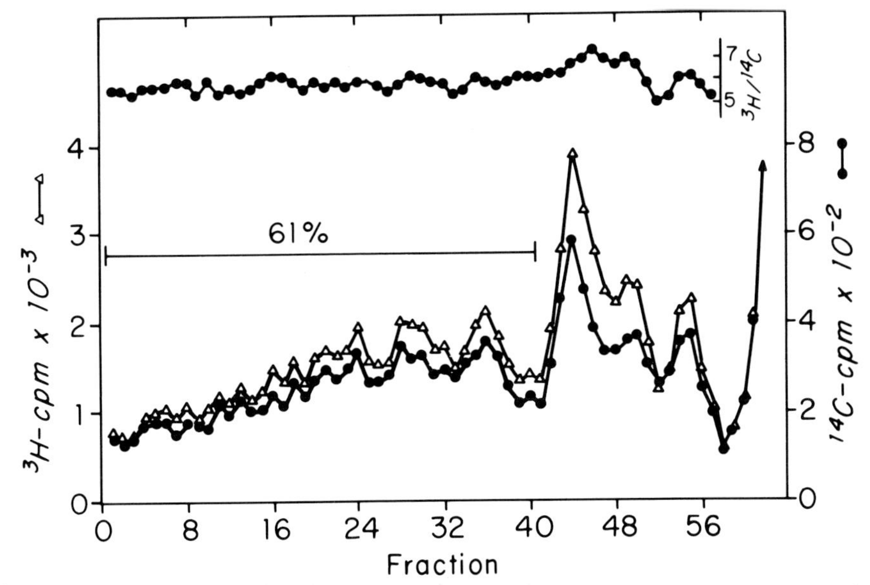

FIG. 2. Polysomes during polyamine starvation. Cells were grown as described in Table 1 for two generations in the presence of 5 μg/ml ^{14}C-uracil (0.012 μC/μg, *putrescine starved*) or ^{3}H-uracil (0.06 μC/μg, *unstarved*). At $A_{540} = 0.1$, cells were harvested and polysomes isolated as described by Forchhammer and Lindahl (5). Polysomes were displayed in a 15 to 30% sucrose-gradient, and samples were counted in a liquid scintillation counter.

particle. To assess the proportion of total ribosomes which were active, sucrose-gradient centrifugation was performed on polysomes extracted from starved and unstarved cells. The sucrose-gradient patterns from a double-label experiment are illustrated in Fig. 2. It can be seen that the proportion of total material in the polysome region, 61%, was independent of putrescine supplementation. In addition, it should be pointed out that there were no apparent differences in the subunit regions of the gradients from the starved and unstarved cells, indicating the absence of an unusual accumulation of ribosomal precursor material or aberrant particles during polyamine starvation. Assuming that half of the monosome region represented active ribosomes (5), we arrived at the conclusion that 70% of the total ribosomes were active in protein synthesis. Using this value, the data of Table 1, and the fact that 87% of the total RNA was ribosomal (3), we calculated polypeptide chain growth rates from the following equation (5):

$$\frac{\text{amino acids}}{\text{ribosome} \times \text{second}} = \frac{\frac{\text{protein}}{120} \times \mu \times \ln 2}{\frac{\text{rRNA}}{1.6 \times 10^{6}} \times 0.70}$$

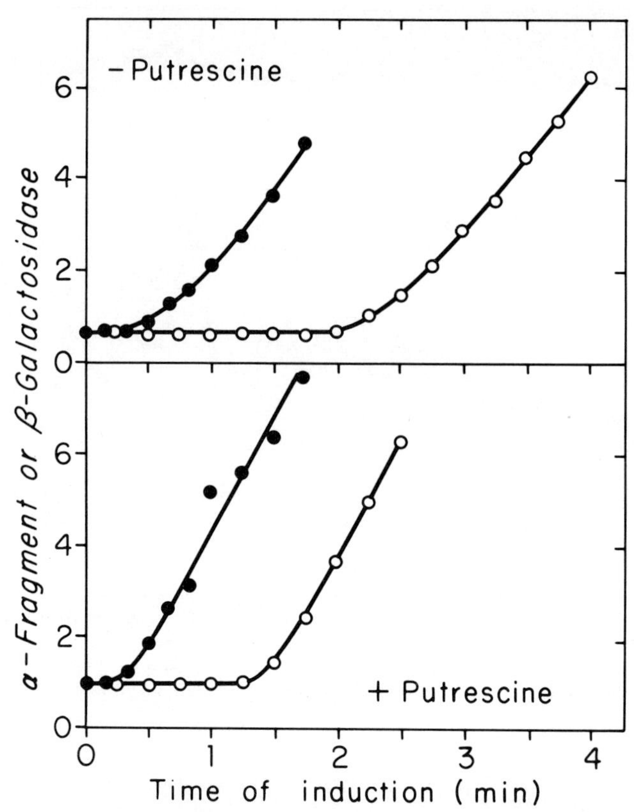

FIG. 3. Kinetics of β-galactosidase and auto-α formation. Cells were grown as described in Table 1 to $A_{540} = 0.3$ and at zero time, isopropylthiogalactoside (IPTG) was added to a final concentration of 10^{-3} M. Auto-α (●) was measured by a modification of the method of Morrison and Zipser (6) (M. T. Hansen, *in preparation*). For the measurement of β-galactosidase (○), 0.5-ml samples of the cultures were taken into tubes containing 20 μl toluene and 40 μl chloramphenicol (2.5 mg/ml). The enzyme was assayed at 28°C using o-nitrophenylgalactoside as substrate (7). Enzyme units were defined as nmoles o-nitrophenol produced per minute. The enzyme units were normalized to 1 ml of culture at $A_{540} = 1.0$. The amount of auto-α produced was converted to β-galactosidase units through normalization of the values obtained with uninduced cultures.

On this basis, polypeptide chain growth rates of 17.5 and 9.5 amino acids per second per ribosome were calculated for unstarved and polyamine-starved cells, respectively. The value of 17.5 amino acids per second is in good agreement with the results of others (5), but the rate of chain elongation is clearly slowed in the polyamine-starved cells.

We have also measured the rate of β-galactosidase chain elongation by the following method. One could assay for a fragment of approximately 60 amino acids at the amino terminal of β-galactosidase (auto-α) by *in vitro* complementation with enzyme from a deletion mutant lacking this fragment (6). It was then possible to estimate the time required for polymerization of the remaining 1,110 amino acids by measuring the time from the first appearance of auto-α to the appearance of complete, active enzyme. Such an experiment is illustrated in Fig. 3. In both starved and unstarved cells, auto-α first began to appear at approximately 15 sec after the addition of isopropylthiogalactoside (IPTG). Complete enzyme began to accumulate at 80 sec in the unstarved cells and at 120 sec in the putrescine-deficient culture. Therefore, it required 65 and 105 sec for completion of β-galactosidase in unstarved and starved cells, respectively. These results yield values for polypeptide chain elongation of 17 amino acids per second in the unstarved cells and 10.5 amino acids per second under conditions of polyamine limitation. These results are in reasonable agreement with the polysome data, and, taken together, these experiments strongly argue that during polyamine limitation the number of active ribosomes is unchanged, but that there is a decrease in the rate of polypeptide chain elongation which is proportional to the growth rate of the cells.

III. RATE OF CHAIN ELONGATION AND STABILITY OF MESSENGER RNA

In order to measure the rate of messenger RNA chain elongation, we have again utilized the β-galactosidase induction system. In this case, we measured the time required from induction to the first appearance of complete Z-gene message. The cells were induced and at various times sampled into the antibiotic streptolydigen. Streptolydigen blocked RNA chain elongation (8), but allowed enzyme synthesis from any completed message present. After allowing time for expression of completed message, the samples were assayed for β-galactosidase. The results are illustrated in Fig. 4. The first appearance of message was observed at 75 and 115 sec in unstarved and starved cells, respectively. If one assumes that the Z-gene consists of 3,500 nucleotides, these results are consistent with elongation rates of 47 and 30 nucleotides per second in unstarved and starved cells, respectively.

In order to validate the above interpretation, it was necessary to demonstrate that there was no lag in inducer action in the polyamine-starved cells. This could be implied from the fact that there was no significant influence of polyamine starvation on the time at which auto-α first appeared (Fig. 3). An additional experiment substantiated this point. Rifampicin, an inhibitor

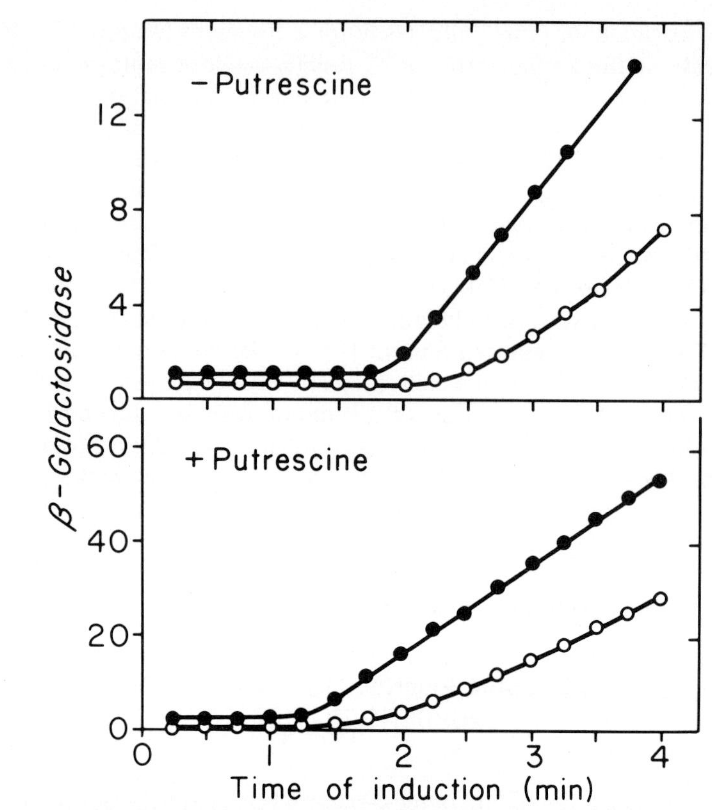

FIG. 4. Kinetics of production of streptolydigen-resistant enzyme-forming capacity. Cultures were grown and induced with IPTG as described in Fig. 3. A portion of the samples (●) were taken into streptolydigen (250 μg/ml), incubated 20 min at 35°C for expression of enzyme-forming capacity, and taken into chloramphenicol and toluene for assay as described in Fig. 3. Other samples (○) were taken directly into chloramphenicol and toluene and assayed.

which blocked initiation but allowed elongation of RNA chains (reviewed in ref. 9), was added at various times relative to the inducer and expression of enzyme-forming capacity was allowed to proceed. This experiment is illustrated in Fig. 5. With both the starved and and unstarved cells, the rifampicin-insensitive enzyme-forming capacity extrapolated back to zero time with no pronounced lag as would be necessary to explain the results of Fig. 4. Thus, the above interpretation that polyamine starvation produced a slowdown in the growth rate of messenger RNA appears warranted.

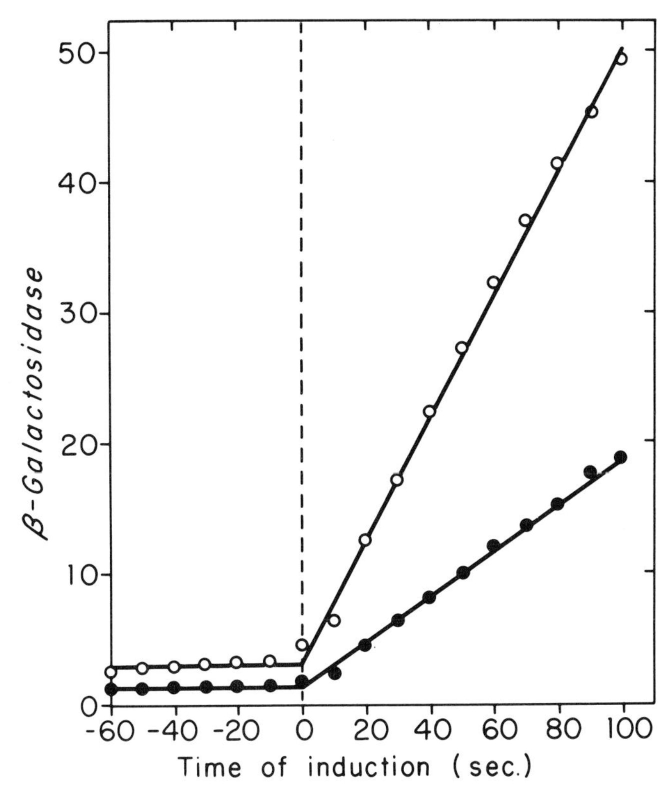

FIG. 5. Kinetics of production of rifampicin-insensitive enzyme-forming capacity. Cells were grown with (○) and without putrescine (●) and induced with IPTG as described in Fig. 3. Rifampicin (60 μg/ml final concentration) was added to samples of the culture at various times relative to zero time, when ITPG was added. After rifampicin addition, the samples were incubated 15 min at 35°C for expression of enzyme-forming capacity, taken into chloramphenicol and toluene, and assayed as described in Fig. 3.

It was also of interest to study the rate of messenger RNA decay in the polyamine-starved cells. β-Galactosidase was induced for 45 sec and the induction period was terminated by the addition of rifampicin. This produced the elementary wave curves shown in Fig. 6. When one plotted the decay of enzyme-forming capacity versus time, half-lives of 60 and 72 sec were obtained for unstarved and starved cells. Thus, polyamine starvation produced, if anything, a slight increase in the stability of β-galactosidase messenger RNA.

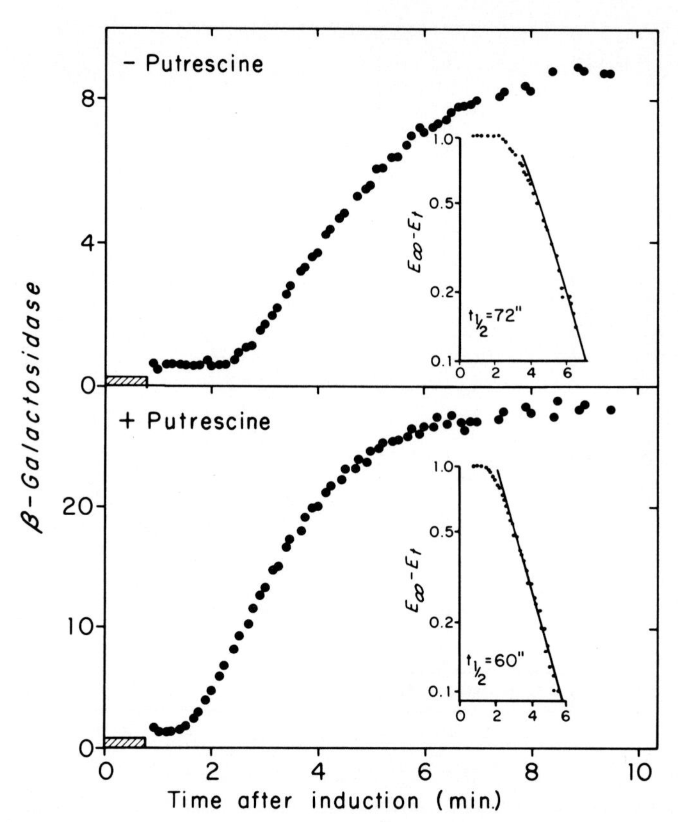

FIG. 6. Elementary wave of β-galactosidase production. Cells were grown and induced with IPTG as described in Fig. 3. At 45 sec after induction, rifampicin (60 μg/ml) was added. Samples of the culture were removed into toluene and chloramphenicol and assayed as described in Fig. 3. The insets are semi-log plots of the remaining enzyme-forming capacity versus time.

IV. DISCUSSION AND CONCLUSIONS

On the basis of the results presented here, it is possible to limit the available models for the *in vivo* action of polyamines in bacteria. For example, an early proposal was that polyamines acted to stabilize cellular RNA (10). This proposal is no longer tenable, either for the normally stable RNA species or for messenger RNA. The fact that the relative rates of RNA synthesis and accumulation fall proportionally during polyamine starvation clearly shows that there can be no massive turnover of transfer and ribosomal RNA. If polyamine starvation shows any effect at all on the half-life of β-galactosidase message, it is to lengthen it slightly. There is no apparent

TABLE 3. *Summary of polypeptide and RNA chain elongation rates*

	+Putrescine	−Putrescine	Unstarved/Starved
Cell growth rate (μ)[a]	2.2	1.3	1.7
Polypeptide elongation rate[b]			
Polysomes	17.5	9.5	1.8
β-Galactosidase	17	10.5	1.6
RNA elongation rate[c]			
β-Galactosidase	47	30	1.6

[a] Doublings/hour
[b] Amino acids/sec/growing chain
[c] Nucleotides/sec/growing chain

role for polyamines in the processing of stable RNA, since no abnormalities in size distribution, precursor accumulation or degree of methylation have been observed during polyamine limitation of growth (3). In addition, no accumulation of ribosomal precursor particles was apparent in the polysome patterns shown here.

The effects of polyamine starvation on the growth rates of protein and RNA chains are summarized in Table 3. It is interesting to note that in both starved and unstarved cells the apparent rate of nucleotide addition to messenger RNA is approximately three times the rate of polypeptide chain growth. This is perhaps not surprising since it is generally considered that translation and transcription occur simultaneously in bacteria. This being the case, it is impossible to say at present whether polyamine starvation produces a separate effect on protein synthesis or whether the decreased polypeptide chain growth rate is due to a jamming of ribosomes behind the slow RNA polymerase. Within experimental error, the decrease in chain elongation rates is proportional to the slowing of the growth rate during polyamine limitation. This suggests, but does not prove, that protein and RNA chain growth is the primary site of polyamine limitation of growth. It should be pointed out that in prokaryotic organisms this is not a typical way to regulate growth. Over a wide variety of growth rates, it appears that the protein and RNA chain elongation rates are nearly constant (5, 12–15). Therefore, it seems unlikely that under normal conditions polyamine levels regulate cell growth in bacteria.

It is attractive to propose that there is a direct effect of polyamines on RNA polymerase. Several groups have shown a two- to threefold stimulation of bacterial RNA polymerases by spermidine at physiological concentrations of potassium ion (150 to 200 mM) (16–18). However, a cautionary note must be introduced. It is entirely possible that the aberrations described here are not primary results of polyamine limitation. For example, there may be a generalized inhibition of nucleic acid synthesis during polyamine starvation, since we have suggested that the rate of DNA chain elongation may be reduced (3). Such a pleiotropic effect could be brought about by structural changes in the cell.

ACKNOWLEDGMENTS

Much of this study was carried out at the University Institute of Microbiology, Copenhagen, Denmark, while I held a fellowship from the John Simon Guggenheim Foundation. I thank Professor Ole Maaløe for his hospitality and stimulating discussions, Mogens Trier Hansen for performing the auto-α measurements, and Kasper von Meyenberg, Nils Fiils, and Lasse Lindahl for their critical and helpful advice. The expert technical assistance of Caroline M. Jorstad is also gratefully acknowledged. This investigation was supported by U.S. Public Health Service grant GM-13957 from the National Institute of General Medical Sciences and by a grant from the American Heart Association (70 646).

REFERENCES

1. Cohen, S. S. (1971) *Introduction to the Polyamines,* Prentice-Hall, Inc., New Jersey.
2. Morris, D. R., and Jorstad, C. M., *J. Bacteriol.,* **101,** 731 (1970).
3. Morris, D. R., and Jorstad, C. M., *J. Bacteriol.,* **113,** 271 (1973).
4. Neuhard, J., *J. Bacteriol.,* **96,** 1519 (1968).
5. Forchhammer, J., and Lindahl, L., *J. Mol. Biol.,* **55,** 563 (1971).
6. Morrison, S. L., and Zipser, D., *J. Mol. Biol.,* **50,** 359 (1970).
7. Pardee, A. B., Jacob, F., and Monod, J., *J. Mol. Biol.,* **1,** 165 (1959).
8. Cassini, G., Burgess, R. R., Goodman, H. M., and Gold, L., *Nature New Biol.,* **23,** 197 (1971).
9. Mosteller, R., and Yanofsky, C., *J. Mol. Biol.,* **48,** 525 (1970).
10. Herbst, E. J., and Doctor, B. P., *J. Biol. Chem.,* **234,** 1497 (1959).
11. Dion, A. S., and Cohen, S. S., *J. Virol.,* **9,** 423 (1972).
12. Coffman, R. L., Norris, T. E., and Koch, A. L., *J. Mol. Biol.,* **60,** 1 (1971).
13. Bremer, H., and Yuan, D., *J. Mol. Biol.,* **38,** 163 (1968).
14. Pato, M. L., and von Meyenberg, K., *Cold Spring Harbor Symp. Quant. Biol.,* **35,** 497 (1970).
15. Rose, J. K., Mosteller, R. D., and Yanofsky, C., *J. Mol. Biol.,* **51,** 541 (1970).
16. So, A. G., Davie, E. W., Epstein, R., and Tissières, A., *Proc. Nat. Acad. Sci. U.S.A.,* **58,** 1739 (1967).
17. Fuchs, E., Millette, R. L., Zillig, W., and Walter, G., *Eur. J. Biochem.,* **3,** 183 (1967).
18. Lee-Huang, S., and Warner, R. C., *J. Biol. Chem.,* **244,** 3793 (1969).

Polyamines in Normal and Neoplastic Growth. Edited by D. H. Russell. Raven Press, New York © 1973.

The Structural and Metabolic Involvement of Polyamines with Herpes Simplex Virus

Wade Gibson and Bernard Roizman

Department of Microbiology, The University of Chicago, Chicago, Illinois 60637

I. INTRODUCTION

Herpes viruses are a group of large, DNA-containing viruses ubiquitous in the animal kingdom. They are composed of at least three major molecular species, including DNA, protein, and lipid, which are organized into three concentric architectural elements, i.e., a core containing DNA, an icosahedral capsid, and an envelope derived from the nuclear membrane (4).

Our interest in the possible relationship between herpes viruses and polyamines stemmed from the following considerations. First, the condensation of viral DNA into a tight configuration able to fit within the nucleocapsid would necessarily require neutralization of its electronegativity. As will be shown here, herpes simplex virus (HSV) does not appear to contain a highly basic core protein that could function in this capacity. Secondly, herpes virus replication is strongly dependent on the amino acid arginine (5, 6). When HSV-infected cells are starved for arginine, nucleocapsids do not form (7). Since arginine can serve as a precursor to polyamines, and since a number of workers have proposed that polyamines may serve to neutralize DNA-phosphate, the question arose whether the observed effect of arginine starvation, namely, absence of capsid assembly, might in fact be due to an effect on polyamine metabolism. With this possibility in mind, we examined the extent to which polyamines are involved with the herpes virion and its replication.

II. RESULTS AND DISCUSSION

A. Presence, specificity, and distribution of polyamines in the virion

The presence of polyamines in the herpes simplex virion was reported elsewhere (1). Pertinent to this chapter is the following summary of these

TABLE 1. *The polyamine content of purified herpes virus*

Exp. No.[a]	μg of DNA[b]	nmoles of polyamine[c] Sp.	Spd.	Ratio Spd./Sp.	Polyamine-N[d]/DNA-P Sp.	Spd.
1	8.64	3.5	5.5	1.6	0.50	0.60
2	9.11	5.0	7.0	1.4	0.70	0.75
3	4.49	2.0	3.0	1.5	0.60	0.65
4	20.30	7.5	13.0	1.7	0.50	0.60
Average ± S.E.				1.6 ± 0.2	0.58 ± 0.17	0.65 ± 0.12

From Gibson and Roizman (1).

[a] Highly purified, enveloped virions were isolated from HSV-1-infected HEp-2 cells according to the procedure of Spear and Roizman (26). Briefly, the infected cells were suspended in 0.01 M sodium phosphate buffer (pH 7.1), ruptured by Dounce homogenization, and made 0.25 M in sucrose. Nuclei were removed from the resulting lysate by centrifugation at 1,500 × g for 10 min, and the clarified lysate layered onto linear, 3 to 30% Dextran-10 (Pharmacia Fine Chemicals, Uppsala, Sweden) gradients made in 0.01 M sodium phosphate buffer, pH 7.1. Following centrifugation at 20,000 rpm for 60 min and 5°C in a Spinco SW 25.3 rotor, a visible band containing enveloped virions was recovered from the gradient. The virions were pelleted from this suspension by centrifugation at 25,000 rpm and 5°C for 2 hr in a Spinco SW 25.3 rotor.

[b] DNA determinations were made on the cold PCA-insoluble pellets by the diphenylamine technique described by Burton (29). Herring sperm DNA (Sigma Chemical Co., St. Louis, Mo.) was used as a standard.

[c] Polyamines were extracted from virions with perchloric acid, dansylated (1-dimethylamino-napthalene-5-sulfonyl chloride, obtained from Pierce Chemical Co., Rockford, Ill.), and separated by thin-layer chromatography on silica gel G plates developed in ethylacetate-cyclohexane, 2:3, essentially by the method of Seiler and Weichmann (27), as modified and described by Dion and Herbst (28). More specific details of the experimental procedures have been previously published (1).

[d] Calculations were based on an average nucleotide weight of 322, and on three nitrogen atoms for spermidine and four nitrogen atoms for spermine.

data: (a) Highly purified preparations of herpes simplex virions were found to contain the polyamines spermidine and spermine in a nearly constant molar ratio of 1.6 ± 0.2. On the basis of the fact that all virions have one molecule of DNA, 100 × 10^6 daltons in molecular weight (8), it can be calculated readily that each virion must contain approximately 40,000 molecules of spermine and about 70,000 of spermidine. As indicated, these amounts either together or separately are adequate to neutralize an appreciable percentage of the viral DNA-phosphate (Table 1). (b) Polyamines are specific structural components of the virion rather than nonspecifically bound host molecules. This conclusion is based on the results of an experiment in which an infected cell lysate containing unlabeled virions was mixed with a similarly prepared lysate of uninfected cells containing labeled polyamines. Virions isolated from the mixture contained polyamines with

TABLE 2. *Characterization of polyamine binding to the virion and subviral components*

Material analyzed[a]	μg of DNA[b]	nmoles of polyamine[b]		Ratio	Polyamine-N[b] / DNA-P	
		Sp.	Spd.	Sp./Spd.	Sp.	Spd.
Purified virions	20.3	7.5	13.0	0.6	0.50	0.60
Purified virions plus NP-40	25.4	9.5	10.5	0.9	0.50	0.40
Purified virions plus NP-40 and urea	17.8	6.0	0.4	15.0	0.45	0.02

From Gibson and Roizman (1).

[a] Details of the experimental procedures were published elsewhere (1). Briefly, a large pool of enveloped virions was prepared as described in the legend to Table 1, and divided into three equal aliquots. The first was pelleted and extracted with perchloric acid; the second was exposed to NP-40 (0.5% v/v final concentration; Shell Oil Co., New York, N.Y.) at 0°C for 15 min prior to pelleting and acid extraction; and the third was similarly treated with NP-40 but pelleted through a barrier of 10% w/w sucrose—0.01 M sodium phosphate (pH 7.1)—2 M urea.

[b] See legend to Table 1 for explanation of experimental techniques and constants used in determining these values.

specific activities less than 10% of those in the initial mixture, indicating that there had been little exchange between unlabeled polyamines bound to the virions and labeled extraneous polyamines. Lastly, (c) it was shown that disruption of the viral envelope with a nonionic detergent and urea removed up to 95% of the viral spermidine but left the amount of spermine essentially unchanged and in a nearly constant ratio with the viral DNA-phosphate. These observations led us to speculate that spermine may function in this virus in a manner similar to that proposed for putrescine and spermidine found in the bacteriophage T-4, namely, to neutralize DNA-phosphate (2, 3). Since spermidine was removed in parallel with the envelope constituents following detergent treatment, it was suggested that this polyamine is specifically associated with the envelope (Table 2).

B. The effect of herpes virus infection on host polyamine metabolism

1. *Interference with Putrescine Synthesis.* These studies were prompted mainly by the observation (1) that ornithine functioned well as a precursor to viral polyamines only when added to the cells prior to infection. To examine the possibility that this effect was the result of herpes virus interference with normal cellular polyamine metabolism, two series of experiments were done. In the first, ^{3}H-ornithine was added to uninfected cells

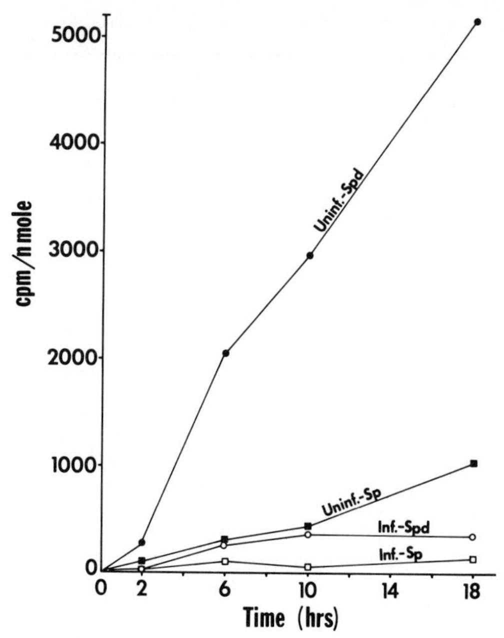

FIG. 1. Changes in polyamine specific activities in infected cells labeled after infection, and uninfected cells labeled during the same interval. Four cultures of HEp-2 cells were infected with HSV-1. One hr after infection, fresh medium containing 5 μC/ml of (3-^{3}H)-D,L-ornithine (2.38 C/mmole, New England Nuclear Corp., Boston, Mass.) was added. At the same time, four uninfected cultures were similarly overlayed with ^{3}H-ornithine-containing medium. At 2, 6, 10, and 18 hr after infection, one culture each of the infected and uninfected cells was collected, as described in the legend to Table 3, and extracted with perchloric acid. Polyamine analyses were done as described in the legend to Table 1. The amount of radioisotope incorporated into each polyamine was determined by scraping its fluorescent spot from the thin-layer plate, drying it at 70°C for 1 hr, suspending it in a toluene-base scintillation fluid, and then measuring the radioactivity in a Packard Tri-Carb scintillation spectrometer. Spd, spermidine; Sp, spermine.

and to 1-hr infected cells, and its conversion to spermine and spermidine then followed in parallel. As indicated in Fig. 1, the specific activities of the infected cell polyamines increased only slightly as compared with those of uninfected cells. At the end of the 18-hr labeling period, the specific activities of spermine and spermidine in the infected cells were only 13 and 6%, respectively, of those in uninfected cells. It is unlikely that the decrease in the utilization of ornithine for polyamine synthesis resulted from an alteration of ornithine uptake by the infected cells, since nearly equal amounts of the radioisotope became incorporated into cold, trichloroacetic acid (7.5%)-precipitable material in both infected and uninfected cells.

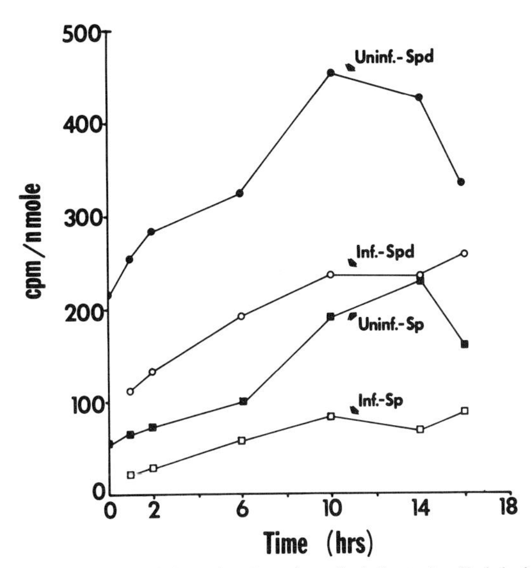

FIG. 2. Changes in the specific activity of polyamines in infected cells labeled before infection, and in uninfected cells labeled during the same time interval. During the 18-hr period immediately prior to infection, HEp-2 cells were incubated in medium containing ^{3}H-ornithine (5 μC/ml, as described in Fig. 1). The labeled medium was then replaced with unlabeled medium, and half of the cultures were infected with HSV-1. At the times indicated, one culture each of infected and uninfected cells was collected, extracted with perchloric acid, and analyzed for its polyamine content and radioactivity as described in Fig. 1. Spd, spermidine; Sp, spermine.

A plausible interpretation of these data is based on the observation that ornithine decarboxylase (ODC) has an extremely rapid rate of turnover (9) and that HSV shuts off host protein synthesis very rapidly after infection (10). Thus, it is possible that ODC would disappear very rapidly from cells after infection and that, therefore, the biosynthesis of putrescine from ornithine would cease.

2. *Interference with Polyamine Metabolism Subsequent to Putrescine Synthesis.* A second experiment was then done to determine whether polyamine metabolism subsequent to the formation of putrescine is affected by herpes virus infection. The cells in this instance were labeled with ^{3}H-ornithine for 18 hr to develop an intracellular pool of labeled putrescine. Half were then infected with HSV and the others left uninfected. As shown in Fig. 2, the specific activities of spermine and spermidine increased in both the infected and uninfected cultures. We do not know why the specific activities of the infected cell polyamines were lower than those of the uninfected cultures. However, one explanation consistent with the idea that ODC may disappear quickly from infected cells is that in the infected cell

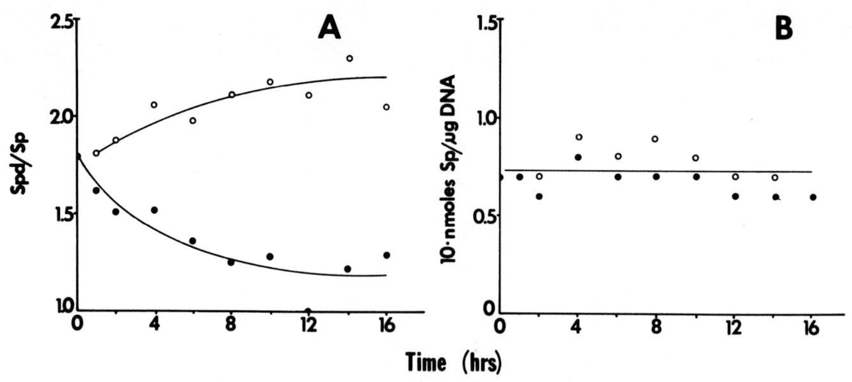

FIG. 3. The ratios of spermidine to spermine and of spermine to DNA in infected and uninfected cells. Spermidine to spermine ratios (Panel A) were calculated from the data obtained from the experiment described in Fig. 2. The amount of spermine relative to DNA was also calculated for both infected and uninfected cells (Panel B). The procedures used to quantitate the amounts of polyamine and DNA were the same as described in the legend to Table 1. Infected cell ratios (○ — ○); uninfected cell ratios (● — ●).

less of the intracellular ornithine becomes "fixed" into the polyamine pathway by conversion to putrescine. A second possibility is that the process of infection may specifically alter either the ornithine or putrescine pool.

Several interesting observations emerged from analyses of the change in the ratio of spermidine to spermine in infected and uninfected cells. Figure 3A shows that this ratio increased slightly in the infected cultures but decreased in the uninfected ones. These changes in the ratios reflect two things. First, as indicated in Fig. 3B, the amount of spermine relative to DNA remained essentially constant in both infected and uninfected cells, even though the amount of DNA increased about 40% in the uninfected cells and nearly doubled in the infected cells. Secondly, while the amount of spermidine remained constant in uninfected cells, it more than doubled in infected cells. Thus, the generally decreasing ratio of spermidine to spermine appears due to an increasing amount of spermine in the presence of a constant amount of spermidine, and the slight increase in the ratio of spermidine and spermine observed in the infected cells appears due to simultaneous increases of both spermine and spermidine, but with spermidine accumulating at a slightly faster rate.

3. *Effect of Arginine Starvation on Cellular Polyamine Metabolism.* In the course of attempts to relate HSV's stringent requirement for arginine with the effect of the virus on host polyamine metabolism, we noticed that if uninfected cells were deprived of arginine their polyamine composition changed from that of normally maintained cells to a composition similar to

that of HSV-infected cells. As indicated in Table 3, the ratio of spermidine to spermine in uninfected cells changed from a value of 0.85 with arginine present to 1.35 in its absence. This second ratio, as shown in the table, is close to that of the polyamines in infected cells maintained either in the presence or absence of arginine.

TABLE 3. *The effect of arginine starvation on the polyamine composition of infected and uninfected cells*[a]

Cells	Ratio SPD/SP[b]	
	Arginine present	Arginine absent
Uninfected	0.85	1.35
Infected	1.56	1.42

[a] Four cultures of HEp-2 cells were used. One was left uninfected and incubated in the presence of fresh, complete medium; a second was infected and then incubated in fresh, complete medium; a third was left uninfected and incubated with fresh medium devoid of the amino acid arginine; and a fourth was infected and also incubated in arginine-deficient medium. After 20 hr, the cells were scraped from the culture bottles, pelleted by centrifugation at 1,500 × g for 10 min at 5°C, and extracted with perchloric acid.

[b] Polyamine analyses were done as described in the legend to Table 1.

Just why the effects of arginine deprivation on the polyamine content of uninfected cells mimic the effects of infection is not clear. It is perhaps noteworthy that Russell (11) and Gfeller and Russell (12) showed that in *Xenopus* cells a close relationship may exist between the nucleolus and polyamine metabolism. As was previously reported (13), in HSV-infected cells the nucleolus disaggregates early after infection. If an association does indeed exist between the nucleolus and polyamine metabolism, then it is perhaps not surprising that conditions such as arginine starvation and infection with HSV, which disrupt nucleolar integrity (14 and 13, respectively), also appear to result in similar changes in the cellular polyamine composition.

C. Possible structural and functional involvement of spermine and spermidine with the herpes virion

Two questions arose from the finding of spermine and spermidine in HSV: what function do these polyamines serve in the virion, and what is the explanation of their apparent compartmentalization within the virion? In

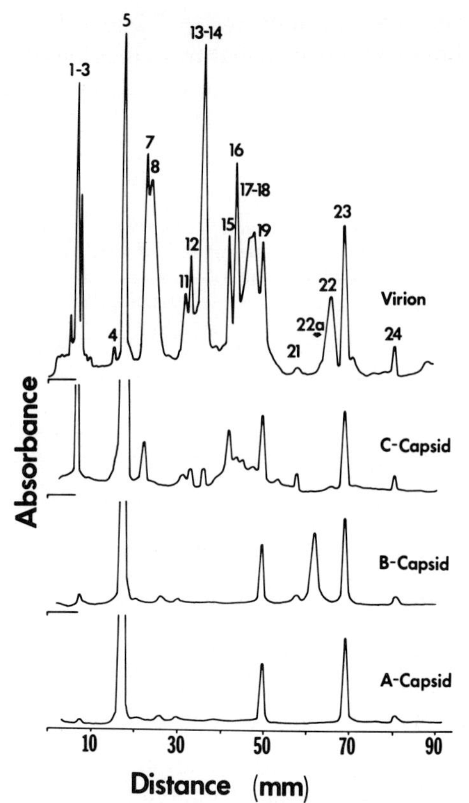

FIG. 4. Electropherograms of the protein components of enveloped HSVs isolated by the procedure of Spear and Roizman (26), C-capsids derived from enveloped virions by detergent treatment, as described by Gibson and Roizman (15), B-capsids containing viral DNA and recovered from the nuclei of infected cells, and A-capsids recovered from the nuclei of infected cells and containing no DNA. The absorbance profiles of the virion, A-capsids, and B-capsids were made from autoradiograms. The C-capsid profile was made from a gel stained with Coomassie Brilliant Blue stain. Details of the techniques used to isolate, solubilize, and electrophorese these virus forms have been presented elsewhere (15).

considering the possible functional role of spermine and spermidine, it is worthwhile examining the composition and structure of the virion in somewhat greater detail. As shown in Fig. 4 the virion consists of 24 structural proteins; of these, only proteins 5, 19, 21, 23, and 24 are capsid constituents (15). The other 19 proteins can be separated from the nucleocapsid (Gibson and Roizman, *unpublished observations*) and have been designated as "noncapsid components." The second panel of Fig. 4 shows the protein composition of C-capsids, a class of particles derived by exposing enveloped virions to detergents (15). Such treatment removes variable amounts of protein from the virion, but invariably results in a particle having a protein composition between that of the virion and that of the C-capsid. The majority of the proteins removed with detergents are glycoprotein constituents of the envelope. Several other proteins which are not components of naked intranuclear capsids are also removed.

Since similar detergent treatment removed spermidine from the virion, but left the amount of spermine essentially unchanged, we proposed that spermidine is an envelope component and spermidine is more intimately associated with the nucleocapsid (1). Although we do not know specifically with which envelope constituents the spermidine is associated, it is interesting to note that at least two non-capsid, virion proteins are phosphorylated (Gibson and Roizman, *unpublished observations*). Spermidine, then, may function to neutralize either these phosphoproteins, or perhaps the phospholipids also known to be present in the envelope (16).

With regard to spermine, we do not have definitive evidence that it is bound to the viral DNA. However, two lines of evidence suggest that if the DNA is indeed neutralized, the species of cation involved is small. First, as we have recently described (15), two types of capsids can be isolated from the nuclei of HSV-infected cells. One type, designated as A-capsids, does not contain DNA; the other, designated as B-capsids, contains normal, intact HSV-DNA molecules. As indicated in Fig. 4, A- and B-capsids also differ with respect to their protein composition. A-capsids are composed of three major and one minor protein (Nos. 5, 19, 23, and 24); B-capsids have two additional components, protein 21 which constitutes about 2% of the total capsid protein, and protein 22a which represents about 20% of the total protein. Protein 22a, however, which might be considered a prime candidate for a core protein, is not a structural component of the mature virion. This was demonstrated by electrophoretically separating the proteins of DNA-containing B-capsids and mature virions, both alone and as a mixture. The bottom panel of Fig. 5 shows that although capsid and virion proteins 5, 19, 21, and 24 co-migrated in the mixture gel, capsid protein 22a does not have a direct counterpart in the virion.

Further, preliminary amino acid analyses indicate that neither capsid protein 22a nor the other major B-capsid proteins have an unusually high percentage of basic residues, although the possibility of modified residues cannot yet be ruled out. Amino acid analyses have not been done on proteins 21 and 24 since together they represent only about 3% of the total capsid protein. It is apparent from these data, then, that the herpes virion does not contain a core protein of the type present in adenoviruses (17, 18) or oncornaviruses (19). In these viruses the core proteins constitute 10 to 20%

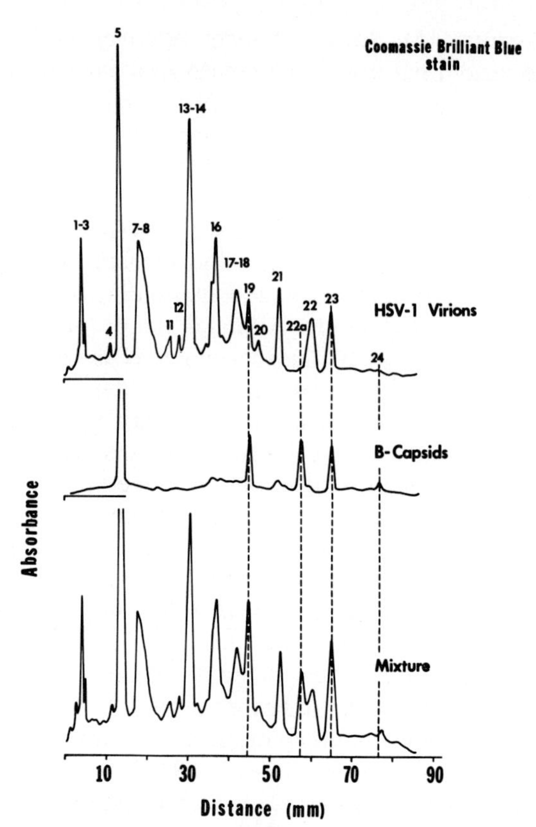

FIG. 5. Electropherogram of B-capsid and virion proteins alone in an artificial mixture. Three companion gels were subjected to electrophoresis, one containing partially purified virions (HSV-1 virions), another containing HSV-1, B-capsid proteins, and a third containing approximately equal amounts of virion and B-capsid proteins (mixture). Absorbance profiles of Coomassie Brilliant Blue stained gels are shown. Details concerning this experiment have been published elsewhere (15).

of the total capsid protein and are composed of nearly 20% arginine residues.

The second line of evidence consistent with the idea that there is no appreciable amount of core protein was recently provided by electron-microscopic analyses of the DNA-containing core structure of the virion (20). Calculations based on the dimensions of the toroidal core, and on the known length of the herpes virus DNA molecule (21), indicate that the core of the virion is just adequate to accommodate the viral DNA. Further, the DNA in negatively stained preparations appears to be very tightly coiled. Thus, the possibility of a basic protein binding to the DNA and serving to neutralize its electronegativity becomes sterically complicated.

Spermine, on the other hand, according to the model of Tsuboi (22) and Liquori et al. (23), can bind in the narrow groove of the DNA and thus neutralize the phosphate charges without presenting similar steric problems.

D. Significance of the compartmentalization of the polyamines within the virion

Whatever their functional involvement, it is clear that the two polyamines are structurally segregated in some manner in the herpes virion. A number of hypotheses have been proposed (1) to explain this segregation; however, the one most amenable to testing and perhaps the most interesting is that available polyamines are compartmentalized within the host cell—spermine inside the nucleus and spermidine outside. Since the HSV nucleocapsid is assembled in the nucleus, spermine alone would be available for binding to the virus during its envelopment at the nuclear membrane and its egress from the infected cells.

One prediction of this hypothesis is that it should be possible to demonstrate directly the compartmentalization of spermine and spermidine within the host cell. This was attempted by using nonionic detergent extraction techniques similar to those used to remove the viral envelope. As shown in Table 4, the distribution of spermine was 70% in the nucleus and 30% in the cytoplasm. No compartmentalization of spermidine was apparent. An explanation of the nearly even distribution of spermidine between the nuclear and cytoplasmic fractions is that the nuclear spermidine may be present in association with a structural component such as the nucleolus or nuclear membrane, and thus partition with the nuclei during fractionation. Consistent with this possibility is the fact that the herpes virion derives its spermidine-containing envelope from the inner lamella of the nuclear mem-

TABLE 4. *Polyamine distribution in the uninfected host cell*[a]

	Spermine		Spermidine	
Cell fraction	nmoles[b]	% of Total	nmoles[b]	% of Total
Nucleus	1.59 ± 0.14	71	1.02 ± 0.06	54
Cytoplasm	0.66 ± 0.20	29	0.87 ± 0.21	46

[a] Uninfected HEp-2 cells were scraped from their culture bottles, pelleted by centrifugation at 1,500 × *g* for 10 min at 5°C, and resuspended in 0.01 M sodium phosphate (pH 7.1) containing NP-40 (0.5% v/v final concentration; Shell Oil Co., New York, N.Y.) for 10 min at 0°C. The nuclei were pelleted by centrifugation at 1,500 × *g* for 10 min at 5°C, and both nuclear and cytoplasmic fractions were extracted with perchloric acid.

[b] Polyamine analyses were done as described in legend to Table 1.

brane (10). This membrane remains morphologically intact during the fractionation procedure.

A second prediction of this hypothesis is that the nucleocapsids of viruses infecting eukaryotic cells should contain largely the polyamine present in the particular host cell compartment in which it is assembled. This prediction is consistent with the observation that turnip yellow mosaic virus and cowpea chlorotic mottle virus, two other eukaryotic viruses known to contain polyamines (24 and 25, respectively), each appears to contain essentially just one of the host polyamines—spermidine. Poxviruses would be interesting to examine in this context since they constitute a class of DNA-containing, enveloped virions that replicate exclusively in the cytoplasm. It might be predicted, therefore, that these viruses would contain only spermidine—both in their core and in their envelope.

III. SUMMARY

Highly purified preparations of enveloped HSV contain the polyamines spermidine and spermine in a nearly constant molar ratio of 1.6 ± 0.2 (1). Spermidine appears confined to the envelope while spermine is more intimately associated with the nucleocapsid. The function of the polyamines in the virion is not known. Analyses of the composition and architecture of the nucleocapsids have revealed that viral DNA is tightly coiled around a central spindle-like structure and that the capsid lacks basic proteins which could neutralize the charges on the DNA and facilitate tight packing. The virion contains sufficient spermine to neutralize a minimum of 40% of its DNA-phosphate. Therefore, the function of spermine could be to neutralize the electronegativity of the DNA in a manner analogous to that postulated in bacteriophage (2, 3). Support for the hypothesis that the compartmentalization of spermine and spermidine observed in the virion reflects the distribution of these polyamines in the host cell was provided by the finding that 70% of the cellular spermine was contained in the nuclei. Results are also presented which suggest that spermine and spermidine synthesis continues after infection whereas conversion of ornithine to putrescine appears to be blocked.

ACKNOWLEDGMENTS

Our work was supported by the U.S. Public Health Service, (CA 08494), the American Cancer Society (VC 103H), and the National Science Foundation (NSF GB 27356X). W. G. was a predoctoral trainee, U.S. Public Health Service (AI 00238). We would like to acknowledge our indebtedness to Drs. H. G. Williams-Ashman and Seymour S. Cohen for their interest and advice during the course of these studies.

REFERENCES

1. Gibson, W., and Roizman, B., *Proc. Nat. Acad. Sci. U.S.A.,* **68,** 2818 (1971).
2. Ames, B. N., Dubin, D. T., and Rosenthal, S. M., *Science,* **127,** 814 (1958).
3. Ames, B. N., and Dubin, D. T., *J. Biol. Chem.,* **235,** 769 (1960).
4. Roizman, B., and Spear, P. G., in Dalton and Haguenau (Editors), *Atlas of viruses,* Academic Press, New York, 1972, p. 79.
5. Tankersley, R. W., Jr., *J. Bacteriol.,* **87,** 609 (1964).
6. Roizman, B., Spring, S. B., and Roane, P. R., Jr., *J. Virol.,* **1,** 181 (1967).
7. Mark, G. E., and Kaplan, A. S., *Virology,* **45,** 53 (1971).
8. Kieff, E. D., Bachenheimer, S. L., and Roizman, B., *J. Virol.,* **8,** 125 (1971).
9. Snyder, S. H., Kreuz, D. S., Medina, V. J., and Russell, D. H., *Ann. N.Y. Acad. Sci.,* **171,** 749 (1970).
10. Roizman, B., in *Current topics in microbiology and immunology,* Vol. 49, Springer-Verlag, Heidelberg, 1969, p. 1.
11. Russell, D. H., *Proc. Nat. Acad. Sci. U.S.A.,* **68,** 523 (1971).
12. Gfeller, E., and Russell, D. H., *Anat. Rec.,* **166,** 306 (1970).
13. Schwartz, J., and Roizman, B., *J. Virol.,* **4,** 879 (1969).
14. Granick, S., and Granick, D., *J. Cell Biol.,* **51,** 636 (1971).
15. Gibson, W., and Roizman, B., *J. Virol.,* **10,** 1044 (1972).
16. Ben-Porat, T., and Kaplan, A. S., *Virology,* **45,** 252 (1971).
17. Laver, W. G., *Virology,* **41,** 488 (1970).
18. Prage, L., and Pettersson, U., *Virology,* **45,** 364 (1971).
19. Fleissner, E., *J. Virol.,* **8,** 778 (1971).
20. Furlong, D., and Roizman, B., *J. Virol.,* **10,** 1071 (1972).
21. Becker, Y., Dym, H., and Sarov, I., *Virology,* **36,** 184 (1968).
22. Tsuboi, M., *Bull. Chem. Soc. Japan,* **37,** 1514 (1964).
23. Liquori, A. M., Constantino, L., Crescenzi, V., Elia, V., Giglio, E., Puliti, R., Savino, D. S., and Vitagliano, V., *J. Mol. Biol.,* **24,** 113 (1967).
24. Beer, S. V., and Kosuge, T., *Virology,* **40,** 930 (1970).
25. Cohen, S. S., *Introduction to the polyamines,* Prentice-Hall, Inc., Englewood Cliffs, N.J., 1971.
26. Spear, P. G., and Roizman, B., *J. Virol.,* **9,** 143 (1971).
27. Seiler, N., and Weichmann, M., *Z. Physiol. Chem.,* **348,** 1285 (1967).
28. Dion, A. S., and Herbst, E. J., *Ann. N.Y. Acad. Sci.,* **171,** 723 (1970).
29. Burton, K., *Biochem. J.,* **62,** 315 (1956).

Polyamines in Normal and Neoplastic Growth. Edited by D. H. Russell. Raven Press, New York © 1973.

Polyamine Metabolism in the Brain

N. Seiler

Unit for Neurochemistry, Max-Planck-Institute for Brain Research, 6000 Frankfurt/M, Germany

I. INTRODUCTION

Although the polyamines seem to be ubiquitously distributed in the living world, occurring in all types of cells, their study in a highly specialized system such as the nervous system may reveal some interesting aspects of their physiological significance and may throw some light on special features of the system. Among the different aspects which are apparently important in connection with the physiological role of the polyamines, their possible interaction with nucleic acids is one on which current interest is focused. The growing fish brain appeared to be a good model for the study of quantitative relations between polyamines and nucleic acids, since brains of many fish species continue to grow parallel to the increase in body size until a very late stage of their lives, whereas in higher vertebrates a period of rapid growth is followed by a period of slow brain growth shortly after birth. The developmental changes occur in these brains within a short time period, and, consequently, errors in age determinations can cause uncertainties in the time course of the quantitative relations.

In addition to the quantitative aspects of polyamine and nucleic acid relationships, some metabolic properties of putrescine in brain will be described, with emphasis on putrescine catabolism. It will be demonstrated that a metabolic pathway from putrescine to γ-aminobutyric acid (GABA) exists not only in bacteria (1, 2) but also in vertebrate tissues. This aspect of putrescine metabolism is especially significant for brain metabolism, since, according to the current view, GABA in vertebrate brain is only produced by decarboxylation of glutamic acid (3).

II. MATERIALS AND METHODS

A. Experimental animals

Rainbow trout (*Salmo irideus,* Gibb.) between 8 and 55 cm long were used. The animals used in the metabolic studies were 12 to 16 cm long. Roaches (*Scardinius erythrophthalmus*) were between 5 and 8 cm long. They were kept in aquariums (125 × 60 × 40 cm) at a water temperature of 12°C, and fed with a standard diet. The albino mice (NMRI, Ges. f. Versuchstierzucht, Hannover) weighed 30 to 35 g. They were also kept under standardized conditions.

B. Radiochemicals and reagents

1,4-^{14}C-Putrescine · 2 HCl (specific activity 20.8 mC/mmole) and 5-^{14}C-D,L-ornithine (specific activity 3.3 mC/mmole) were purchased from New England Nuclear Corporation (Boston). ^{3}H-Spermine (specific activity 39.1 mC/mmole) was labeled in the positions C_1 and C_{10}. It was prepared from 1,10-dicyano-3,8-diazadecane by catalytic tritiation (4). 1-Dimethylamino-naphthalene-5-sulphonylchloride (DANS-Cl) was prepared in our laboratory by the method described previously (5).

C. Intracerebral injections

Twenty μl of the radioactive compounds in distilled water was injected into unanesthetized animals with microsyringes equipped with a needle 0.4 mm in diameter and 4 mm long. The location of the optic lobes, where the solutions were injected, could be observed as light spots dorsal to the eyes. The needle was inserted first into one of these spots and then removed again in order to allow CSF to flow out. After removal of a droplet of CSF with a tampon, the needle was inserted again into the same hole. The solution was now injected. To minimize the escape of radioactive material from the head, the needle was kept in position for 20 sec before it was removed from the skull. Intraventricular injections to mice were carried out with chloral-hydrate anesthetized animals according to published methods (6, 7).

D. Preparation of brain extracts

The animals were decapitated and the brains were removed from the skull within 15 to 20 sec and immediately homogenized with 10 vol. ice-cold 0.2 N perchloric acid. (Brains, which were removed at −15°C from animals killed by freezing with liquid nitrogen, yielded analytical data almost identical to those obtained from brains removed from the skull at room temperature.) After an hour of storage at 3°C, the homogenates were centrifuged for 20 min at approximately 800 × *g*. The supernatants were used as the perchloric acid extracts.

E. Ion-exchange chromatography

Two ml of the perchloric acid extract was neutralized with 2 N NaOH. The final volume was made up to 5 ml, and 2-ml samples were mixed with 1 ml of 6 N HCl and heated at 70°C for 2 hr in order to hydrolyze the amide groups. The samples were then evaporated to dryness in a stream of air. The residue was stored in a desiccator over KOH. Finally it was taken up in 2 ml citric acid buffer, pH 3.20. The 1.5-ml portions of this solution were applied to the ion-exchange columns of an amino acid analyzer (Unichrom, Beckman Instruments, Munich) to which a Packard flow detector (Model 3041) was directly attached. Separation of the acidic and neutral amino acids, including GABA, was achieved by elution first with citrate buffer, pH 3.20, and, after the appearance of the alanine peak, with citrate buffer, pH 4.80 (8).

F. Determination of the polyamines

Spermidine, spermine, and putrescine were determined by fluorometry of their DANS-derivatives. In the case of the mouse brains, the methods described in detail earlier (9, 10) were applied. In the case of fish brains, portions (0.8 ml) of the perchloric acid extracts were reacted with excess DANS-Cl under the usual reaction conditions (5). The DANS-amides were extracted with 5 ml of toluene (or benzene). The toluene extract was evaporated to dryness and the residue taken up in 0.5 ml of toluene. Aliquots (40 μl) of this solution were applied to 20 × 20 cm plates with a 200-μ thick silica gel G (E. Merck, Darmstadt) layer. The plates were placed in a dry ammonia atmosphere for 60 min in order to transform N-DANS-2-oxo-pyrrolidine, the reaction product of GABA with DANS-Cl (11), into DANS-γ-aminobutyramide. This compound remains near the origin, whereas N-DANS-pyrrolidine exhibits an R_F value similar to that of bis-DANS-putrescine in the solvent system used for the separation of the DANS-derivatives of the polyamines. Without reactivation of the layer, the chromatogram was first developed with cyclohexane plus ethylacetate (50 + 50) and then twice with cyclohexane plus ethylacetate (70 + 50) in the same direction. The fluorescent spots corresponding to bis-DANS-putrescine, tri-DANS-spermidine, and tetra-DANS-spermine were scraped out. The DANS-amides were extracted with 5 ml of methanol plus concentrated ammonia (95 + 5) (12).

Fluorescence measurements were performed at 520 nm (fluorescence activation, 365 nm) after sedimentation of the silica gel by centrifugation (for detailed descriptions of the method, see 5, 12, 13). In order to ensure the homogeneity of the *bis*-DANS-putrescine spot, some samples were separated bidimensionally with tetrachloromethane plus methanol (90 + 6) (five runs in the first dimension) and chloroform plus triethylamine (100 + 10) in the second dimension, or with cyclohexane plus ethylacetate (70 + 50) (three runs in the first dimension) and tetrachloromethane plus methanol (90 + 6) (two runs in the second dimension) (14). In addition some putrescine determinations were carried out by a specific mass spectrometric method (15) after dansylation and thin-layer chromatography (TLC). Since all three methods yielded the same results, it was concluded that in the case of the fish brains the *bis*-DANS-putrescine spot was sufficiently well separated for quantitative estimation by the above-mentioned, unidimensional thin-layer chromatographic system.

Standardization of the determinations was achieved by addition of putrescine · 2 HCl to some tissue samples in amounts which approximately corresponded to the amine content of the samples or by dansylation of a mixture of putrescine · 2 HCl, spermidine phosphate, and spermine phosphate under identical reaction conditions. In the unidimensional chromatograms, tissue samples and standard samples were chromatographed on the same plate. From each brain two samples were dansylated and two samples were applied to the thin-layer plate from each reaction mixture.

For the determination of the specific activity of the different polyamines the total amount of the dansylated amine mixture was separated unidimensionally with the above-mentioned solvents. After extraction of the different DANS-amine derivatives with dioxane and of DANS-γ-aminobutyramide with methanol plus concentrated ammonia (95 + 5), they were rechromatographed in the following solvents:

DANS-spermidine and DANS-spermine: benzene plus triethylamine (100 + 20);

DANS-putrescine (1st dimension): trichloroethylene plus methanol (95 + 5), and benzene plus methanol (95 + 5), and (2nd dimension): chloroform plus triethylamine (100 + 10);

DANS-γ-aminobutyramide (1st dimension): methylacetate plus isopropanol plus concentrated ammonia (90 + 50 + 5), (2nd dimension): chloroform plus ethylacetate plus methanol plus acetic acid (60 + 100 + 40 + 2) (16).

The rechromatographed compounds were extracted from the layer with 5 ml of dioxane. After fluorescence measurement, 4-ml aliquots of these solutions were used for radioactivity determination. To each sample, 10 ml of liquid scintillator (5 g PPO per liter toluene) was added.

G. Thin-layer electrophoresis

Aliquots of the brain extracts were applied to thin-layer plates in 3-cm streaks after neutralization with $KHCO_3$ and precipitation of $KClO_4$ by addition of ethanol. The plates were sprayed with pyridine–acetic acid–citric acid buffer, pH 4.8 (17), and then submitted to electrophoresis at 20 V/cm in a Camag TLC chamber for 2 hr at 2°C. The air-dried electrophoretograms were scanned for a first survey with the radiochromatogram scanner LB 2720 (Berthold, Erlangen). Then 40 × 5 mm zones were scraped out and the silica gel submitted to radioactivity measurement. Each counting vial contained 1 ml methanol, 20 μl water, and 10 ml liquid scintillator.

In the case of the isolation of N-acetyl putrescine, a 15-cm streak of the brain extract was applied to the thin-layer plate and the zone corresponding in its electrophoretic mobility to N-acetyl putrescine was scraped out and subdued to dansylation.

H. Separation of DANS-N-acetyl putrescine

The DANS-derivatives in the N-acetyl putrescine zone of the electrophoretogram were separated bidimensionally in the solvents benzene plus methanol (90 + 10) (1st dimension) and chloroform plus triethylamine (100 + 20) (2nd dimension).

I. Nucleic acids

Nucleic acid determinations were carried out in the perchloric acid-insoluble residue of the brain homogenates. RNA was determined by a modified Schmidt-Thannhauser procedure according to the recommendations of Fleck and Munro (18–20). DNA was determined by a modification (21) of the diphenylamine reaction (22).

J. Mass spectrometry

Mass spectra were prepared from the eluted spots of the DANS-derivatives with a Varian MAT CH5 single focusing mass spectrometer at an electron beam energy of 70 eV (concerning mass spectrometry of DANS-derivatives, see 23, 24).

III. QUANTITATIVE RELATIONSHIPS BETWEEN POLYAMINES AND NUCLEIC ACIDS

The data in Table 1 demonstrate that RNA and spermidine concentrations in fish and mouse brains are approximately the same. Spermine concentrations are lower and DNA and especially putrescine concentrations are considerably higher in fish than in mouse brain. The relatively low spermine content and the extremely high concentration of putrescine in fish brain may reflect the thin neuronal layers in comparison to the extensive fibrous zones in fish brain. High spermine concentrations are characteristic for brain areas rich in nerve cells, whereas white matter and peripheral nerves preferentially contain spermidine and putrescine (25–28).

TABLE 1. *Polyamine and nucleic acid concentration in trout, roach, and mouse brain*

Animal	Brain weight (mg)	Putrescine (μmole/g)	Spermidine (μmole/g)	Spermine (μmole/g)	RNA-phosphate (μmole/g)	DNA-phosphate (μmole/g)
Trout (*Salmo irideus*, Gibb.)	70–130	0.62 ±0.01	0.44 ±0.01	0.17 ±0.003	7.52 ±0.15	7.18 ±0.39
Roach (*Scardinius erythrophthalmus*)	20–30	–	0.52 ±0.035	0.24 ±0.03	6.41 ±0.57	7.18 ±0.55
Mouse (albino, NMRI)	430–450	0.011 ±0.0012	0.44 ±0.016	0.35 ±0.025	6.38 ±0.19	4.57 ±0.13

During the embryological development of mouse brain, a decrease of spermidine, spermine, and DNA concentrations was observed. The DNA decrease preceded that of the polyamines (29). In the growing trout brain, however, constant concentrations of RNA, putrescine, and spermidine were measured, and, with increasing brain weight, decreasing concentrations of spermine and DNA were measured. In other words, the total content of RNA, putrescine, and spermidine was in a linear correlation with the brain weight, at least in brains weighing between 60 and 800 mg, whereas the total increase of spermine and DNA was lower than the increase in brain size (30). The analytical data obtained from the individual animals are compiled in the diagrams of Fig. 1. It can be derived from these data that a linear correlation exists between spermine and DNA ($r = 0.87$) in trout brain (30). An increase of 1 μmole spermine is accompanied by a

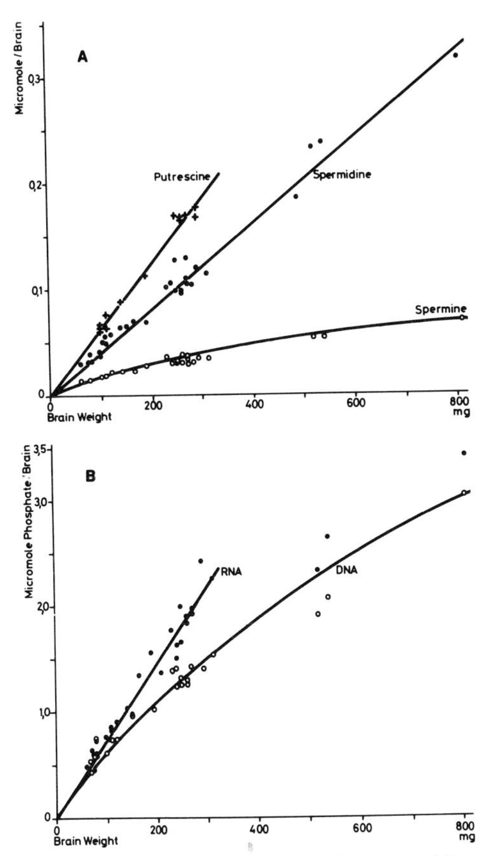

FIG. 1. Relationship between total amounts of putrescine, spermidine, spermine, RNA and DNA, and brain weight of the rainbow trout (*Salmo irideus*, Gibb.). Each point represents the mean value of duplicate estimations.

change of approximately 41 μmoles of DNA-phosphate. The net increase of DNA in fish brain indicates that during the growth process the absolute number of cells is increased. Mitotic activity in fish brain was observed until a very late stage of development with different methods (31).

Although the cell population in the brain is inhomogeneous and the amount of DNA per cell is not the same for the different cell types, the ratio between the concentration of a cell constituent and the concentration of DNA may nevertheless be taken as a measure for the amount of the constituent per cell. Table 2 demonstrates the average amount of polyamines and RNA per cell. These data were calculated from the data of Table 1 and Fig. 1. The differences between putrescine and spermine concentrations

TABLE 2. *Polyamine and RNA amounts/cell in trout, roach, and mouse brain*

Animal	Brain weight (mg)	Putrescine	Spermidine	Spermine	RNA-phosphate
		(μmole/μmole DNA-phosphate)			
Trout (*Salmo irideus*, Gibb.)	70–130	0.086	0.061	0.024	1.05
	150–300	0.12	0.082	0.026	1.45
	500–800	–	0.11	0.025	–
Roach (*Scardinius erythrophthalmus*)	20–30	–	0.073	0.033	0.89
Mouse (albino, NMRI)	430–450	0.0024	0.096	0.076	1.40

Mean values of cell constituent/DNA-ratios are calculated from the data of Table 1 and Fig. 1.

per cell in fish and mouse brain are even more striking than the concentrations of these substances based on tissue wet weight.

A more important fact which can be derived from the data of Table 2 is the following: whereas the concentrations of a typical cytoplasmic constituent, the RNA [and also of protein (30)], is increased per cell, and, concomitantly with RNA, the concentrations of putrescine and spermidine, the average concentration per cell of spermine remains constant throughout life in the fish brain. In other words, we observed in the growing fish brain a constant proportion between spermidine (and putrescine) and RNA, and between spermine and DNA.

Polyamines are known to bind to nucleic acids and to other polyanionic macromolecules (32). One among several possible functions may be the neutralization of a part of the phosphate groups of nucleic acids, which are known to be partially neutralized by basic proteins (33, 34). However, the localization of the polyamines in certain cellular compartments is still an unsolved problem. Its solution would facilitate the interpretation of their functional role. Several lines of evidence have been gathered for their localization within the cell.

Autoradiographic studies after the administration of radioactive putrescine cannot distinguish between the different putrescine metabolites. Even more than 20 days after its administration, radioactivity is still found in the putrescine of fish (Fig. 2) and mouse brain (35). Furthermore, spermidine, spermine, and the nonsoluble acid fraction are labeled to a considerable extent. Shortly after the putrescine injection, a part of the radioactivity is present in the form of N-acetyl putrescine and various putrescine degradation products, among which GABA has been identified (see below).

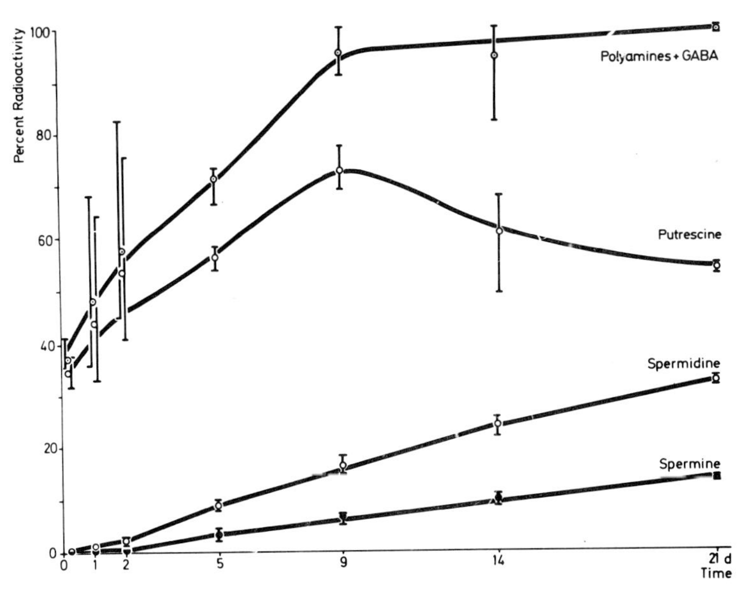

FIG. 2. Distribution of radioactivity among putrescine, spermidine, and spermine in percent of the radioactivity present in the perchloric acid extract of trout brain after the intracerebral injection of 10 μC 1,4-^{14}C-putrescine·2HCl. Each point represents the mean value of triplicate determinations; the bars indicate maximal deviations.

In mouse and fish brain, accumulation of radioactivity derived from putrescine was, with some exceptions, preferential in the areas rich in nerve cells, i.e., in the areas with maximal protein and RNA synthesis (35, 36), radioactivity being demonstrable both in the cytoplasm and in the nuclei of different cell types (37, 38).

The usual methods used for the separation of subcellular elements (centrifugation or electrophoresis of tissue homogenates in water containing media) do not rule out secondary redistribution of the polyamines during tissue processing, since polyamines are presumably bound to polyanions by reversible (ionic) bonds. In bacterial tRNA and ribosomes, however, nonexchangeable spermidine has been detected (39, 40). These observations are in agreement with the presumed structural relationship between these intrinsic cell constituents.

Although the evidence is rather indirect, constant proportions between certain polyamines and nucleic acids under varying physiological conditions can be assumed to be expressions of functional or even structural relationships. In bacteria (41, 42), liver (43–50), peripheral nerve (26, 27) and during the embryological development of several species (29, 52–55), concomitant changes of polyamines and nucleic acids have been observed under varying conditions. From these, and from numerous other findings, which cannot be mentioned here in detail (see, however, 32, 56–60), a preferential interrelation seems to be probable between spermidine and RNA, although it must be assumed that functional aims other than the different types of RNA may exist for spermidine.

It may be inserted here that we were not able to confirm reports describing RNA changes in brain after the administration of Mg-Pemoline (61) (see also 62), 1,1,3-tricyano-2-amino-1-propene (63), or in rats or mice taught to walk a wire (64), so that the comparison of RNA changes with assumed polyamine changes in brain was not possible.

The structural or functional partners of spermine are much less clear than those of spermidine. They may be identical with those of spermidine in certain cells; however, they may be different in others. No clear-cut quantitative relationships have been observed between spermine and other intrinsic cell constituents, although the spermine content is normally elevated with increasing amounts of DNA. During the drug-induced liver growth, spermine increase precedes that of DNA (50). In this connection the constant proportion between spermine and DNA in the growing fish brain should be emphasized.

In fish brain an increase of the cell volume seems to parallel the increasing cell number during brain growth, because RNA (Table 2) and protein (30) content per cell is higher in the brains of old animals than in the brains of young animals. The constancy of the spermine content per cell indicates that the cellular compartment of spermine is apparently unaffected by the cell growth. One of the cellular compartments, which presumably is not changed by the increase of the cell volume, is the cell nucleus. Nevertheless, the evidence that spermine is preferentially localized in the cell nucleus is rather indirect. Such an assumption can only be justified if we also consider other observations concerned with the compartmentalization of spermine.

The most important evidence which suggested that spermine but not spermidine is concentrated in the cell nucleus is the high spermine content of nucleated blood cells in comparison to enucleated ones (65). However, cell nuclei of rat liver and calf thymus, prepared in nonpolar organic solvents, contained both spermidine and spermine (66). In favor of a specific function for spermine in the cell nucleus is the following observation: a small but significant amount of spermine was bound to DNA by intact cells of *Escherichia coli* (67), whereas reversibly bound spermine was found both in the DNA and in the RNA fraction. Good evidence, at least for the localization of spermidine and spermine in different cellular compartments, was found by observing a proportional decrease of spermidine and cytoplasmic RNA in the liver of mice deprived of food. Although in these experiments the RNA content of liver was decreased to approximately 50% of the normal content, no decrease of the spermine (and DNA) content was observed (49). Different compartments of spermine and spermidine were recently reported in herpes simplex virions. Spermine is present within the nucleocapsid and assumed to be associated with DNA (68).

IV. POLYAMINE SYNTHESIS

Twenty-four hours after the intraventricular injection of 2.5 μC of 5-^{14}C-D,L-ornithine, the following amounts of radioactivity were found in the trout brain: putrescine, 138,000 dpm/μmole; spermine, 39,500 dpm/μmole, spermidine, 78,000 dpm/μmole; and GABA, 8,400 dpm/μmole. These data demonstrate that putrescine and the polyamines can be formed in fish brain from ornithine. The incorporation of intracerebrally injected radioactive putrescine into spermidine and spermine suggests that the polyamines in fish brain are synthesized along the pathway, which in principle seems to be the same for all types of cells (32, 69, 70).

As compared with mammalian brain, the rate of polyamine synthesis is low in fish brain. Twenty-one days after the administration of ^{14}C-putrescine approximately 50% of the radioactivity which is present in an acid-soluble form is still represented by putrescine (Fig. 2), whereas in rat and mouse brain the main part of radioactivity is found in spermidine and spermine shortly after the putrescine administration (35, 71).

The endogenous concentration of a substance is controlled by the rates of synthesis and degradation. In trout brain putrescine concentration is high. According to preliminary results, the ornithine decarboxylase activity is, however, low in this organ. In addition, the concentration of ornithine is considerably lower than the concentration of putrescine (30). Therefore it would not be surprising if putrescine is formed in trout brain not only from ornithine, but also from other precursors, for instance, agmatine. Since, however, the degradative and synthetic processes starting with putrescine are also slow, a second precursor for putrescine is not necessarily to be assumed.

Radioactivity is eliminated from trout brain in an exponential rate-time course after the intracerebral injection of ^{14}C-putrescine, even if the injected amounts of putrescine were above physiological concentration. A half-life of 8 days was calculated from the slope of the elimination curve for putrescine in trout brain (Fig. 3). Spermine and spermidine reach their maximal specific activities approximately 51 days after the ^{14}C-putrescine injection (Fig. 4); i.e., their synthesis rate is slow. Furthermore this figure shows that the maximal specific activities of spermidine and spermine remain constant

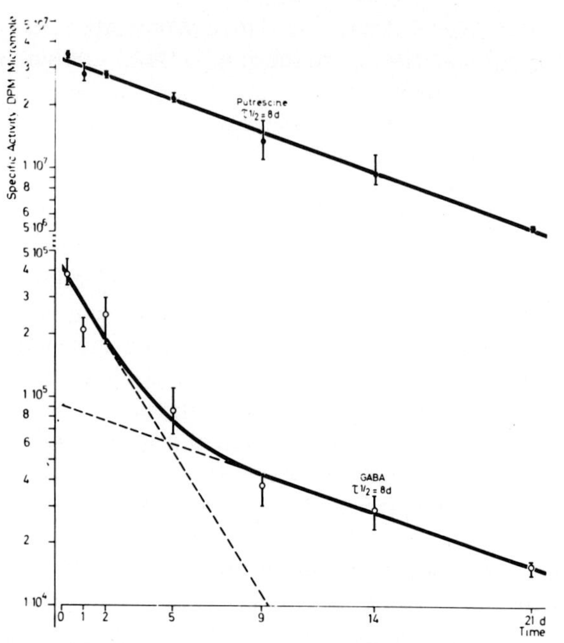

FIG. 3. Decline of specific activity of ^{14}C-putrescine and ^{14}C-GABA in trout brain after the intracerebral administration of 10 μC 1,4-^{14}C-putrescine·2HCl. Each point represents the mean ± S.E.M. for three fish. (Semilogarithmic scale)

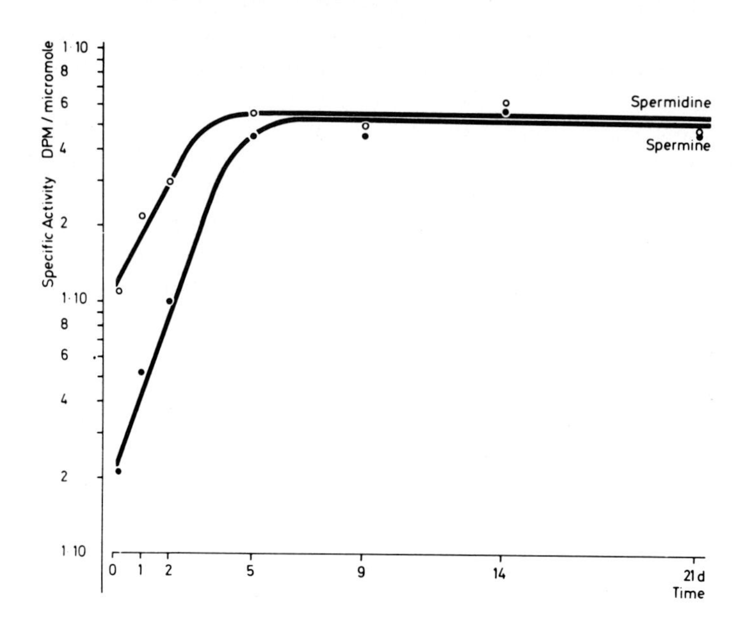

at least for a period of 2 weeks, so that we were unable to determine the biological half-life of the polyamines in trout brain. Low turnover rates are already known for spermine in the mammalian brain. For spermidine, however, a biological half-life of 5 days was found in the rat brain (71).

V. POLYAMINE DEGRADATION

Polyamine degradation in vertebrate tissues has scarcely been studied. A diamine oxidase (72) may be responsible for the catabolism of putrescine, and plasma polyamine oxidases (73) may play some role in the degradation of spermidine and spermine in some vertebrate species, especially in ruminants. The oxidative deamination of spermine in chicken brain is believed to be catalyzed by a spermine oxidase (74). Kakimoto and co-workers (75) discovered N-(4-aminobutyl)-3-aminopropionic acid (putreanine) in mammalian CNS, a possible decomposition product of polyamines. The deamination of putrescine to γ-aminobutyraldehyde and the dehydrogenation of this aldehyde to GABA comprise a pathway that is realized in different bacterial strains (1, 2). Spermine is another precursor of GABA in microorganisms (76). In vertebrates the only precursor of GABA is glutamic acid according to current view (3), and GABA formation is believed to be restricted to the nervous system and to related structures.

We found GABA, although in low concentrations, in a variety of nonneural mammalian tissues (77). At that time not even low glutamate decarboxylase activities were known in visceral organs (78, 79). Therefore we were looking for a GABA precursor which was not restricted to the nervous system, but which occurred rather ubiquitously.

In a first series of experiments, the incorporation of putrescine carbon into the GABA skeleton was demonstrated (80) in liver and brain of rats. These experiments gave some evidence for the existence of a direct pathway from putrescine to GABA, a pathway which does not include glutamic acid as an intermediate. Formation of GABA from glutamic acid was inhibited by thiosemicarbizide to a larger proportion than the formation of GABA from putrescine (81). This observation is further evidence for the existence of the direct pathway. Since, however, both series of experiments were performed with the same method, namely the dansylation procedure (5), amino acid separations were carried out in the present work by ion-exchange chroma-

---

←

FIG. 4. Specific activities of spermidine (open circles) and spermine (dots) in the brain of trout (*Salmo irideus*, Gibb.) after the intracerebral administration of 10 μC 1,4-^{14}C-putrescine·2HCl. Each point represents the mean value of triplicate determinations. (Semilogarithmic diagram)

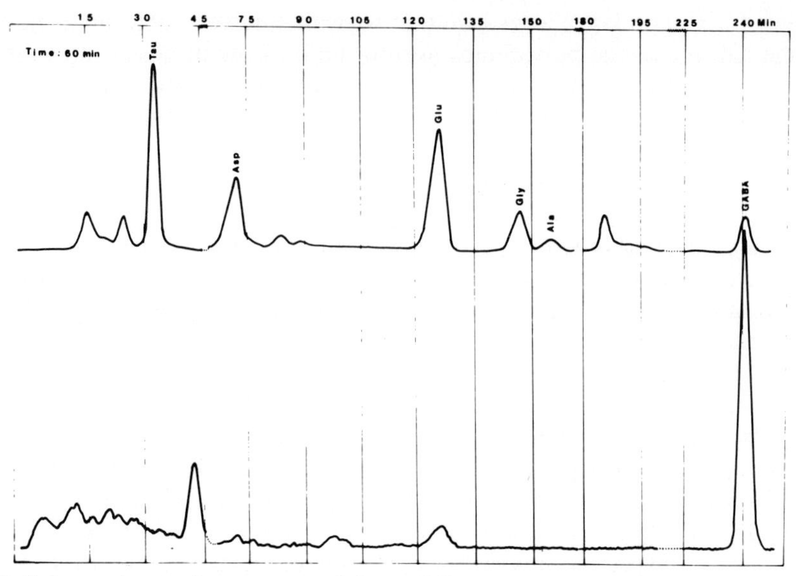

FIG. 5. Ion exchange chromatogram of the partially hydrolyzed perchloric acid extract of a mouse brain (see Materials and Methods), 60 min after the intraventricular injection of 20 μC of 1,4-^{14}C-putrescine·2HCl. Upper scan: colorimetry of the ninhydrine positive compounds; lower scan: radioactivity scan obtained with a flow counter, which was directly coupled to the ion-exchange column according (8). Ordinate, arbitrary units of radioactivity and transmittance, respectively; abscissa, elution time.

tography and radioactivity measurements with a flow counter, which was directly attached to the ion-exchange column. With regard to amino acids, radioactivity was found mainly in the brain GABA, after the intraventricular injection of 20 μC of 1,4-^{14}C-putrescine to mice (Fig. 5). Glutamic acid was the only other common amino acid labeled to a measurable degree. As is apparent from Fig. 5, the specific activity of glutamic acid was considerably lower than the specific activity of GABA. [In the case of 5-^{14}C-ornithine injections, glutamic acid was preferentially labeled. This is not surprising, however, since the mutual conversion of these amino acids, with glutamic acid semialdehyde as an intermediate, is well known (82).] The other radioactive peaks which appear in the chromatogram have not been identified. Essentially the same results have been obtained using trout as experimental animals. Even after intraperitoneal injections of ^{14}C-putrescine, the specific activity of GABA was considerably higher than the specific activity of glutamic acid during the entire experimental period. Sixty min after the i.p. injections of 10 μC ^{14}C-putrescine a specific activity of 2.7 μC/mmole was found for GABA and 0.14 μC/mmole for glutamic acid.

If we compare the decline of the specific activities of putrescine and GABA in the brain of trout after the intracerebral injection of ^{14}C-putrescine, we observe initially a relatively high specific activity of GABA which decreases rapidly (Fig. 3). After approximately 9 days, the slope of the GABA curve exactly parallels the slope of the putrescine elimination curve. The ratio between the specific activities of GABA and putrescine during this phase is 0.003. It can be assumed from this observation that a relatively large proportion of GABA is produced from exogenous putrescine, until it is equilibrated with the endogenous pool. A constant, although small, degradation rate of putrescine to GABA is observed during physiological conditions.

Since there is no evidence for a small glutamic acid pool in the brain which is very actively synthesizing GABA and is thus responsible for the formation of GABA with a higher specific activity than that of glutamic acid, the observations reported here confirm our previous statement about the existence of a pathway from putrescine to GABA in the vertebrate organism, which does not include glutamate decarboxylation as an intermediary step. Of course, most brain GABA is formed from glutamic acid, and it is likely that in other tissues glutamate decarboxylation comprises a considerable part of the GABA formation as well. The small proportion of GABA which is formed in brain from putrescine as compared to glutamate can be assessed from the low specific activity of GABA in comparison to putrescine (Fig. 3).

It is likely that the pathway from putrescine to GABA is a part of one of the general physiological routes for the degradation of putrescine to CO_2. The partial oxidation of exogenous putrescine to CO_2 in mammals (82) and of GABA in non-neural tissue (83) is well established. The existence of this pathway is a good explanation for the occurrence of the GABA-degrading enzymes GABA-α-ketoglutarate aminotransferase (GABAT) and succinate semialdehyde dehydrogenase (SSADH) not only in neural but also in non-neural tissue. The reactions presumably involved in putrescine metabolism are schematically compiled in Fig. 6. The quantitative aspects of putrescine metabolism in brain are scarcely understood. In addition to polyamine synthesis and degradation to CO_2 via GABA, a certain part of putrescine is acetylated. We have demonstrated the occurrence of ^{14}C-N-acetylputrescine in trout brain after the intracerebral injection of radioactive putrescine by electrophoretic separation, subsequent dansylation, and mass spectrometry of the DANS-derivative. The endogenous concentration of this putrescine metabolite is unknown.

We are not only ignorant about some of the important quantitative aspects of putrescine metabolism in brain but also about the enzyme reactions responsible for the deamination of putrescine. Since it is known that diamine

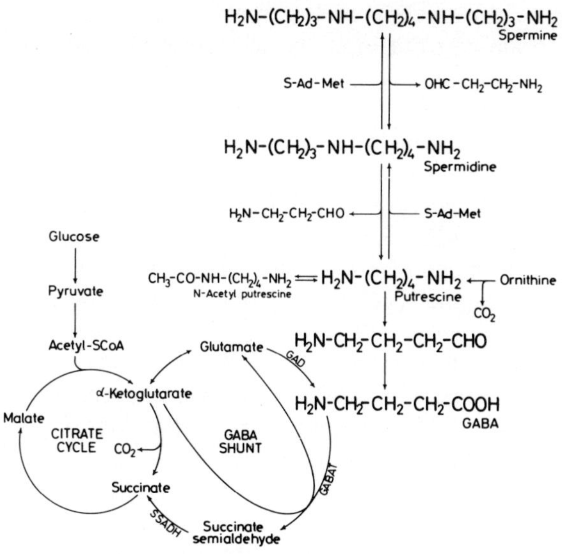

FIG. 6. Reactions involved in putrescine metabolism.

oxidase activities determined in brain homogenates are low (72), transamination reactions have to be taken into consideration as the initial step of putrescine deamination.

After a single intraventricular ^{3}H-putrescine dose, we were able to demonstrate radioactive putrescine in the mouse brain over a period of more than 28 days (35). We regard this observation as a hint for the existence of a degradative pathway of the polyamines spermidine and spermine to putrescine, in analogy to that observed in bacteria (76) (Fig. 6). Since suitably labeled polyamines are not available at present, we were only able to demonstrate the degradation of spermine to spermidine in trout brain. The electrophoretograms are shown in Fig. 7. They have been obtained from brain extracts at different times after the injection of ^{3}H-spermine, which was labeled in the positions C_1 and C_{10}. In addition to spermidine some other radioactive compounds seem to be formed from spermine, which may have been derived from the aldehyde, presumably split off by the oxidase reaction (see Fig. 6). Since spermine was not labeled in the putrescine moiety, it was not possible to demonstrate the formation of putrescine or GABA in this experiment. However, we were able to show by autoradiography that spermine is rapidly penetrating the cell membranes (36). If it should turn out that putrescine is formed from spermine and spermidine under physiological conditions in the vertebrate organism, the pathway outlined above would not only serve as a general degradative route for putrescine, but could

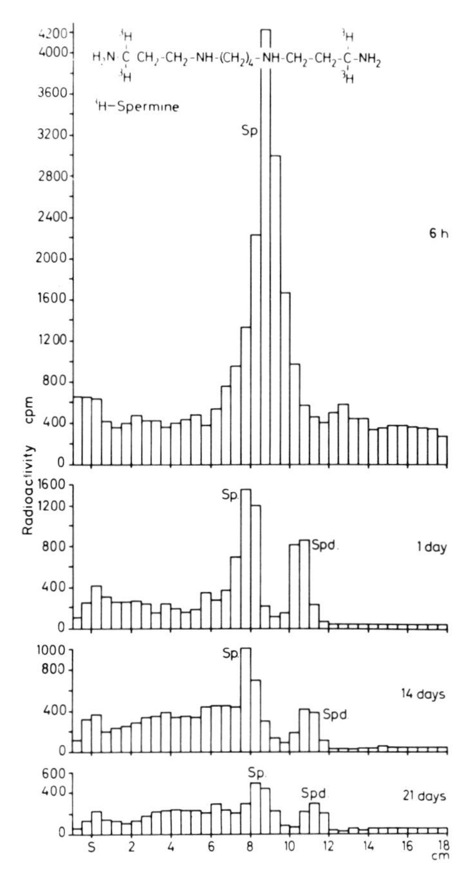

FIG. 7. Thin-layer electrophoretograms of perchloric acid extracts of trout brains after the intracerebral injection of 8 μC of ^{3}H-spermine. Ordinate: radioactivity units (cpm); abscissa: distance from the origin. Electrophoretic conditions: pyridine—acetic acid—citric acid buffer, pH 4.8 (17); 2 hr, 20 V/cm; 2°C.

comprise the whole spectrum of polyamines, as a possible regulating system for these intrinsic cell constituents.

VI. SUMMARY

Fish brains continue to grow until a very late stage of development. They seemed, therefore, to be a good model for the study of quantitative relations between polyamines and nucleic acids under physiological conditions. In the brain of the rainbow trout (*Salmo irideus,* Gibb.), constant proportions were observed between the cytoplasmic constituents RNA and protein and spermidine and putrescine throughout brain growth on one side, and between DNA and spermine on the other side. Only the presumably cytoplasmic constituents were in a linear correlation with the brain size, whereas the total increase of DNA and spermine was lower than the increase of brain weight.

Polyamine synthesis from radioactive putrescine and the degradation of spermidine and spermine are considerably lower in fish brain than in mammalian brain. By determining the specific activities of GABA and glutamic acid after the intracerebral injection of 1,4-^{14}C-putrescine, further evidence was gathered for the existence of a pathway from putrescine to GABA in vertebrate brain, a pathway which does not include glutamate decarboxylation as an intermediary step.

REFERENCES

1. Jakoby, W. B., and Fredericks, J., *J. Biol. Chem.*, **234,** 2145 (1959).
2. Kim, K., and Tchen, T. T., *Biochem. Biophys. Res. Comm.*, **9,** 99 (1962).
3. Baxter, C. F., in A. Lajtha (Editor), *Handbook of neurochemistry*, Vol. 3, Plenum Press, New York, 1970, p. 289.
4. Fischer, H. A., unpublished.
5. Seiler, N., *Meth. Biochem. Anal.*, **18,** 259 (1970).
6. Haley, T. J., and McCormick, W. G., *Brit. J. Pharmacol. Chemotherapy*, **12,** 12 (1957).
7. Noble, E. P., Wurtman, R. J., and Axelrod, J., *Life Sci.*, **6,** 281 (1967).
8. Möller, H., Seiler, N., and Werner, G., *Z. Analyt. Chem.*, **243,** 126 (1968).
9. Seiler, N., and Wiechmann, M., *Hoppe-Seyler's Z. Physiol. Chem.*, **348,** 1285 (1967).
10. Seiler, N., and Askar, A., *J. Chromatog.*, **62,** 121 (1971).
11. Seiler, N., and Wiechmann, M., *Hoppe-Seyler's Z. Physiol. Chem.*, **349,** 588 (1968).
12. Seiler, N., and Wiechmann, M., *Z. Analyt. Chem.*, **220,** 109 (1966).
13. Seiler, N., and Wiechmann, M., in A. Niederwieser and G. Pataki (Editors), *Progress in thin-layer chromatography and related methods*, Vol. 1, Ann Arbor–Humphrey Science Publishers, Ann Arbor, London, 1970, p. 94.
14. Seiler, N., and Wiechmann, M., *J. Chromatog.*, **28,** 351 (1967).
15. Seiler, N., and Knödgen, B., *Organic mass spectrometry*, **7,** 97 (1973).
16. Seiler, N., and Wiechmann, J., *Experientia*, **20,** 559 (1964).
17. Fischer, F. G., and Bohn, H., *Hoppe-Seyler's Z. Physiol. Chem.*, **308,** 108 (1957).
18. Fleck, A., and Munro, H. N., *Biochim. Biophys. Acta*, **55,** 571 (1962).
19. Munro, H. N., and Fleck, A., *Meth. Biochem. Anal.*, **14,** 113 (1966).
20. Munro, H. N., and Fleck, A., *Analyst*, **91,** 78 (1966).
21. Croft, D. N., and Lubran, M., *Biochem. J.*, **95,** 612 (1965).
22. Burton, K., *Biochem. J.*, **62,** 315 (1956).
23. Seiler, N., Schneider, H., and Sonnenberg, K.-D., *Z. Analyt. Chem.*, **252,** 127 (1970).
24. Seiler, N., Schneider, H. H., and Sonnenberg, K.-D., *Anal. Biochem.*, **44,** 451 (1971).
25. Shimizu, H., Kakimoto, Y., and Sano, I., *J. Pharmacol. Exp. Therapeut.*, **143,** 199 (1964).
26. Kremzner, L. T., Barrett, R. E., and Terrano, M. J., *Ann. N.Y. Acad. Sci.*, **171,** 735 (1970).
27. Seiler, N., and Schröder, J. M., *Brain Res.*, **22,** 81 (1970).
28. Seiler, N., and Schmidt-Glenewinkel, T., in preparation.
29. Shimizu, H., Kakimoto, Y., and Sano, I., *Nature*, **207,** 1196 (1965).
30. Seiler, N., and Lamberty, U., *J. Neurochem.*, **20** (1973).
31. Kirsche, W., *Z. Mikrosk.-Anat. Forsch.*, **77,** 313 (1967).

32. Tabor, H., and Tabor, C. W., *Pharmacol. Rev.,* **16,** 245 (1964).
33. Clark, R. J., and Felsenfeld, G., *Nature New Biol.,* **229,** 101 (1971).
34. Zimmermann, E., *Angew. Chem.,* **84,** 451 (1972).
35. Fischer, H. A., Korr, H., Seiler, N., and Werner, G., *Brain Res.,* **39,** 197 (1972).
36. Erdmann, G., Fischer, H. A., Seiler, N., and Werner, G., in preparation.
37. Gfeller, E., and Russell, D. H., *Z. Zellforsch.,* **120,** 321 (1971).
38. Fischer, H. A., Schröder, J. M., and Seiler, N., *Z. Zellforsch.,* **128,** 393 (1972).
39. Cohen, S. S., *Ann. N.Y. Acad. Sci.,* **171,** 869 (1970).
40. Cohen, S. S., and Lichtenstein, J., *J. Biol. Chem.,* **235,** 2112 (1960).
41. Raina, A., Jansen, M., and Cohen, S. S., *J. Bact.,* **94,** 1684 (1967).
42. Cohen, S. S., and Raina, A., in H. J. Vogel, J. O. Lampen, and V. Bryson (Editors), *Organizational biosynthesis,* Academic Press, New York, 1967, p. 157.
43. Dykstra, W. G., and Herbst, E. J., *Science,* **149,** 428 (1965).
44. Raina, A., Janne, J., and Simes, M., *Biochim. Biophys. Acta,* **123,** 197 (1966).
45. Raina, A., Jänne, J., Hannonen, P., and Hölttä, E., *Ann. N.Y. Acad. Sci.,* **171,** 697 (1970).
46. Heby, O., and Lewan, L., *Virchows Arch. Abt. B. Zellpath.,* **8,** 58 (1971).
47. Neish, W. P., and Key, L., *J. Cancer,* **2,** 69 (1967).
48. Andersson, G., and Heby, O., *J. Nat. Cancer Inst.,* **48,** 165 (1972).
49. Seiler, N., Werner, G., Fischer, H. A., Knötgen, B., and Hinz, H., *Hoppe-Seyler's Z. Physiol. Chem.,* **350,** 676 (1969).
50. Seiler, N., and Askar, A., *Hoppe-Seyler's Z. Physiol. Chem.,* **353,** 623 (1972).
51. Dion, A. S., and Herbst, E. J., *Ann. N.Y. Acad. Sci.,* **171,** 723 (1970).
52. Caldarera, C. M., Giorgi, P. P., and Casti, A., *Boll. Soc. Ital. Biol. Sper.,* **45,** 460 (1969).
53. Russell, D. H., Snyder, S. H., and Medina, V. J., *Life Sci.,* **8,** 1247 (1969).
54. Caldarera, C. M., Barbiroli, B., and Moruzzi, G., *Biochem. J.,* **97,** 84 (1965).
55. Caldarera, C. M., and Moruzzi, G., *Ann. N.Y. Acad. Sci.,* **171,** 709 (1970).
56. Bachrach, U., *Ann. Rev. Microbiol.,* **24,** 109 (1970).
57. Cohen, S. S., *Introduction to the polyamines,* Prentice-Hall, Englewood Cliffs, 1971.
58. Herbst, E. J., and Tanguay, R. B., *Progr. Mol. Subcell. Biol.,* **2,** 166 (1970).
59. Raina, A., and Jänne, J., *Fed. Proc.,* **29,** 1568 (1970).
60. Stevens, L., *Biol. Rev.,* **45,** 1 (1970).
61. Simon, L. N., and Glasky, A. J., *Life Sci.,* **7,** 197 (1968).
62. Ellis, D. B., Sawyer, J. L., and Brink, J. J., *Life Sci.,* **7,** II, 1259 (1968).
63. Hyden, H., *The neuron,* Elsevier Publishing Company, Amsterdam, 1967, p. 179.
64. Dellweg, H., Gerner, R., and Wacker, A., *J. Neurochem.,* **15,** 1109 (1968).
65. Shimizu, H., Kakimoto, Y., and Sano, I., *Arch. Biochem. Biophys.,* **110,** 368 (1965).
66. Stevens, L., *Biochem. J.,* **99,** 11P (1966).
67. Johnson, H. G., and Bach, M. K., *Arch. Biochem. Biophys.,* **128,** 1984 (1968).
68. Gibson, W., and Roizman, B., *Proc. Nat. Acad. Sci. U.S.A.,* **68,** 2818 (1971).
69. Williams-Ashman, H. G., and Lockwood, D. H., *Ann. N.Y. Acad. Sci.,* **171,** 882 (1970).
70. Pegg, A. E., *Ann. N.Y. Acad. Sci.,* **171,** 977 (1970).
71. Russell, D. H., Medina, V. J., and Snyder, S. H., *J. Biol. Chem.,* **245,** 6732 (1970).
72. Zeller, E. A., in P. D. Boyer, H. Lardy, and K. Myrbäck (Editors), *The enzymes,* Vol. 8, Academic Press, New York, 1963, p. 313.
73. Blaschko, H., *Adv. Comp. Physiol. Biochem.,* **1,** 67 (1962).
74. Caldarera, C. M., Moruzzi, M. S., Rossoni, C., and Barbiroli, C., *J. Neurochem.,* **16,** 309 (1969).
75. Kakimoto, Y., Nakajima, T., Kumon, A., Matsuoka, Y., Imaoka, N., Sano, I., and Kanazawa, A., *J. Biol. Chem.,* **244,** 6003 (1969).
76. Razin, S., Gery, I., and Bachrach, U., *Biochem. J.,* **71,** 551 (1959).
77. Seiler, N., and Wiechmann, M., *Hoppe-Seyler's Z. Physiol. Chem.,* **350,** 1493 (1969).
78. Whelan, D. T., Scriver, C. R., and Mohyuddin, F., *Nature,* **224,** 916 (1969).
79. Haber, B., Kuriyama, K., and Roberts, E., *Biochem. Pharmacol.,* **19,** 1119 (1970).
80. Seiler, N., Wiechmann, M., Fischer, H. A., and Werner, G., *Brain Res.,* **28,** 317 (1971).

81. Seiler, N., and Knödgen, B., *Hoppe-Seyler's Z. Physiol. Chem.*, **352,** 97 (1971).
82. Meister, A., *Biochemistry of the amino acids,* Vol 1, Academic Press, New York, 1965, p. 355.
83. Jänne, J., *Acta Physiol. Scand. Suppl.,* **300,** 7 (1967).
84. Wilson, W. E., Hill, R. J., and Koeppe, R. E., *J. Biol. Chem.*, **234,** 347 (1959).

Polyamines in Normal and Neoplastic Growth. Edited by D. H. Russell. Raven Press, New York © 1973.

Anti-Polyamine Antibodies and Growth

G. Quash,* Luce Gresland,** E. Delain,*** and J. Huppert**

*Department of Biochemistry, University of the West Indies, Mona, Kingston 7, Jamaica, West Indies, **Unité de Biologie Moléculaire, Groupe de Recherche N°8 du C.N.R.S., Institut Gustave Roussy, ***Laboratoire de Microscopie Electronique, Institut Gustave Roussy, 94, Villejuif, France

Many roles have been proposed for the polyamines spermine, spermidine, and putrescine in certain physiological processes, i.e., tRNA methylation (1), aminoacylation of tRNA (2), RNA synthesis (3), bacterial cell division (4), and the growth of animal cells in tissue culture (5, 6).

In vivo the experiments which have led to the above-mentioned conclusions have involved the measurement of putrescine, spermidine, and spermine levels, and concomitantly those of RNA in response to specific stimuli, e.g., administration of growth hormone to hypophysectomized rats (7), during tissue regeneration (8), or during embryonic development (9).

It has been reported (6) that alterations occur when polyamines are added *in vitro* in a limited concentration range to the biosynthetic systems for RNA or to cells growing in tissue culture. If the roles defined by the *in vitro* approach are a true reflection of what occurs *in vivo,* then the functioning of *in vitro* systems containing *physiological* concentrations of polyamines should be altered if the amines are selectively removed. Of the *in vitro* systems mentioned, we chose to study growth, in view of the early rise and subsequent fall in polyamine levels in rapidly growing normal tissue as opposed to their accumulation in malignant cells (10). This coincidence between elevated polyamine levels and malignant growth has more recently been substantiated by Russell and her colleagues (11) who have shown that there is a significant rise in polyamine levels in the urine of patients suffering from malignant diseases.

In order to affect growth by selectively sequestering the polyamines, it is first necessary to develop a reagent(s) capable of reacting specifically with these compounds. To achieve this specificity under physiological conditions, we decided to use the immunological approach.

I. PREPARATION AND CHARACTERIZATION OF ANTI-POLYAMINE ANTIBODIES

A. Principle of the immunological approach

As previously described (12), our approach consisted of using a polyamine as a hapten linked to a protein (e.g., bovine serum albumin or lysozyme) and injecting this complex mixed with Freund's complete adjuvant into experimental animals by the technique of Agustin (13). Boosters were given intravenously, and the animals were bled 5 days after the last injection.

B. Evidence for the presence of antibodies to polyamines

To avoid cross-reaction with the antibodies directed against the protein carrier, the sera were tested with poly-L–glutamic acid–spermine (poly-G-spermine). With this antigen the following results were obtained. (1) γ-Globulins prepared from the immunoserum gave a characteristic antigen–antibody precipitation curve with dissolution of the precipitate in excess antigen. (2) The amount of protein precipitated at the equivalence point increased as immunization proceeded. Typical values were as follows:

Serum	μg protein precipitated
normal pre-immune serum	48
immune serum after 4 weeks	720
" " " 6 "	1,252
" " " 8 "	1,293

(3) The complement fixation test was positive with the antiserum diluted 1/100 and 1.7 μg of poly-G-spermine. (4) Sheep red blood cells (SRBC) sensitized with poly-G-spermine at a concentration of 20 μg/ml gave positive hemagglutination tests with the serum diluted up to 1/1000.

From the foregoing we concluded that the injection of a spermine–protein complex to rabbits resulted in the formation of antibodies to spermine. The next step was to determine if there was cross-reaction with the different polyamines and, if so, to what extent.

C. The cross-reactions of anti-polyamine antibodies

This was investigated by adding varying concentrations of free spermine, spermidine, putrescine, cadaverine, and lysine individually to the antiserum 4 hr before poly-G-spermine was added. After further incubation for 16 hr at 4°C, the protein precipitated was estimated. If the free amines added combine with the antibodies present in the antiserum, there should be a diminution in the amount of protein precipitated. This is precisely what was found, as shown below.

Amine added	% maximum inhibition
spermine	95
spermidine	85
putrescine	65
diaminopropane	65
cadaverine	50
lysine	0

These results showed definitely that cross-reaction was taking place between the different polyamines but that there was no interaction with lysine. There were two possible explanations: (1) the presence in the antiserum of either one antibody type having different affinities for the various polyamines, or (2) different types of antibodies each with its specificity for the tetramine spermine, the triamine spermidine, and the diamines putrescine and diaminopropane.

D. Evidence for the existence of antibodies with different specificities

To distinguish between the two possibilities given above, four antigens were made: poly L-glut–spermine, poly-L-glut–spermidine, poly-L-glut–putrescine, and poly-L-glut–diaminopropane. Precipitin reactions between these antigens and the antiserum resulted in, as previously described (14), typical antigen–antibody–precipitin curves with values for protein precipitated at the equivalence point of the following:

Antigen	μg protein precipitated
poly-L-glut–spermine	1,750
poly-L-glut–spermidine	875
poly-L-glut–putrescine	320
poly-L-glut–diaminopropane	320

These results were indicative of different antibody types; however, to prove this conclusively we decided to eliminate all the anti-diamine antibodies by absorbing the antiserum completely with poly-L-glut–diamino-

TABLE 1.

Antigen	Serum dilutions $\frac{1}{50}$	$\frac{1}{100}$	$\frac{1}{200}$	$\frac{1}{400}$	$\frac{1}{800}$	$\frac{1}{1600}$	Sensitized red blood cells	Serum blank
Poly-G-spermine								
Absorption (1)	++++	++++	+++	+++	+++	++	–	–
" (2)	++++	++++	+++	+++	+++	++		–
" (3)	++++	++++	+++	+++	+++	++		–
" (4)	++++	++++	+++	+++	++	+		–
" (5)	++++	+++	+++	+++	++	+		–
Poly-G-DAP								
Absorption (1)	+++	+++	+++	+	–	–	–	–
" (2)	+++	++	+	–	–	–		–
" (3)	+++	+	–	–	–	–		–
" (4)	+	–	–	–	–	–		–
" (5)	–	–	–	–	–	–		–

propane (poly-G-DAP). If in fact the antiserum contains more than one antibody type, then antibodies still capable of reacting with spermidine and spermine should remain after this absorption.

Thus, after each addition of poly-L–glut–DAP and elimination of the precipitate, the antibodies remaining in the serum were titred by the hemagglutination technique. As can be seen in Table 1, the addition of poly-L–glut–DAP brought about a successive diminution in the antidiamine titer, whereas there was no diminution in the anti-tetramine titer. If there were only one antibody type with different affinities, there should have been a corresponding decrease in the antibodies reacting with spermine. This, therefore, is evidence for the existence of distinct antibody types with specificities for diamines, and also for triamines and tetramines, since this same serum gave characteristic antigen–antibody precipitation curves with poly-L–glut–spermidine and poly-L–glut–spermine (14).

From the results presented above it was concluded that antibodies to polyamines can be formed by injecting a spermine–protein complex to experimental animals and that the antibodies so formed show distinct specificities for diamines, triamines, and tetramines. These specific immunological tools could now be used to investigate directly the role of polyamines in growth.

II. EFFECT OF THE ANTISERUM ON GROWTH

A comparative investigation was undertaken to ascertain the effect of the antiserum on the growth of normal mouse fibroblasts, L cells (spontaneously transformed mouse fibroblasts), and baby hamster kidney cells transformed by polyoma virus (BHK Py) growing in tissue culture.

All cells in monolayer were grown in Eagle's medium supplemented with glutamine and bicarbonate. Waymouth medium was used for suspension cultures of transformed BHK cells. Rabbit or calf serum was used at a concentration of 10% or as specified. Cells were grown in incubators at 37°C in a humid atmosphere of air and 5% Co_2. Further experimental details and the results obtained have already been reported (15).

A. Cell growth as a function of antiserum concentration

To determine the effect of antiserum on growth, cells were counted 4 days after the seeding of 4.5×10^5 cells in plastic Petri dishes containing medium plus varying concentrations of rabbit antiserum. Control cultures were seeded similarly but contained, instead, medium supplemented with pre-immune serum from the *same* rabbit. Calf serum was added to the cultures, both experimental and control, to obtain a final serum concentration of 10%.

At the end of 4 days, the period chosen to maximize any differences in cell growth, cells were still in their exponential phase and approached the stationary phase on day 5.

The results presented in Table 2 show that at 8% antiserum (0.5 ml) there is a significant effect on cell number in the case of both L cells and normal mouse fibroblasts. With pre-immune serum at the same concentration, the cells grew at the same rate as those grown in the presence of calf serum. The fact that at lower serum concentrations the drop in cell number was not as significant could indicate that a threshold amount of antiserum is necessary to produce an effect on a given number of cells. Another point that emerges is that the final cell count after 4 days in the case of both L cells and normal mouse fibroblasts was less than 4.5×10^5, i.e., the number

TABLE 2. *Viable cell count after 4 days*

Type of cell	Vol. pre-immune or antiserum	Pre-immune serum 253	Antiserum 253
L cells	0.5	7.1×10^6	5.1×10^4
	0.3	8.0×10^6	6.3×10^6
	0.1	7.6×10^6	6.8×10^6
Mouse fibroblasts	0.5	1.2×10^6	2.5×10^4
	0.3	1.4×10^6	4.5×10^5
	0.1	1.7×10^6	1.7×10^6

Dishes were seeded with 4.5×10^5 cells in 5.4 ml of Eagle's Medium to which was added the indicated amount of rabbit serum completed to a total volume of 6.0 ml with calf serum.

of cells originally seeded. This means that cytolysis or cytostasis followed by cell death was taking place.

This second alternative can be ruled out immediately from the results obtained with BHK Py growing in suspension in the presence of antiserum. Here, after an initial 50% decrease in the number of cells on the first day, the cell count remained constant for the next 3 days.

These results were obtained with an antiserum containing complement and antibodies to the protein carrier–lysozyme and to the hapten–spermine. It now had to be determined whether the cytolysis observed was due to the presence in the antiserum of antibodies, anti-lysozyme, and/or anti-spermine, as well as complement.

B. Anti-polyamine antibodies—the cytolytic agents

This was investigated by growing BHK Py in suspension culture in the presence of antiserum and also in antiserum from which the antibodies had been eliminated by successive absorptions with an insoluble poly-L–glut–spermine complex. The absence of anti-polyamine antibodies was verified by hemagglutination as described (15).

The results showed that, although there was a drop in the number of viable cells in the presence of antiserum, in the presence of absorbed serum cells grew as well as those growing in the presence of pre-immune serum. Of 1.2×10^5 cells/ml seeded on day 1, there remained, on day 5, 5×10^4 cells/ml in the presence of antiserum, and only 3.8×10^5 cells/ml with absorbed antiserum.

These results provide evidence that it is the anti-polyamine antibodies which are implicated in cytolysis.

C. The need for complement in cytolysis

A comparison was made of the rate of growth of cells in pre-immune serum and in antiserum decomplemented by heating at 56°C for 30 min. The rates were identical for the two sera and for all cell types examined, i.e., normal mouse fibroblasts, L cells, normal chick embryo fibroblasts, and BHK Py. Thus, complement is a necessary factor for cytolysis.

Although the growth rates showed no significant differences, a microscopic examination of L cells revealed that in the presence of decomplemented antiserum these cells became rounded and no longer adhered to the plastic surface (15). With normal mouse fibroblasts, no such effect was observed; the cells retained their typical elongated shape and adhered to the cell surface.

These microscopic modifications of L cells growing in decomplemented antiserum were further substantiated by electron micrographs. At a magnification of 20,000, all the cell components were normal except for the lysosomes which were numerous. In cells treated with non-decomplemented antiserum, there was, in addition to the lysosome increase, an accumulation of large lipid droplets (15).

The observations are similar to those obtained by other authors (16) using a decomplemented antiserum to mouse ascites cells; here, an increased lysosome content was also found. Although the mechanism for this accumulation of lysosomes and lipids is not evident, the observation that this occurs with L cells (i.e., transformed mouse fibroblasts) but not with normal mouse fibroblasts prompted the next step in the investigation, i.e., to try to determine whether this difference in reactivity of cells is due to quantitative differences in the same amine(s) or to qualitative differences between normal fibroblasts and L cells. To do this, the identity of the amines involved in cytolysis had to be determined.

D. Identification of the amines involved in cytolysis

The easiest and most rapid way of accomplishing this was, as has been shown (17), to determine whether precytolytic events could be arrested by the separate addition of either putrescine, spermidine, or spermine. This approach had two advantages. (1) It should provide further evidence that it is the anti-polyamine antibodies which are involved in cytolysis and not the antibodies directed against the protein carrier. (2) An antipolyamine antiserum contains antibodies specific for putrescine, spermidine, and spermine. Therefore, if it is the diamines which take part in cytolysis, then the addition of putrescine should prevent cytolysis by competition of the putrescine for antibodies reacting with putrescine-containing sites; if it is spermidine, then the addition of free spermidine but not of putrescine should arrest precytolytic events. Finally, by the same reasoning, for spermine-containing sites only spermine but not spermidine or putrescine should prevent cytolysis.

Before this could be done, the time taken for cytolysis to occur had to be determined.

E. The sequence of events in precytolysis

Cells (2×10^4) were seeded into Falcon micro test II tissue culture plates and allowed to grow for 24 hr. The medium was then aspirated off and replaced by medium containing antiserum; microscopic examination was carried out at 15-min intervals thereafter. The 24-hr interval between trypsinization, seeding, and addition of the antiserum was introduced to eliminate the possibility of the antibodies reacting with cryptic sites (18) revealed by trypsin treatment.

Microscopic observation showed that already at 30 min there was an effect on cell morphology, which became more marked at 45 min; by 60 min, there were cells with distended membranes. Lysis was complete by 24 hr.

F. Putrescine inhibition of cytolysis

After a cell antiserum contact time of 30 or 60 min, spermine, spermidine, or putrescine was added. Putrescine was the only amine which prevented cytolysis at 60 min; however, when it was added at 30 min, not only was cytolysis prevented but the cells actually recovered. Therefore, one of the amines involved in cytolysis is putrescine. Since identical results were obtained with BHK Py transformed by polyoma virus, it would seem that, regardless of whether malignant transformation is of spontaneous or viral origin, the polyamine involved in cytolysis is putrescine, and probably membrane-bound putrescine, for the following reasons. (1) Both light and electron microscopy have shown that precytolytic events involve the cell membrane which is modified and distended (17). (2) Cytolysis requires complement (15) thus indicating a surface phenomenon. (3) If cytolysis was the result of the interaction of the antibodies with medium diamines, the addition of free putrescine should have increased rather than decreased cytolysis.

Furthermore, we can also deduce that the antibodies involved in cytolysis must be *anti-diamines,* since from the earlier immunochemical work it is known that putrescine reacts with anti-diamines only; spermidine reacts with anti-diamines and anti-triamines; spermine reacts with anti-diamines, anti-triamines, and anti-tetramines. We can therefore conclude that the cytolysis of L cells and BHK Py is brought about by the interaction of anti-diamine antibodies with putrescine-containing sites on the cell membrane.

Further investigations on growth are being directed to the identification of the amine(s) involved in the cytolysis of normal cells, their quantitative estimation, and a comparison of the number and distribution of amine sites on normal and malignant cells. In addition to their effect on growth, anti-polyamine antibodies are at present being used to investigate the role of polyamines in viral infections and in developing a quantitative specific assay for polyamines in biological fluids which would be less tedious than the qualitative immunoelectrophoretic technique previously published (19).

ACKNOWLEDGMENTS

One of the authors (G. Q.) acknowledges the financial support of: The Standing Advisory Committee for Medical Research in the British Caribbean; The French Foreign Ministry for a "bourse d'études" at Institut Pasteur, Paris, France; The International Union Against Cancer for an Eleanor Roosevelt International Cancer Fellowship at Institut Gustave Roussy, Villejuif, France.

Drs. M. Wilson, J. Jonard, and R. Wahl collaborated in the developmental work on the immunochemistry of polyamines.

REFERENCES

1. Leboy, P. S., *Biochemistry,* **9,** 1577 (1970).
2. Doctor, B. P., Fournier, M. J., and Thornsvard, C., *Ann. N.Y. Acad. Sci.,* **171,** 863 (1970).
3. Gumport, R. I., and Weiss, S. B., *Biochemistry,* **8,** 3618 (1969).
4. Inouye, M., and Pardee, A., *J. Bacteriol.,* **101,** 770 (1970).
5. Ham, R. G., *Biochem. Biophys. Res. Comm.,* **14,** 34 (1964).
6. Pohjampelto, P., and Raina, A., *Nature,* **235,** 247 (1972).
7. Kostyo, J., *Biochem. Biophys. Res. Comm.,* **23,** 150 (1966).
8. Raina, A., Jänne, J., and Siimes, M., *Biochim. Biophys. Acta,* **123,** 197 (1966).
9. Russell, D., and Snyder, S. H., *Proc. Nat. Acad. Sci. U.S.A.,* **60,** 1420 (1968).
10. Siimes, M., and Jänne, J., *Acta Chem. Scand.,* **21,** 815 (1967).
11. Russell, D., Levy, C., Schimpff, S., and Hawk, I., *Cancer Res.,* **31,** 1555 (1971).
12. Quash, G., and Jonard, J., *Compt. Rend. Acad. Sci. Paris,* **265D,** 934 (1967).
13. Agustin, R., *Immunology,* **2,** 1 (1959).
14. Jonard, J., Quash, G., and Wahl, R., *Compt. Rend. Acad. Sci. Paris,* **265D,** 1099 (1967).
15. Quash, G., Delain, E., and Huppert, J., *Exp. Cell Res.,* **66,** 426 (1971).
16. Bitensky, L. *Lysosomes,* Ciba Foundation Symp., Churchill, London, 1963, p. 371.
17. Quash, G., Gresland, L., Delain, E., and Huppert, J., *Exp. Cell Res.,* **75,** 363, 1972.
18. Burger, M., and Noonan, K., *Nature,* **228,** 512 (1970).
19. Quash, G., and Wilson, M., *B. West Ind. Med. J.,* **16,** 81 (1967).

Polyamines in Normal and Neoplastic Growth. Edited by
D. H. Russell. Raven Press, New York © 1973.

Polyamine-Synthesizing Enzymes in Regenerating Liver and in Experimental Granuloma

Aarne Raina, Juhani Jänne*, Pekka Hannonen*, Erkki Hölttä*, and Juhani Ahonen**

*Department of Biochemistry, University of Kuopio, Kuopio, Finland, *Department of Medical Chemistry, University of Helsinki, Helsinki, Finland, and **Department of Medical Chemistry, University of Turku, Turku, Finland*

The first observations in 1964–1965 which indicated that the synthesis of spermidine was increased in regenerating rat liver (1–3) opened a widening area of studies concerning polyamine metabolism in animal tissues undergoing rapid growth. A number of hypotheses have been presented that suggest important functions for putrescine, spermidine, and spermine, especially in rapidly growing animal tissues. Although we now know much about the metabolism of the polyamines, their physiological function has not yet been solved. Even without knowledge of the exact function(s) of these compounds in mammalian tissues, we are beginning to understand some of the mechanisms involved in the regulation of polyamine synthesis in various systems.

This chapter is designed to describe the biosynthesis of polyamines in two rapidly growing tissues, regenerating rat liver and experimental granuloma.

I. EXPERIMENTAL PROCEDURES

A. Animals

Female rats of the Wistar strain were used in all experiments.

B. Partial hepatectomies and production of granulation tissue

Partial hepatectomies were performed under light ether anesthesia as described by Higgins and Anderson (4). Granulomas were produced by the method of Viljanto (5) by implanting six pieces of sterilized viscose cellulose sponges (10 × 10 × 20 mm, dry weight 70 to 80 mg) under the dorsal skin of the rats. The rats were killed 4 to 42 days after implantation.

C. Preparation of tissue extracts

Livers were removed after decapitation and immediately homogenized with two volumes of ice-cold 0.25 M sucrose containing 1 mM 2-mercaptoethanol and 0.3 mM EDTA (or 100 mM KCl–0.3 mM EDTA in the lyophilization experiments). The granulomas were rapidly removed and freed from adjoining tissue. Each granuloma was divided into two equal pieces to be used for chemical analyses or enzyme assays. The piece of granuloma used for enzyme assays was homogenized with two volumes of 0.25 M sucrose–5mM dithiothreitol with an Ultra-Turrax homogenizer (Janke et Kunkel) three times for 5 sec. The homogenates were centrifuged for 20 min at 20,000 × g_{max}, and the resulting supernatant fraction was further centrifuged for 60 min at 105,000 × g_{max}. The final supernatant fraction was used for the enzyme assays as such or after dialysis for 16 hr against 100 volumes of 5 mM potassium phosphate (pH 7.2) containing 50 mM KCl, 1 mM EDTA, 2 mM 2-mercaptoethanol, and 0.5 mM dithiothreitol.

D. Chemicals

Unlabeled and labeled S-adenosylmethionine were synthesized essentially as described by Pegg and Williams-Ashman (6). Decarboxylated S-adenosylmethionine was synthesized and purified as described earlier (7, 8). DL-Ornithine-1-^{14}C (specific activity 37 mC/mmole) was purchased from the Radiochemical Center (Amersham) and treated before use as described earlier (9, 10). Putrescine-1, 4-^{14}C-dihydrochloride (specific activity 17.5 mC/mmole), spermidine trihydrochloride (specific activity 10.22 mC/mmole), DL-methionine-2-^{14}C (specific activity 4.08 mC/mmole), and DL-methionine-1-^{14}C (specific activity 3.54 mC/mmole) were purchased from the New England Nuclear Corporation. The radioactive putrescine was purified before use on a Dowex 50 (H^+) column (11).

E. Enzyme assays

Ornithine decarboxylase activity was assayed from undialyzed cytosol fractions as described earlier (10). Due to the low enzyme activity in granuloma tissues, ornithine-1-^{14}C of high specific activity was used as the substrate (0.05 mM). The activity of S-adenosylmethionine decarboxylase, spermidine, and spermine synthases and the synthesis of spermidine from S-adenosylmethionine-2-^{14}C were assayed as described by Hannonen et al. (12). The lyophilization experiments and the enzyme assays were performed as described by Hölttä and Jänne (13).

F. Chemical analyses

Polyamines were determined from the acid tissue extracts by the method of Raina et al. (14). Nucleic acids were assayed in a hot trichloroacetic acid extract as described by Ashwell (15). Protein was determined by the method of Lowry et al. (16).

II. BIOSYNTHESIS OF POLYAMINES IN REGENERATING RAT LIVER

The earliest change observed in polyamine metabolism after partial hepatectomy is the intensive stimulation of ornithine decarboxylase activity (17–20), resulting in an accumulation of putrescine (18, 21). At later stages of regeneration, there is an accumulation of spermidine concomitantly with the accumulation of RNA (1–3). In the present work we try to characterize further the factors leading to an increased synthesis of spermidine in regenerating rat liver.

A. Polyamine-synthesizing enzymes in regenerating liver

Spermidine and spermine are synthesized in mammalian tissues by at least three different enzymes: (1) a putrescine-activated S-adenosylmethionine decarboxylase, (2) spermidine synthase, and (3) spermine synthase. These three enzymes are separable proteins and probably also function as independent enzymes (22–24) although the existence of an enzyme complex has been suggested (25, 26). There are several reasons to believe that the accumulation of putrescine is primarily responsible for the increased synthesis and accumulation of spermidine in regenerating liver whereas at later stages of regeneration the stimulation of spermidine-synthesizing enzymes may also contribute to the accumulation of spermidine (12, 13, 19, 21, 27). Figure 1 shows the pattern of individual enzyme activities needed for the synthesis of spermidine and spermine in regenerating rat liver. As shown in Fig. 1, all three activities increased in a relatively parallel manner after partial hepatectomy. After the initial parallel increase, it appeared that S-adenosylmethionine decarboxylase returned more rapidly toward the control level than did spermidine synthase activity. As can be seen in Fig. 1, the peak activities occurred on day 2 or later after partial hepatectomy, as

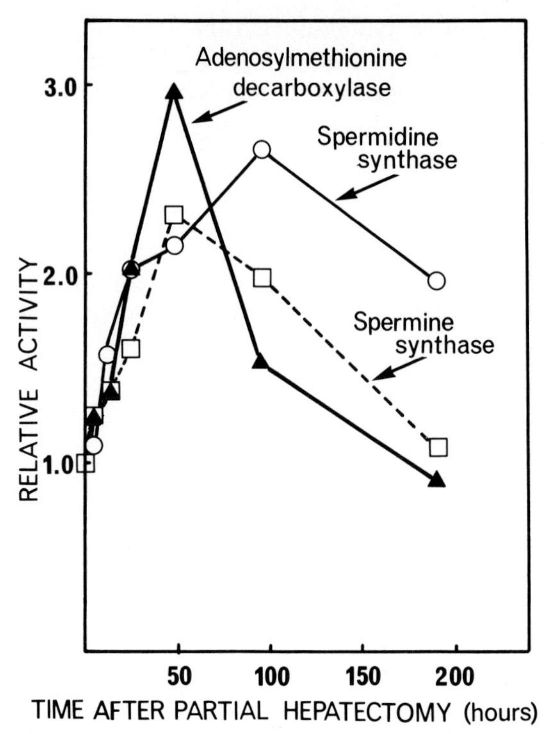

FIG. 1. Stimulation of the enzyme activities involved in the synthesis of spermidine and spermine after partial hepatectomy. The enzyme activities were measured as described under Experimental Procedures. The curves were replotted, using relative enzyme activities, from the data given by Hannonen et al. (12).

compared to the early stimulation of ornithine decarboxylase activity at 4 hr after the operation (11, 17). The parallel initial increase in the three enzyme activities might indicate a coordinate regulation of the enzyme activities after partial hepatectomy. However, the later time courses of the enzyme activities do not support this kind of hypothesis. The interpretation of the data is difficult especially in the light that these three enzymes appear to have marked differences in their apparent turnover rates. Figure 2 shows the decay patterns of the three enzymes in regenerating rat liver after administration of cycloheximide. S-Adenosylmethionine decarboxylase appeared to have a very short half-life—only 35 min—whereas spermidine and spermine synthases seemed to be rather stable. A somewhat longer half-life for S-adenosylmethionine decarboxylase from normal liver was reported by Russell and Taylor (28). It is worthy of mention that there was no decay

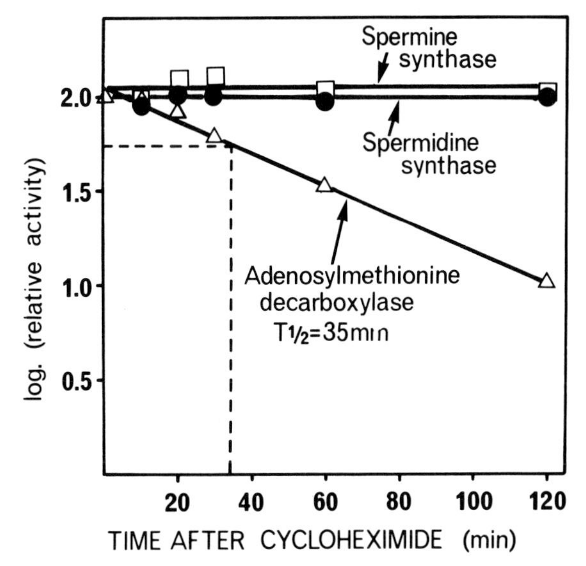

FIG. 2. Effect of cycloheximide on S-adenosylmethionine decarboxylase and spermidine and spermine synthase activities from regenerating liver. The animals were partially hepatectomized 48 hr before sacrifice. Cycloheximide (8 mg/kg) or saline was administered at times indicated. Each point represents a value obtained from a pooled sample of five to six rats. The lines were plotted by computer by the least squares method.

in spermidine and spermine synthases even 5 hr after the administration of cycloheximide.

B. Stimulation of S-adenosylmethionine decarboxylase and spermidine synthesis by putrescine

Both the observed differences in the total activities of individual enzymes (12, 22) and the decay rate of spermidine synthesis from S-adenosylmethionine and putrescine after injection of cycloheximide (11) suggest that the rate-limiting enzyme in the synthesis of spermidine in rat liver is the S-adenosylmethionine decarboxylase activity and not spermidine synthase activity. Considering the relatively late stimulation of S-adenosylmethionine decarboxylase after partial hepatectomy, it appears unlikely that this increase is responsible for the early stimulation of spermidine synthesis *in vivo* from labeled precursors (3, 21) and spermidine accumulation in the liver (1–3). It appears to us that the primary changes leading to the increased synthesis of spermidine in regenerating rat liver are the intensive stimulation of ornithine decarboxylase activity and concomitant accumulation of putres-

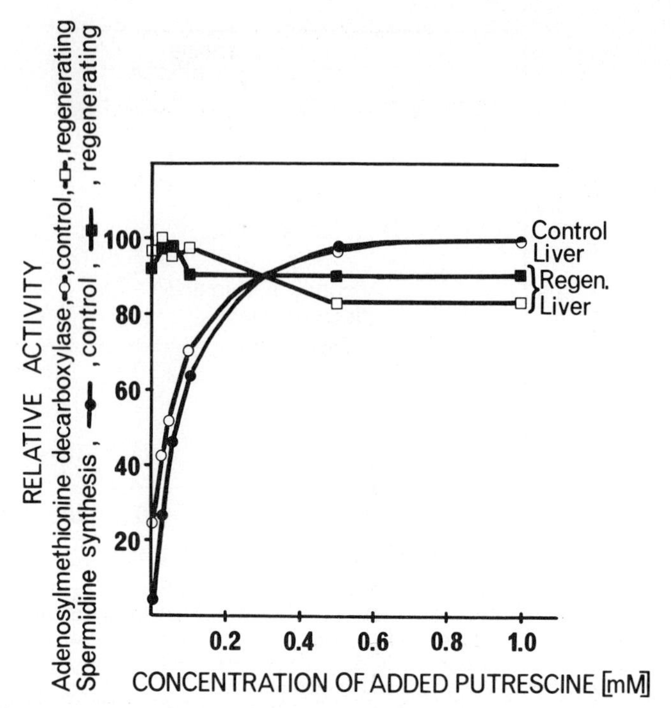

FIG. 3. Effect of exogenous putrescine on S-adenosylmethionine decarboxylase activity and spermidine synthesis from S-adenosylmethionine-2-^{14}C *in vitro*. Liver cytosol fractions were prepared from normal or 12-hr regenerating liver and lyophilized as described in the text. S-adenosylmethionine decarboxylase activity was assayed under normal assay conditions (see Experimental Procedures).

cine in the liver remnant (11, 12). Figure 3 represents an attempt to demonstrate the critical role of putrescine in the synthesis of spermidine. When a cytosol fraction from normal rat liver was lyophilized and dissolved in a specified volume of buffer so that in the final incubation mixture all the constituents were at the same concentration as in the living cell (13), the S-adenosylmethionine decarboxylase activity and the synthesis of spermidine from S-adenosylmethionine were strongly stimulated by exogenous putrescine. This may mean that in the liver cell S-adenosylmethionine decarboxylase is not saturated by endogenous putrescine. However, under the same conditions the cytosol fraction obtained from 12-hr regenerating liver showed practically no dependence on exogenous putrescine as far as the activity of S-adenosylmethionine decarboxylase and the synthesis of spermidine from radioactive S-adenosylmethionine were concerned. Both systems seemed to be saturated by the endogenous putrescine. It should also be mentioned that an extensive dialysis of the cytosol fraction from regenerating liver restored its response to putrescine. The experimentally measured K_m for putrescine as the activator of S-adenosylmethionine decarboxylase showed no difference between dialyzed cytosol fractions from either normal or regenerating liver.

After the resolution of the enzyme activities needed for the synthesis of spermidine and spermine in mammalian tissues, the role of putrescine has become more obvious. First, very low concentrations of putrescine activate the S-adenosylmethionine decarboxylase from several animal tissues (6, 8, 22, 23, 29). Second, putrescine serves as a substrate for spermidine synthase reaction, and, finally, putrescine acts as a competitive inhibitor of spermine synthase (30, 31).

C. Stimulation of ornithine decarboxylase after partial hepatectomy

The mechanism of the stimulation of ornithine decarboxylase activity and thus the accumulation of putrescine in regenerating rat liver is still unsolved. Ornithine decarboxylase has been purified extensively from rat ventral prostate (10) and rat liver (32). So far it has not been possible to find any low molecular weight effector for this enzyme. The compounds studied include at least those participating in the neighboring reactions, e.g. urea cycle, several nucleotides, etc. (10). It has been reported that partially purified ornithine decarboxylase from rat liver was inhibited by some nucleotides (32). However, using somewhat cruder enzyme preparations we have not been able to confirm this observation. The effect of nucleotides on the ornithine decarboxylase activity would be interesting especially in the light of the observation that ornithine decarboxylase from *Escherichia coli* is stimulated by a variety of nucleotides, especially by guanosine triphosphate. Certain thiols appear to be the only exceptions of low molecular weight compounds which can stimulate *in vitro* ornithine decarboxylase from animal sources, most probably by altering its physical behavior (10, 33). Thus it appears that the activity of animal ornithine decarboxylase is regulated mainly by the rate of synthesis and degradation of the enzyme protein. The extremely short apparent half-life of the enzyme is in good agreement with this kind of hypothesis (11, 34).

In addition to partial hepatectomy, treatment with several different hormones also results in an intensive stimulation of ornithine decarboxylase in their target organs (35–38); the growth hormone is especially effective in increasing the hepatic ornithine decarboxylase activity (39–41). It is possible that several factors, including humoral, are involved in the stimulation of ornithine decarboxylase activity after partial hepatectomy. In fact, we have

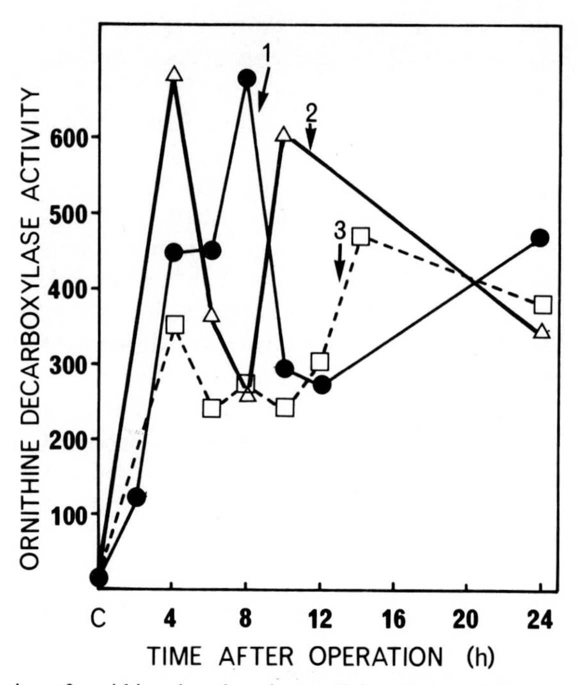

FIG. 4. Stimulation of ornithine decarboxylase activity after partial hepatectomy in rats of different ages. Ornithine decarboxylase activity was assayed under normal assay conditions. Each value represents a pooled sample obtained from three to four rats. The weights of the animals were as follows: Curve 1, 105 to 120 g; curve 2, 140 to 160 g; curve 3, 161 to 180 g. Data from Hölttä and Jänne (13).

done a series of experiments to study the ornithine decarboxylase activity at short time intervals after partial hepatectomy using animals of different ages. As shown in Fig. 4, there are certain indications that the stimulation of ornithine decarboxylase activity after partial hepatectomy occurs in several phases. There was a peak of ornithine decarboxylase activity invariably at 4 hr postoperatively; a second peak occurred later, apparently depending on the age of the animal. In young animals the second peak occurred very soon after the first peak (Curve 1), whereas in older animals there was a delay of several hours between the two peaks (Curves 2 and 3). We tentatively call the first phase a humoral phase and the second phase the regenerative response of the liver tissue itself. The involvement of a humoral phase in stimulation of ornithine decarboxylase activity is supported by the observation that a prior hypophysectomy of the rat delays but does not prevent the stimulation of ornithine decarboxylase activity after partial hepatectomy (42). Furthermore, the stimulation of ornithine decarboxylase activity after a single injection of growth hormone occurs at about 4 hr after the injection (40) independently of the age of the animal (41). The possibility for rapid changes with different phases in the ornithine decarboxylase activity is understandable in the light of the very short half-life of the enzyme. However, the definitive mechanism of the stimulation of ornithine decarboxylase activity at molecular level is still to be discovered.

III. POLYAMINE SYNTHESIS IN EXPERIMENTAL GRANULOMA

The formation of granulation tissue is a basic biological phenomenon in all mammalian repair processes. Granulation tissue can be induced experimentally by several methods, e.g., by implanting viscose cellulose sponges subcutaneously in the rat (5, 43). Three phases can be distinguished in the development of granulation tissue formation thus induced: (a) an intensive cellular proliferation during the first week, (b) rapid synthesis of collagen during the second and third weeks, and (c) finally a phase of involution (43). Experimental granuloma has been used as a model not only for studies on connective tissue formation and metabolism but also for studies on many fundamental biological problems, e.g., on differentiation of protein synthesis and on development and aging (44). Our earlier studies on granulation tissue have been directed to reveal changes in nucleic acid metabolism during different developmental phases (43). This work extends these studies to polyamine metabolism. A comparison is also made between the accumulation of polyamines and nucleic acids at various phases of granuloma development.

A. Accumulation of polyamines and nucleic acids during the development of granuloma

We reported earlier that there is a remarkable parallelism in the synthesis and accumulation between polyamines and ribonucleic acid in various systems studied so far (45). The same appears to be true also in the experimental granuloma, as shown in Fig. 5. Spermidine, spermine, and RNA accumulated in a relatively parallel way up to 21 days after implantation. Thereafter the amount of polyamines and RNA remained relatively constant, or showed slight decrease up to 42 days. The concentration of spermidine was about twice that of spermine. The putrescine content of granuloma was rather low at all phases of granuloma development.

The constant ratio of polyamine to RNA is also demonstrated in Table 1. This holds true especially of spermidine. On the other hand, the ratio of

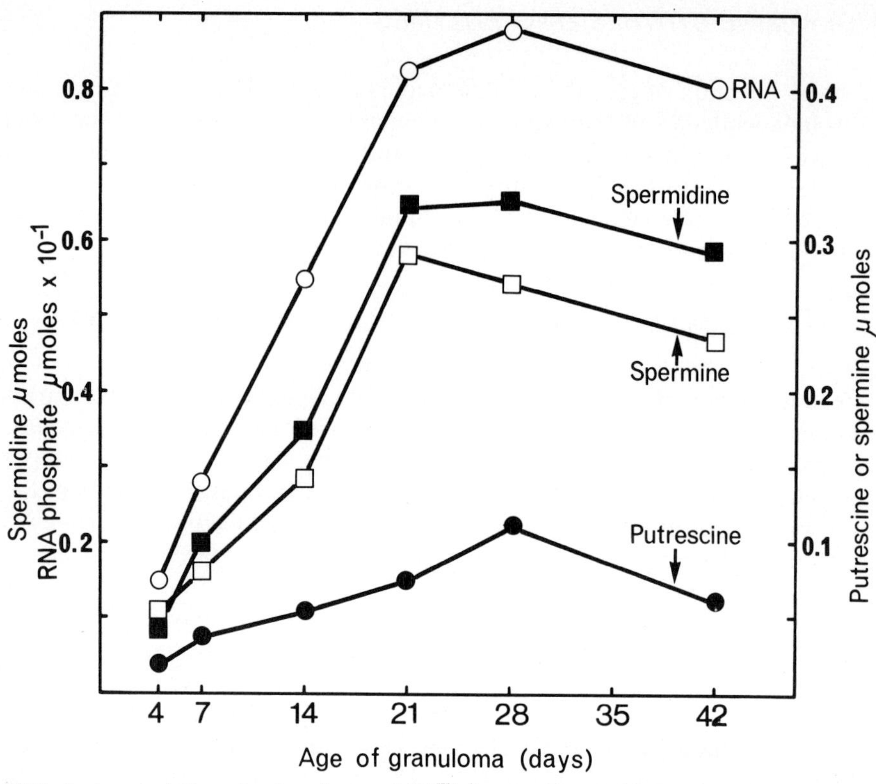

FIG. 5. Accumulation of polyamines and RNA in experimental granuloma. The values are given as μmoles of polyamines or RNA phosphate per (single) granuloma.

TABLE 1. *The ratio of polyamines to RNA and DNA phosphate in granuloma*

Age (days)	nmoles per μmole RNA phosphate			nmoles per μmole DNA phosphate		
	Pu	Spd	Sp	Pu	Spd	Sp
4	12.4 (10.7–15.1)	56.0 (40.4–64.3)	34.8 (30.8.–37.1)	6.9 (5.6–8.2)	30.7 (26.1–34.2)	19.3 (18.0–20.1)
7	13.1 (11.4–16.0)	71.5 (63.7–80.8)	28.6 (27.5–29.2)	9.7 (7.3–12.4)	52.3 (49.8–54.3)	21.1 (17.9–24.2)
14	9.7 (9.0–11.1)	62.9 (56.1–69.6)	25.5 (21.9–27.4)	15.6 (12.9–19.1)	100.4 (89.0–120.0)	40.5 (36.1–47.2)
21	9.1 (8.4–10.0)	78.4 (73.6–82.4)	35.4 (32.3–37.9)	14.5 (11.9–17.5)	123.6 (115.6–138.1)	55.8 (50.7–63.0)
28	12.5 (10.4–14.6)	74.2 (68.5–78.2)	31.1 (27.6–34.3)	24.2 (16.4–29.0)	142.6 (107.7–169.5)	59.8 (43.4–68.1)
42	7.6 (6.9–8.7)	72.3 (69.2–75.4)	28.9 (27.4–30.2)	13.7 (8.8–22.3)	124.4 (96.4–178.5)	49.7 (37.4–70.7)

For experimental details see Fig. 5.

polyamine to DNA (Table 1) and to protein (not tabulated) revealed much larger variations.

B. Polyamine-synthesizing enzymes in granulation tissue at different phases of development

Experimental granuloma as a model of differentiating tissue seemed to offer an interesting tool for studying the changes in polyamine-synthesizing enzyme activities at various phases of the development, although the material used for the induction of granulation tissue, i.e., viscose cellulose sponge, is not ideal for preparation of tissue extracts for enzyme assays. Special attention was therefore paid to careful homogenization of tissue samples (see also Experimental Procedures).

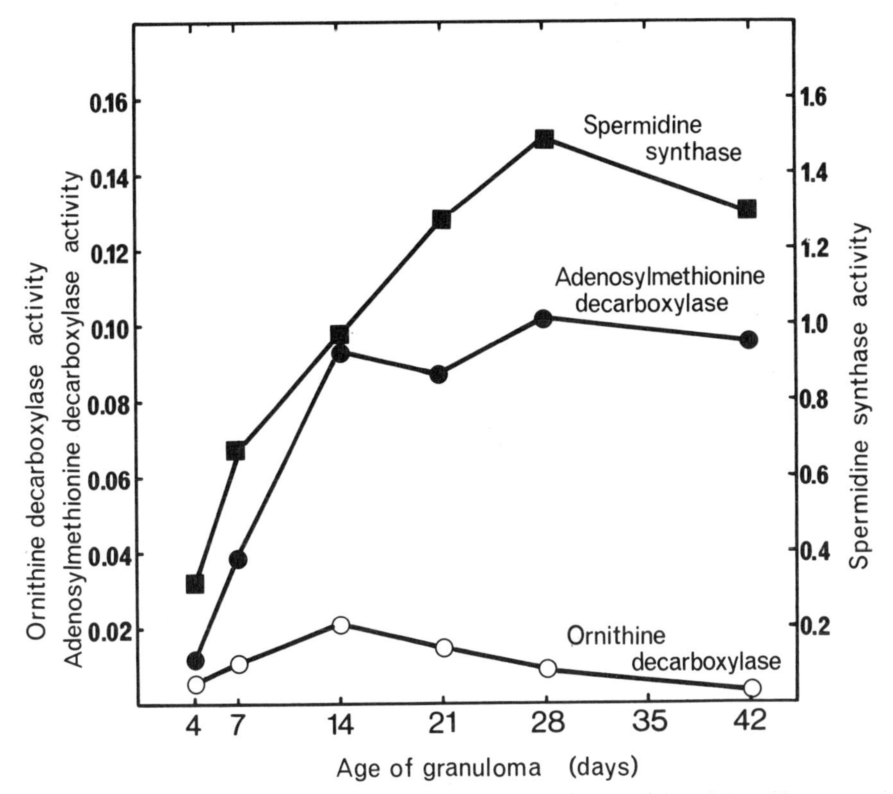

FIG. 6. Polyamine-synthesizing enzyme activities in experimental granuloma. The enzyme activities are expressed as nmoles of radioactive CO_2 liberated from ornithine-^{14}COOH (ornithine decarboxylase) and from S-adenosylmethionine-^{14}COOH (S-adenosylmethionine decarboxylase) or nmoles of spermidine synthesized from putrescine-^{14}C and decarboxylated S-adenosylmethionine (spermidine synthase) per mg protein per 30 min.

Ornithine decarboxylase, S-adenosylmethionine decarboxylase, and spermidine synthase activities could be assayed in the granulation tissue, whereas the activity of spermine synthase was so low that no reliable measurements could be made. Figure 6 shows the pattern of polyamine-synthesizing enzyme activities in the experimental granuloma during development. The activity of all three enzymes increased up to 14 days after implantation. Thereafter S-adenosylmethionine decarboxylase and ornithine decarboxylase remained constant or decreased slowly, whereas spermidine synthase activity continued to increase up to 28 days. The enzyme pattern shows an obvious resemblance to that observed in the regenerating liver (Fig. 1) as far as the sequence of the events is concerned. As in the regenerating rat liver, spermidine synthase activity in granulation tissue was 10 times higher (in terms of specific activity) than the activity of S-adenosylmethionine decarboxylase (Fig. 6).

It is noteworthy that after the initial increase in the enzyme activities, and also in the polyamine and RNA contents, there was very little change during the period of involution of the granulation tissue.

IV. GENERAL CONCLUSIONS

Regenerating rodent liver is a useful tool for studying the relationship of polyamine metabolism to rapid growth. It also serves as a basic model for the sequence of the stimulation of polyamine synthesis during the period of rapid growth. Basically the same sequence (i.e., a primary stimulation of ornithine decarboxylase and concomitant accumulation of putrescine, and the stimulation of spermidine synthesis first by an activation of S-adenosylmethionine decarboxylase by putrescine and later a stimulation of spermidine and spermine-synthesizing enzymes) can be seen in liver after growth hormone treatment (39, 40). The postulated sequence described is a mixture of coarse and fine regulation of various enzyme activities. At least in liver putrescine seems to play a crucial role in the regulation of spermidine synthesis.

The situation is generally very similar in the experimental granuloma. The same sequence can be seen in the stimulation of the different enzyme activities during the development of the granulation tissue.

It is not known how long the fascinating problem of the physiological function of mammalian polyamines will remain unsolved. The observation of the close parallelism for the accumulation of polyamines and RNA can be regarded as further evidence pointing to the importance of polyamines in cell physiology and metabolism. We feel confident that these compounds are some way related to the growth and/or its regulation in animal tissues.

ACKNOWLEDGMENTS

The studies presented here were supported by grants from the National Research Council for Medical Sciences, Finland, the Sigrid Jusèlius Foundation and the Orion Scientific Research Foundation. The skillful technical assistance of Mrs. Riitta Sinervirta, Mrs. Sirkka Kanerva, and Mrs. Terttu Jompero is gratefully acknowledged.

REFERENCES

1. Raina, A., Jänne, J., and Siimes, M., *Abstracts of 2nd FEBS Meeting,* Vienna, 1965.
2. Dykstra, W. G., Jr., and Herbst, E. J., *Science,* **149,** 428 (1965).
3. Raina, A., Jänne, J., and Siimes, M., *Biochim. Biophys. Acta,* **123,** 197 (1966).
4. Higgins, G. H., and Anderson, R. M., *Arch. Pathol.,* **12,** 186 (1931).
5. Viljanto, J., *Acta Chir. Scand.,* Suppl. **333,** 1 (1964).
6. Pegg, A. E., and Williams-Ashman, H. G., *J. Biol. Chem.,* **244,** 682 (1969).
7. Jänne, J., Williams-Ashman, H. G., and Schenone, A., *Biochem. Biophys. Res. Comm.,* **43,** 1362 (1971).
8. Raina, A., and Hannonen, P., *FEBS Letters,* **16,** 1 (1971).
9. Raina, A., and Jänne, J., *Acta Chem. Scand.,* **22,** 2375 (1968).
10. Jänne, J., and Williams-Ashman, H. G., *J. Biol. Chem.,* **246,** 1725 (1971).
11. Raina, A., Jänne, J., Hannonen, P., and Hölttä, E., *Ann. N.Y. Acad. Sci.,* **171,** 697 (1970).
12. Hannonen, P., Raina, A., and Jänne, J., *Biochim. Biophys. Acta,* **273,** 84 (1972).
13. Hölttä, E., and Jänne, J., *FEBS Letters,* **23,** 117 (1972).
14. Raina, A., Jansen, M., and Cohen, S. S., *J. Bacteriol.,* **94,** 1684 (1967).
15. Ashwell, G., in S. P. Colowick and N. O. Kaplan (Editors), *Methods in enzymology,* Vol. III, Academic Press, New York, 1957, p. 73.
16. Lowry, O. H., Rosebrough, N. J., Farr, A. L., and Randall, R. J., *J. Biol. Chem.,* **193,** 265 (1951).
17. Russell, D. H., and Snyder, S. H., *Proc. Nat. Acad. Sci. U.S.A.,* **60,** 1420 (1968).
18. Jänne, J., and Raina, A., *Acta Chem. Scand.,* **22,** 1349 (1968).
19. Schrock, T. R., Oakman, N. J., and Bucher, N. L. R., *Biochim. Biophys. Acta,* **204,** 564 (1970).
20. Fausto, N., *Biochim. Biophys. Acta,* **190,** 193 (1969).
21. Jänne, J., *Acta Physiol. Scand.,* Suppl. **300,** 1 (1967).
22. Hannonen, P., Jänne, J., and Raina, A., *Biochem. Biophys. Res. Comm.,* **46,** 341 (1972).
23. Jänne, J., and Williams-Ashman, H. G., *Biochem. Biophys. Res. Comm.,* **42,** 222 (1971).
24. Jänne, J., Schenone, A., and Williams-Ashman, H. G., *Biochem. Biophys. Res. Comm.,* **42,** 758 (1971).
25. Feldman, M. J., Levy, C. C., and Russell, D. H., *Biochem. Biophys. Res. Comm.,* **44,** 675 (1971).
26. Feldman, M. J., Levy, C. C., and Russell, D. H., *Biochemistry,* **11,** 671 (1972).
27. Russell, D. H., and Lombardini, J. B., *Biochim. Biophys. Acta,* **240,** 273 (1971).
28. Russell, D. H., and Taylor, R. L., *Endocrinology,* **88,** 1397 (1971).
29. Coppoc, G. L., Kallio, P., and Williams-Ashman, H. G., *Int. J. Biochem.,* **2,** 673 (1971).
30. Pegg, A. E., and Williams-Ashman, H. G., *Arch. Biochem. Biophys.,* **137,** 156 (1970).
31. Hannonen, P., Jänne, J., and Raina, A., *Biochim. Biophys. Acta,* **289,** 225 (1972).
32. Friedman, S. J., Halpern, K. V., and Canellakis, E. S., *Biochim. Biophys. Acta,* **261,** 181 (1972).
33. Jänne, J., and Williams-Ashman, H. G., *Biochem. J.,* **119,** 595 (1970).

34. Russell, D. H., and Snyder, S. H., *Mol. Pharmacol.,* **5,** 253 (1969).
35. Pegg, A. E., Lockwood, D. H., and Williams-Ashman, H. G., *Biochem. J.,* **117,** 17 (1970).
36. Kobayashi, Y., Kupelian, J., and Maudsley, D. V., *Science,* **172,** 379 (1971).
37. Cohen, S., O'Malley, B. W., and Stastny, M., *Science,* **170,** 336 (1970).
38. Williams-Ashman, H. G., Jänne, J., Coppoc, G. L., Geroch, M. E., and Schenone, A., in G. Weber (Editor), *Advances in enzyme regulation,* Vol. X, Pergamon Press, New York, 1972, p. 225.
39. Jänne, J., Raina, A., and Siimes, M., *Biochim. Biophys. Acta,* **166,** 419 (1968).
40. Jänne, J., and Raina, A., *Biochim. Biophys. Acta,* **174,** 769 (1969).
41. Russell, D. H., Snyder, S. H., and Medina, V. J., *Endocrinology,* **86,** 1414 (1970).
42. Snyder, S. H., and Russell, D. H., *Fed. Proc.,* **29,** 1575 (1970).
43. Ahonen, J., *Acta Physiol. Scand.,* Suppl. **315,** 1 (1968).
44. Kulonen, E., in E. A. Balazs (Editor), *Chemistry and biology of intercellular matrix,* Vol. III, Academic Press, London, 1970, p. 1811.
45. Raina, A., and Jänne, J., *Fed. Proc.,* **29,** 1568 (1970).

Polyamines in Normal and Neoplastic Growth. Edited by D. H. Russell. Raven Press, New York

Aspects of Polyamine Biosynthesis in Normal and Malignant Eukaryotic Cells

H. G. Williams-Ashman, G. L. Coppoc*, Amelia Schenone, and George Weber**

The Ben May Laboratory for Cancer Research, University of Chicago, Chicago, Illinois 60637; Department of Veterinary Physiology and Pharmacology, Purdue University, Lafayette, Indiana 47907; and Department of Pharmacology**, Indiana University School of Medicine, Indianapolis, Indiana 46202*

The increasing attention now being given to the roles of aliphatic polyamines in cancerous cells is an important and timely development. Among the reasons for further studies on the biochemistry of spermidine and spermine in malignant tissues are the following. First, these minimolecular bases can, in cell-free systems, profoundly influence the synthesis, degradation, and functions of nucleic acids, and also can affect many other types of biochemical processes that may be crucial for the growth and differentiation of cells. Secondly, striking alterations in the concentrations and biosynthesis of polyamines have been shown to accompany various phases of the growth and/or differentiation of many animal tissues. And thirdly, there are indications that some anticancer drugs can either directly or indirectly alter the levels of polyamines in sensitive malignant cells; certain of these substances exert selective inhibitory actions on polyamine biosynthetic enzymes *in vitro.*

In this chapter we will attempt to consider in perspective two specialized features of polyamine biochemistry: (a) the nature and regulation of enzymes responsible for the synthesis of spermidine and spermine in normal and malignant cells, and (b) aliphatic polyamine formation and turnover in relation to the speed of growth of certain animal tumors. No exhaustive coverage of the relevant literature will be provided because these fields are treated comprehensively in a number of recent reviews (1–9).

I. PATHWAYS FOR POLYAMINE PRODUCTION IN ANIMAL TISSUES

Currently there is consensus that in animal cells the diaminobutane moiety of spermidine and spermine is derived from putrescine which in turn originates from ornithine, and that decarboxylated S-adenosyl methionine (Ado-met) provides propylamino groups for the enzymic synthesis of polyamines. At present, only one function for putrescine in animal tissues is well established, that is, to serve as a precursor for spermidine and spermine. Nevertheless, other possible functions of putrescine should not be ignored. Putrescine can be incorporated into peptide linkages (almost certainly in union with the γ-carboxyl group of glutamate derived from glutaminyl groups in polypeptide chains) by the action of transamidating enzymes (10, 11), but it has not been studied if such reactions occur in living cells and, if they do, if they are of physiological importance. Putrescine can also serve as a substrate for diamine oxidases (7, 12, 13); however, it is far from clear if the aldehyde products derived from putrescine in this manner are of any metabolic importance.

The propylamino moiety of decarboxylated Ado-met has not, up to now, been shown to serve as a precursor in animal cells for any substances other than spermidine and spermine. This raises three considerations. First, it should be kept in mind that other but as yet undetected propylamino group transfer reactions may conceivably occur in living cells, e.g., the formation of propylaminated derivatives of proteins or nucleic acid bases. Secondly, no biological substance other than decarboxylated Ado-met has yet been found to contain an "active propylamino" group. Thirdly, it is possible that there is more than one physiological role for the spermidine and spermine synthase reactions, which can be represented by the following equations:

$$\text{decarboxylated Ado-met} + \text{putrescine} \rightarrow \text{spermidine} + \text{MTA} + H^+$$
$$\text{decarboxylated Ado-met} + \text{spermidine} \rightarrow \text{spermine} + \text{MTA} + H^+.$$

These equations are usually considered as routes to the synthesis of spermidine and spermine, and it is incontrovertible that this represents their main biological function. However, it is noteworthy that in both of these synthase reactions the formation of polyamines is stoichiometric with the production of 5′-methylthioadenosine (MTA). It is customary to consider MTA as a "garbage product" of the polyamine synthase reactions, which in some animal tissues is rapidly removed by the action of an enzyme that phosphorolytically cleaves MTA to yield adenine and what is probably 1-phospho-5-methylthioribose (14). Nevertheless, it could be imagined that MTA could itself act in some situations as a regulator of enzyme systems or as a precursor of physiologically important substances, and that in certain cells the primary function of the spermidine and spermine synthase reactions would be to produce MTA rather than polyamines. Noteworthy in this context is the fact that, although some bacteria and yeasts contain enzymes that form MTA by cleavage of S-adenosyl methionine (15, 16), such a splitting reaction has not been described in animal cells; therefore, the spermidine and spermine synthase reactions are the only known reactions that can result in production of MTA in animal tissues.

A. Ornithine decarboxylase

In all animal tissues that have been examined heretofore, a pyridoxal phosphate-dependent enzyme that specifically catalyzes the decarboxylation of L-ornithine represents the only mechanism for the formation of putrescine (1, 2, 8, 17). With the exception of a claim that a considerable fraction of the total ornithine decarboxylase of chick embryo homogenates is associated with the nuclear fraction (18), the enzyme appears to be localized almost completely in the cytosol of animal cells. No definitive evidence suggesting the existence of isoenzymes of animal ornithine decarboxylases has been advanced, although this eventuality should be explored more closely. The K_m for L-ornithine for the decarboxylase in a number of animal tissues is of the order of 0.1 mM at pH 7.4 (19). Evaluation of reports of ornithine decarboxylase activities in various mammalian tissues should take into account recent observations that, with respect to tissues like rat ventral prostate (19, 20) and liver (20, 21), there are marked variations in the degree of activation of carbon dioxide release that can be induced by addition of different thiol compounds.

Mammalian ornithine decarboxylases exhibit Michaelis-Menten kinetics and in general seem to be rather unsusceptible to allosteric-like regulation by nucleotides, S-adenosyl methionine and its derivatives, and many other types of small molecules (8, 19, 21). Ornithine decarboxylase in some animal tissues is inhibited in a competitive fashion by putrescine and to a lesser extent by spermidine and spermine (17, 22); however, the K_i values for these substances are sufficiently high that these inhibitions by polyamines are of doubtful physiological significance. In the case of some tissues, such as rat ventral prostate, dialysis of centrifuged homogenates or passage of the extracts through gel filtration columns increases their ornithine decarboxylase activities (17); this may in part be due to removal of polyamines (which are present in exceptionally high concentrations in rat ventral prostate) but might also reflect removal of other inhibitors of the enzyme. The extent to which various specific or nonspecific inhibitors of ornithine decarboxylase may be present in various animal tissues and subcellular fractions thereof needs to be examined much more thoroughly.

The activity of ornithine decarboxylases that are denizens of various normal (20, 23–25) and malignant (26, 27) animal tissues falls very rapidly after administration of inhibitors of protein biosynthesis such as cyclohexi-

mide, which does not directly inhibit the enzyme even in high concentrations. From these observations, it has been concluded that ornithine decarboxylase in many animal tissues has a half-life of considerably less than 30 min. This is probably a correct interpretation of such experiments. Nevertheless, the conclusion that mammalian ornithine decarboxylases turn over very rapidly must remain somewhat tentative until (a) the actual amounts, rather than activities, of the enzyme are determined under appropriate conditions by suitable radioimmunological or other procedures, and (b) it is proven that protein synthesis inhibitors such as cycloheximide do not alter the normal rate of intracellular degradation of the enzyme.

Under some conditions, the enhancement of tissue ornithine decarboxylase that results from, for example, partial hepatectomy, unilateral nephrectomy, or administration of various hormones and drugs is diminished or abolished by treatment of the animals with inhibitors of DNA-directed RNA synthesis (such as actinomycin D) as well as with inhibitors of protein conditions. Under other circumstances, actinomycin D does not prevent increases in ornithine decarboxylase due to growth, hormonal, and pharmacological stimuli. Brandt et al. (28) postulated that regulation of the synthesis of ornithine decarboxylase can occur at both transcriptional and post-transcriptional levels. Evidence for post-transcriptional control of ornithine decarboxylase synthesis in cultures of rat hepatoma cells has also been advanced by Hogan (27). Very recently, Beck et al. (29) reported that hepatic ornithine decarboxylase is increased many fold within 3 hr after injection of dibutyryl cyclic AMP; the effect of the cyclic nucleotide was inhibited by actinomycin D given simultaneously, but not when the antibiotic was administered 1 hr after dibutyryl cyclic AMP. Simultaneous treatment of the animals with actinomycin D had considerably less inhibitory effect on the increase in hepatic ornithine decarboxylase following injection of dexamethasone. These results hint that cyclic AMP might be involved in the stimulation of hepatic ornithine decarboxylase activity by a variety of agents. Additional complexities of dietary and other factors that influence swift oscillations in liver ornithine decarboxylase activities are discussed below.

B. Formation of decarboxylated S-adenosyl methionine

The enzyme S-adenosyl methionine decarboxylase was discovered by Herbert and Celia Tabor in *Escherichia coli;* this enzyme has been purified to a state of homogeneity and shown to contain covalently bound pyruvate as a prosthetic group (30). The *E. coli* Ado-met decarboxylase and a comparable enzyme in *Azotobacter vinelandii* (31) require Mg^{++} and are not stimulated by aliphatic polyamines. The enzymic decarboxylation of Ado-met by soluble extracts of various animal tissues was first demonstrated by Pegg and Williams-Ashman (32, 33). Unlike the bacterial enzymes, Ado-met decarboxylases from various invertebrate and vertebrate animal tissues do not require any metal ion activator, but are, on the contrary, markedly stimulated by putrescine, spermidine (33), and a few other amines (8, 34). Extracts of baker's yeast contain a very active Ado-met decarboxylase that is not enhanced by metal ions but shows an almost complete requirement for putrescine (35). In contrast, an enzyme from mung bean that catalyzes release of carbon dioxide from Ado-met is activated by Mg^{++} but not by putrescine (31). In a preliminary communication, Mitchell and Rusch (36) state that the Ado-met decarboxylase of *Physarum polycephalum* is not enhanced by putrescine, no mention being made of any metal ion requirement. From these findings, it is evident that Ado-met decarboxylases from various sources may require no dissociable cofactors, may be activated by Mg^{++}, or may be enhanced by putrescine and some other simple amines. A dual requirement for both Mg^{++} and an amine activator has not been reported for any Ado-met decarboxylase described up to now. The differential activations by Mg^{++} or putrescine of Ado-met decarboxylases from different organisms does not seem to accord with any simple evolutionary patterns.

The nature of the prosthetic group, if any, of Ado-met decarboxylases of higher animal tissues is uncertain. Pegg and Williams-Ashman (33) reported in 1969 that the rat ventral prostate enzyme, even after extensive purification, was not activated by exogenous pyridoxal phosphate. Decarboxylation of Ado-met by the 80-fold purified enzyme was, however, strongly inhibited by 4-bromo-3-hydroxy-benzyloxyamine (NSD-1055), and the inhibitory effect of NSD-1055 was reversed by addition of pyridoxal phosphate, but not by unsubstituted pyridoxal. More recently, Feldman et al. (37, 38) stated that a purified preparation of rat liver Ado-met decarboxylase is inhibited by NSD-1055 and that the inhibition is reversed by pyridoxal phosphate. Feldman et al. (38) also stated that after dialysis of the semipurified liver enzyme against phosphate buffer of pH 7.5, the liver enzyme activity is increased by 10 μM pyridoxal phosphate. The latter observation does not accord with reports that highly purified and extensively dialyzed preparations of the rat ventral prostate (32) and baker's yeast (35) Ado-met decarboxylases are not enhanced by pyridoxal phosphate in the absence of inhibitors such as NSD-1055. The enzymes from the two latter sources are powerfully inhibited by preincubation with sodium borohydride in the absence of putrescine and Ado-met (8). Noteworthy in this connection is the pronounced inhibition of *E. coli* Ado-met decarboxylase, which contains protein-bound pyruvate as a prosthetic group, by sodium borohydride (30).

The question as to whether tightly bound pyridoxal phosphate is really the prosthetic group of these enzymes remains to be firmly established. Unequivocal proof that pyridoxal phosphate is the prosthetic group of mammalian Ado-met decarboxylases must await isolation of the enzymes in a homogenous form and direct demonstration of the presence of bound pyridoxal phosphate molecules that participate in the removal of CO_2 from Ado-met.

High concentrations of pyridoxal phosphate cause a rapid release of CO_2 from Ado-met in the absence of any enzymes. The rates of these nonenzymic decarboxylations depend, among other things, on (a) the initial concentrations of Ado-met and pyridoxal phosphate, (b) temperature, (c) the nature of the buffer employed at neutral pH, and (d) the pH of the reaction mixture (39). Metal ions, particularly Mn^{++}, markedly stimulate the decarboxylation of Ado-met by pyridoxal phosphate. Without added divalent metal ions, the nonenzymic reaction is inhibited markedly by dithiothreitol when orthophosphate is employed as buffer, but is much less so with diglycine or Tris buffers at pH 7.2. At equimolar initial concentrations, Ado-met is decarboxylated by pyridoxal phosphate at a faster rate than is methionine. Pyridoxal phosphate is more effective than is unsubstituted pyridoxal in promoting loss of CO_2 from Ado-met, whereas pyridoxamine and pyridoxamine phosphate are hardly active in this regard. These facts underscore the need for setting up extensive and suitable controls to correct for nonenzymic reactions in studies on effects of pyridoxal phosphate on Ado-met decarboxylases. Noteworthy in this connection is the observation of Mazelis (40) that nonenzymic release of CO_2 from methionine, induced by Mn^{++} plus pyridoxal phosphate, is greatly enhanced by addition of horseradish peroxidase. We have found (39) that similar peroxidase-accelerated reactions occur with Ado-met as substrate, indicating that measurement of Ado-met decarboxylase activities of tissues that are rich in peroxidases must be interpreted with caution.

Rat ventral prostate Ado-met decarboxylase is activated not only by putrescine and spermidine, but also by some other amines (8, 34), including cadaverine (1,5-diaminopentane), 1,3-diaminopropane, and N-(2-aminoethylpropane)-1,3-diamine [the latter substance was kindly provided by Drs. E. J. Modest and M. Israel (41)]. Baker's yeast Ado-met decarboxylase is not enhanced by cadaverine or 1,3-diaminopropane (34). Many other substances, including spermine, 3-aminobutyronitrile, 4-amino-1-butanol, synthalin, various ribo- and deoxyribo-polynucleotides as well as various other compounds listed elsewhere (8), do not enhance prostate Ado-met decarboxylase. It is worth noting that highly refined preparations of rat prostate (33) and liver (38) Ado-met decarboxylase invariably exhibit appreciable activity in the absence of activating amines, in marked contrast to the yeast enzyme (35). Enhancement of animal tissue Ado-met decarboxylases by putrescine depends critically on (a) pH, (b) temperature, and (c) the initial levels of Ado-met and the amine activator. Not only the K_m for activation but also maximal velocities at saturating levels of activating amines vary markedly among different substances such as putrescine, cadaverine, and spermidine (8, 34).

C. Relationship of mammalian Ado-met decarboxylases to spermine and spermidine synthases

When the activation of rat tissue Ado-met decarboxylase by putrescine and spermidine was first discovered (32, 33), it was found that with partially purified preparations of the enzyme in the presence of putrescine, release of carbon dioxide from Ado-met was stoichiometric with synthesis of spermidine and MTA. The same enzyme preparations from rat prostate also catalyzed the synthesis of spermidine from putrescine and added decarboxylated Ado-met (enzymically synthesized with the *E. coli* enzyme) at more than twice the rate observed with Ado-met itself. The first attempts to separate prostatic Ado-met decarboxylase from spermidine synthase were unsuccessful (33). A survey of the properties of prostatic spermine synthase, which catalyzes the formation of spermine from spermidine and decarboxylated Ado-met, was undertaken (42) soon after the first announcement (33) of the discovery of this enzyme. Early attempts to separate prostate spermidine synthase from spermine synthase by chromatography, or on the basis of differential heat stabilities, were also unfruitful (42). However, the fact that Ado-met decarboxylase, spermidine synthase, and spermine synthase reactions in mammalian tissues may be catalyzed by enzymes that are completely separable from one another is now indicated by the following lines of evidence. (a) Spermidine synthase can be fractioned free from putrescine-activated Ado-met decarboxylase and vice versa in rat ventral prostate (43, 44), brain (45), and liver (46), as well as from baker's yeast (35) and *E. coli* (30). The claim of Feldman et al. (37, 38) that Ado-met decarboxylase cannot be separated from spermidine synthase in rat liver does not accord with a study by Hannonen et al. (46). In rat prostate (44) and even more so in liver (46), the absolute activity of spermidine synthase is many times greater than that of the decarboxylase. (b) Combination of spermidine synthase preparations with little or no Ado-met decarboxylase activity with synthase-free preparations of Ado-met decarboxylase from prostate allows synthesis of spermidine to occur from added Ado-met and putrescine (44); similarly, reestablishment of the coupling between Ado-met decarboxylase and spermine synthase has been demonstrated with purified enzymes from rat brain (45). (c) Spermidine and spermine synthases have been separated from one another using prostate (43, 44), liver (46), and brain (45) extracts as starting material. (d) The ratios of the activities of the decarboxylase and the two synthases in crude extracts of different rat tissues are not constant (45); (e) Methylglyoxal *bis*(guanylhydrazone) (MGBG), an extremely potent inhibitor of putrescine-activated Ado-met decarboxylases from rat tissues and yeast (34) does not inhibit rat prostate spermidine synthase as determined with decarboxylated Ado-met as substrate, but does, as expected, prevent the synthesis of spermidine from Ado-met and putrescine by enzyme preparations that contain an active decarboxylase and an excess of spermidine synthase (47). (f) Experiments involving administration of cycloheximide suggest that in regenerating rat liver the half-life of Ado-met decarboxylase is very short, whereas the half-lives of spermidine and spermine synthases seem to be very much longer (48). (g) Changes in the activity of Ado-met decarboxylase and spermidine synthase during liver regeneration do not completely parallel one another (48).

When both enzymes are present in suitable amounts, but not necessarily with spermidine synthase in great excess with respect to Ado-met decarboxylase, the synthesis of spermidine from Ado-met and putrescine can occur at a linear rate with respect to time and without accumulation of more than negligible amounts of the decarboxylated Ado-met intermediate (33, 43). Among the factors that facilitate the tight and efficient coupling between the spermidine synthase and Ado-met decarboxylases in the overall formation of spermidine is the very low K_m (roughly 20 μM) for decarboxylated Ado-met in the synthase reaction at pH values near neutrality (8, 44). A comprehensive understanding of mechanisms involved in the coupling between the Ado-met decarboxylases and spermidine synthases of animal tissues must await, among other things, accurate assessment of the equilibrium constants for these reactions.

It might be thought that separation of mammalian Ado-met decarboxylase from spermidine and spermine synthases would be of interest only as an academic exercise in enzymology, and that in living cells these enzymes are held together in the cytosol by noncovalent linkages so as to form a multienzyme complex (49) that is exceptionally efficient in promoting the synthesis of spermidine and spermine. However, many of the aforementioned experimental results do not lend any strong support to this contention. Furthermore, gel filtration experiments with crude preparations of rat liver Ado-met decarboxylase (containing considerable spermidine synthase activity) suggested that the size (molecular Stokes radius) of the enzyme was roughly the same as that of highly purified Ado-met decarboxylase (utterly devoid of spermidine synthase) from the same source, and in addition, the Stokes radius of liver spermidine synthase appeared to be about the same as that of the Ado-met decarboxylase (46). The latter facts hint that there does not exist any specific complex between Ado-met decarboxylase and spermidine synthase in crude liver extracts, because, if this were the case, one might expect that the gel filtration behavior of the enzymes would be different after their separation from one another by suitable purification procedures (46). It may be added that the clear-cut separation of Ado-met decarboxylase from spermidine and spermine synthases in a variety of mammalian tissues does not support the contention of Russell and Potyraj (50) that measurements of the spermidine-activated Ado-met decarboxylase activity of rat uterine extracts provide a reliable index of the rate of enzymic synthesis of spermine by this organ, because putrescine enhances the enzyme to a greater extent than spermidine.

Four other facets of the mechanism and regulation of polyamine biosynthesis in animal cells may now be mentioned. First, the absolute activity of the known biosynthetic enzymes in many normal resting animal tissues is sufficiently feeble that it may well be that the levels of these enzymes, rather than the concentrations of substrates, is often rate-limiting to the overall processes of spermidine and spermine formation. Secondly, it is now quite clear that putrescine can control the rate of polyamine biosynthesis by means other than serving as substrate for the spermidine synthase reaction. This is because putrescine enhances animal cell Ado-met decarboxylases (8, 31) and inhibits spermine synthase by competing with the spermidine substrate (42). The concentrations of putrescine that exert the latter regulatory actions on the isolated enzyme systems involved are entirely commensurate with the rather low steady-state levels of putrescine found in many animal cells. Thirdly, it is conceivable that other, and as yet unknown, pathways for polyamine formation may be operative in animal tissues; one reaction that may be worth searching for in this connection is a putative dismutation between two molecules of spermidine to yield one molecule of spermine and one molecule of putrescine (2, 8). Fourthly, it must be remembered that the steady-state concentrations of polyamines in any tissue must obviously depend on a balance between their rate of production or uptake and their rate of metabolic transformation within cells and their possible leakage out of cells into body fluids. With the exception of prostatic secretion in some mammalian species, spermidine and spermine do not seem to leach out of animal cells very readily under normal circumstances (8). Nevertheless, pathways for the intracellular degradation or conjugation of polyamines in animal cells need to be examined more thoroughly than has been accomplished up to now (1, 2, 7, 8).

D. Specific inhibitors of polyamine biosynthetic enzymes

Insufficient experimental attention has been paid to the possible presence in animal cells of inhibitors of various polyamine biosynthetic enzymes. It is of interest, however, that neither S-adenosyl homocysteine nor MTA exerts any powerful inhibitory actions toward ornithine decarboxylase, Ado-met decarboxylase, and spermidine synthase (8). This is in contrast to powerful inhibitions of many enzymic transmethylations by S-adenosyl homocysteine (which is formed after Ado-met loses its methyl group from the sulfonium pole) (51, 52) and depression of the action of some transmethylases by MTA, a product of both the spermidine and spermine synthase reactions (51).

A number of derivatives of Ado-met that would be interesting to test as inhibitors of Ado-met decarboxylase and polyamine synthases have recently been synthesized chemically in Lederer's laboratory (53). As pointed out elsewhere (8), interference with both the substrate and the regulatory functions of putrescine with respect to the various biosynthetic enzymes may represent fruitful targets for the design of drugs that could specifically block polyamine formation *in vivo*. Recently, MGBG was shown to inhibit yeast and animal Ado-met decarboxylases at strikingly low concentrations (34). At levels of a few micromolar, the drug inhibited decarboxylase activity to a greater extent in the presence of the putrescine activator. Thus it appears that there may be some preferential interaction of MGBG with a putative putrescine-binding site on animal tissue Ado-met decarboxylase molecules. However, at higher concentrations, all decarboxylase activity in the presence of putrescine is abolished by MGBG, which in concentrations approaching the millimolar range also depresses the action of *E. coli* Ado-met decarboxylase (34). MGBG is without influence on ornithine decarboxylase and spermidine synthase of rat prostate in concentrations as high as 0.5 mM. Whether this drug can selectively decrease polyamine levels in living normal and malignant animal cells by virtue of inhibiting the key biosynthetic enzyme Ado-met decarboxylase is currently under investigation in collaboration with Drs. E. Mihich and C. Dave of the Roswell Park Memorial Institute (Buffalo, New York).

Russell and Levy (54) have reported that a few drugs that increase the survival time of mice bearing L1210 leukemia apparently lower the levels of putrescine, spermidine, and spermine in the tumor. These authors state: "Preliminary studies of ornithine and Ado-met decarboxylase activities in the tumors of the drug-treated mice indicate that these enzyme activities are directly affected by the drugs administered and that the decreased polyamine content is due to inhibition of the polyamine biosynthetic pathway." It would be interesting to know whether the drugs that Russell and Levy found to decrease polyamine concentrations in L1210 leukemia directly influence various polyamine biosynthetic enzymes *in vitro* in any specific fashion, especially at concentrations commensurate with their actions of tumor polyamine levels in living animals.

II. SOME FEATURES OF POLYAMINE FORMATION IN MALIGNANT TISSUES

The current upsurge of interest in the aliphatic polyamines of cancerous cells has been sparked by a variety of findings. These include: (a) correlations between polyamine and nucleic acid concentrations during various phases of growth of animal tumors (for extensive bibliographies, see references 1, 8, 26, 55); (b) reports that the urinary excretion of polyamines may increase dramatically in certain cancer patients (56); (c) the increases in polyamines and some of their biosynthetic enzymes that occur in many normal tissues in states of rapid growth induced by a wide variety of stimuli (for references see 1, 2, 8, 26); and (d) alterations in polyamine concentrations in some tumors that may either directly or indirectly result from treatments with certain antineoplastic drugs (54).

We have recently examined the activity of polyamine biosynthetic decarboxylases and the concentrations of putrescine, spermidine, and spermine in a spectrum of Morris hepatomas of vastly different growth rates and degrees of differentiation (26). This family of neoplasms have well-standardized growth rates, exhibit remarkable constancies of some of their metabolic characteristics, in many instances have well-defined karyotypes, and in general maintain their unique biological and biochemical properties through a large number of generations of transplants. Thus, the Morris rat hepatomas may serve as particularly useful test objects for study of tumor growth rates in relation to the production, turnover, and possible regulatory functions of aliphatic polyamines.

Under conditions where the corresponding control animals were of the same age, sex, and strain, and subject to the same dietary regimen of *ad libitum* feeding with all animals being killed at the same time each day, it was found that the putrescine content of all but one (hepatoma 7800) of the seven tumor strains examined was significantly greater than normal (26). Putrescine levels tended to be highest in the most rapidly growing Morris tumors, and all of the hepatomas that showed greatly enhanced putrescine concentrations had elevated ornithine decarboxylase activities. Nevertheless, there was no strict parallelism between ornithine decarboxylase activities and the tissue putrescine concentrations. In contrast, the putrescine-activated Ado-met decarboxylases of all but one of the Morris hepatomas studied was significantly lower than that of the normal control livers, (the exception being hepatoma 7777 which had a significantly increased Ado-met decarboxylase activity). Thus, in every instance, the ratio of ornithine decarboxylase/Ado-met decarboxylase was greater in the hepatomas as compared with normal liver controls (26). A lowering of putrescine-activated Ado-met decarboxylase activities, under conditions where the latter may be rate-limiting to operation of the spermidine synthase reaction, would tend, of course, to facilitate enhancement of steady-state tissue putrescine concentrations, that may also reflect heightened ornithine decarboxylase action. In all of the Morris hepatoma lines examined, as in many normal rat tissues, the Ado-met decarboxylase of soluble extracts was greatly enhanced by exogenous putrescine; as might also be expected, the degree of activation of carbon dioxide release from Ado-met elicited by exogenous putrescine was considerably diminished with extracts of those liver tumors that exhibited abnormally high putrescine concentrations (26).

Hölttä and Jänne (58) have recently reported that during the first 10 hr after partial hepatectomy, there is a significant lowering in the liver remnant of Ado-met decarboxylase activity determined in the presence of an excess of putrescine, *pari passu* with very large increases in ornithine decarboxylase activity. Thus, in these early phases of liver regeneration, the ratio of ornithine to Ado-met decarboxylase activities is tremendously enhanced. In the same study (58), the extent of putrescine accumulation in the early phases of liver regeneration corresponded fairly closely with the increased ornithine decarboxylase, which is in contrast to the measurements of putrescine levels and ornithine decarboxylase activities in Morris hepatomas of varying growth rates (26). Noteworthy in this connection are reports that putrescine-activated Ado-met decarboxylase activities rise considerably during the later phases of rat liver regeneration, attaining a peak at about 48 hr after partial hepatectomy (48, 58, 59).

As already mentioned, the concentration of putrescine was increased in all except one of the Morris hepatoma lines examined in our studies, and the putrescine levels in the most rapidly growing tumors were five to 10 times those observed in the normal livers of control rats. The general trend in spermidine concentrations with respect to tumor growth rates was in the same direction, although smaller in magnitude, in five of the seven hepatoma lines investigated. In contrast, spermine concentrations did not appear to alter in any systematic way the growth rates of the tumors. According to the Molecular Correlation Concept proposed by Weber (65), the behavior of the putrescine and spermidine levels falls into class 2, in which are grouped biochemical parameters that are altered in the same direction in the hepatoma spectrum, whereas the spermine values may be grouped in class 3, which subsumes parameters that exhibit no correlation with tumor growth rates.

The spermidine/spermine quotient was significantly increased in five out of the seven hepatoma lines in comparison with the normal liver controls. Although no simple correlation was found in this series between increased spermidine/spermine ratios and the growth rates of the various hepatomas, the ratios in comparison with normal liver controls tended to be greatest in the most rapidly growing and dedifferentiated tumors (26).

These observations underscore other reports (e.g. 13, 54, 55, 60) that there never seems to be any marked decline or even loss of aliphatic polyamines and their biosynthetic enzymes in many different types of fast growing and very anaplastic malignant animal tumors.

There is evidence (26, 27) to suggest that the turnover of ornithine and Ado-met decarboxylases in rat hepatomas may be extraordinarily rapid, as has been found regarding these enzymes in many normal animal tissues (18, 20, 50). Apart from obvious considerations of the possible presence in crude tissue extracts of substances that may directly inhibit the decarboxylases of polyamine biosynthesis (cf. 61), the present lack of specific radioimmunological or other procedures for estimation of the actual concentrations of these enzyme molecules makes it mandatory to exercise considerable caution in the interpretation of most published studies on fluctuations in the activities of ornithine and Ado-met decarboxylases in various tissues that are concomitances of altered states of tissue growth. That normal liver ornithine decarboxylase activities may vary widely during a given day, and in relation to restricted periods of food intake, is evident from the experiments of Hayashi et al. (62). (It is conceivable that swift endocrine changes, such as rapid alterations in growth hormone output, are partially involved in the latter phenomena.) The recent findings of Katunuma and his colleagues (63) that animal tissues contain group-specific proteinases that initiate the degradation of various pyridoxal phosphate-requiring enzymes, and also inhibitors of these enzyme-degrading proteinases, are of obvious pertinence to the mechanisms of regulation of ornithine decarboxylase levels in different animal tissues.

Cavia and Webb (64) reported that the spermidine and especially the spermine concentrations in two Morris hepatoma lines (7800 and 5123D) were considerably lower than in normal liver controls, although these tumors exhibited increased spermidine/spermine ratios, and markedly heightened ornithine decarboxylase activities. Williams-Ashman et al. (26), who did not examine the hepatoma 5123D line, found that Morris hepatoma 7800 did not show an increased ornithine decarboxylase activity, although the spermidine/spermine quotient was slightly increased, with only marginal declines in the concentrations of spermidine and spermine. Although Cavia and Webb (64) and Williams-Ashman et al. (26) made measurements on Morris hepatoma 7800 transplants in rats of the same (Buffalo) strain and sex and of approximately the same age and body weight, the disparity in the findings between the two groups may well be due to factors such as dietary intake or the endocrine status of the host animals. It must be emphasized that the concentration of polyamines in malignant tumors as compared with their normal tissues of origin, when expressed in terms of unit wet weights of tissue, may also depend critically on differences in tissue water content, on the proportion of stromal cells of host origin in the solid tumor specimens, and on the amounts of necrotic tissue that may contaminate the neoplasms under investigation. It may also be appropriate to mention here that the differences in aliphatic polyamine concentrations and decarboxylase activities in various Morris hepatomas as compared with normal liver controls were of the same order of magnitude in the studies of Williams-Ashman et al. (26) when the results were expressed in terms either of wet weight of tissue or on a per cell basis, as determined by counts of cell nuclei.

Weber et al. (57) have shown that in a series of Morris hepatomas of increasing growth rates and degrees of dedifferentiation, there is a dramatic loss of ornithine carbamyltransferase activity, which is likely to be an important factor in the loss of a functional urea cycle in the more anaplastic tumors of this class. In the same study (57), it was pointed out that the decrease in ornithine carbamyltransferase is accompanied by an increase in ornithine decarboxylase, so that the ratio of ornithine decarboxylase to ornithine carbamyltransferase activities may increase in the more rapidly growing hepatomas by a factor of ten thousand or more. In the faster growing tumors, this might facilitate utilization of ornithine for polyamine production. However, in normal rat liver the activity of ornithine carbamyltransferase is many hundreds of thousands times greater than that of ornithine decarboxylase when both enzymes are measured under optimal conditions, and the two enzymes are quite different with regard to both their intracellular localizations and the magnitude of the Michaelis constants for their various substrates. Thus, at present it is difficult to assess the degree to which the activities of ornithine decarboxylase and ornithine carbamyltransferase in normal liver vs. various hepatomas may reflect the *in vivo* channeling of ornithine into putrescine and polyamine biosynthesis on the one hand, and, on the other hand, into the formation of citrulline. In addition, the extent to which the action of the active mitochondrial enzyme ornithine transaminase may also contribute to the utilization of ornithine in normal liver and various hepatomas remains to be completely clarified.

III. POSSIBLE BIOLOGICAL SIGNIFICANCE OF ALTERATIONS IN ORNITHINE AND POLYAMINE METABOLISM IN MORRIS HEPATOMAS

It seems that most biochemical characteristics which are maintained or increased in tumor cells can be related to a biologically selective advantage conferred on the cancer cells by these alterations (65). It is, therefore, of particular interest that in the Morris hepatoma spectrum the polyamine levels are preferentially maintained, and the total concentrations of spermidine plus spermine plus putrescine tend to be raised with the increase in tumor growth rate (26). This preferential retention of polyamines stands in sharp contrast with a decline or loss of some macromolecules, including various enzymes, from the faster growing tumors of this series. For instance, there is a marked decrease in the concentrations of total cellular and mitochondrial protein and of total phospholipides in the hepatomas; moreover, there is a progressive loss of the four key gluconeogenic enzymes: glucose-6-phosphatase, fructose-1,6-diphosphatase, phosphoenolpyruvate carboxykinase, and pyruvate carboxylase. There is also a decline of respiration, and in the activities of glucokinase and liver-type pyruvate kinase, a decrease in the activities of enzymes of pyrimidine catabolism (dihydrothymine and dihydrouracil dehydrogenases, thymidine phosphorylase), and the various enzymes that catabolize purines and amino acids (65). Since in the Morris hepatoma spectrum it is chiefly the enzymes involved in growth processes (i.e., those involved in glycolysis, and the synthesis of DNA, RNA, and proteins) that are retained or increased, the preferential retention or even increase in the levels of aliphatic polyamines, and of ornithine decarboxylase which is the first enzyme of the polyamine biosynthetic sequence, might be cautiously interpreted as evidence for a link between the polyamines and neoplastic growth rates.

The preferential maintenance and increase of ornithine decarboxylase in the hepatoma spectrum is especially striking when considered from the standpoint of competing pathways for ornithine utilization. In the Morris rat hepatoma spectrum, out of three pathways of ornithine utilization, only ornithine decarboxylase is retained and increased in the various hepatoma lines, whereas ornithine carbamyltransferase (57) and ornithine transaminase (N. Katunuma, I. Tomino, and G. Weber, to be published) are decreased in parallel with an increase in tumor growth rates.

In the hepatoma spectrum with the increase in tumor growth rate, a progressive metabolic imbalance was discovered which was manifest in the imbalance in opposing key enzymes and metabolic pathways in carbohydrate, pyrimidine, and DNA metabolism (65). To these growth rate-linked alterations may be added the recently reported imbalance in enzymes of ornithine utilization (26, 57). The selectivity, specificity, and functional significance of the changes in polyamine formation with regard to the fundamental nature of the neoplastic transformation, however, require a great deal more investigation in depth. The well-known interactions of aliphatic polyamines with polynucleotides of one form or another, and with enzyme systems that assemble and degrade nucleic acids, are most germane to these considerations. Nevertheless, much further experimentation will be necessary before it can be decided if the alterations in polyamine levels and ornithine decarboxylase that correlate with the growth rates of various Morris hepatomas are in any functional way related to the turnover and functions of nucleic acids in these neoplasms.

ACKNOWLEDGMENTS

Studies from the authors' laboratories were supported by U.S. Public Health Service grants HD-04592, CA-13526, and CA-05034, and by grants from the American Cancer Society and the Damon Runyon Memorial Fund, Inc.

REFERENCES

1. Cohen, S. S., *Introduction to the polyamines,* Prentice-Hall, Englewood Cliffs, N.J., 1971.
2. Williams-Ashman, H. G., Pegg, A. E., and Lockwood, D. H., *Adv. Enzyme Regul.,* **7,** 291 (1969).
3. Pegg, A. E., *Ann. N.Y. Acad. Sci.,* **171,** 977 (1970).
4. Stevens, L., *Biol. Rev.,* **45,** 1 (1970).
5. Raina, A., and Jänne, J., *Federation Proc.,* **29,** 1568 (1970).
6. Smith, T. A., *Endeavour,* **31,** 22 (1972).
7. Bachrach, U., *Ann. Rev. Microbiol.,* **24,** 110 (1970).
8. Williams-Ashman, H. G., Jänne, J., Coppoc, G. L., Geroch, M. E., and Schenone, A., *Adv. Enzyme Regul.,* **10,** 225 (1972).
9. Caldarera, C. M., and Moruzzi, G., *Ann. N.Y. Acad. Sci.,* **171,** 709 (1970).
10. Clarke, D. D., Mycek, M. J., Neidle, A., and Waelsch, H., *Arch. Biochem. Biophys.,* **79,** 338 (1959).
11. Lorand, L., and Campbell, L. K., *Anal. Biochem.,* **44,** 207 (1971).
12. Zeller, E. A., in P. D. Boyer, H. Lardy, and K. Myrbäck (Editors), *The enzymes,* Vol. 8, Academic Press, New York, 1963, p. 313.
13. Tabor, H., and Tabor, C. W., *Pharmacol. Rev.,* **16,** 245 (1964).
14. Pegg, A. E., and Williams-Ashman, H. G., *Biochem. J.,* **115,** 241 (1969).
15. Mudd, S. H., *J. Biol. Chem.,* **234,** 87 (1951).
16. Lombardini, J. B., and Talalay, P., *Adv. Enzyme Regul.,* **9,** 349 (1971).
17. Pegg, A. E., and Williams-Ashman, H. G., *Biochem. J.,* **108,** 533 (1968).
18. Snyder, S. H., Kreuz, D. S., Medina, V. J., and Russell, D. H., *Ann. N.Y. Acad. Sci.,* **171,** 749 (1970).
19. Jänne, J., and Williams-Ashman, H. G., *J. Biol. Chem.,* **246,** 1725 (1971).
20. Jänne, J., and Williams-Ashman, H. G., *Biochem. J.,* **119,** 595 (1970).

21. Friedman, S. J., Halpern, K. V., and Canellakis, E. S., *Biochim. Biophys. Acta*, **261**, 181 (1972).
22. Schrock, T. R., Oakman, N. J., and Bucher, N. L. R., *Biochim. Biophys. Acta*, **204**, 564 (1970).
23. Russell, D. H., and Snyder, S. H., *Mol. Pharmacol.*, **5**, 253 (1969).
24. Kaye, A. M., Icekson, I., and Lindner, H. R., *Biochim. Biophys. Acta*, **252**, 150 (1971).
25. Raina, A., Jänne, J., Hannonen, P., and Höltä, E., *Ann. N.Y. Acad. Sci.*, **171**, 693 (1970).
26. Williams-Ashman, H. G., Coppoc, G. L., and Weber, G., *Cancer Res.*, **32**, 1924 (1972).
27. Hogan, B. L. M., *Biochem. Biophys. Res. Comm.*, **45**, 301 (1971).
28. Brandt, J. T., Pierce, D. A., and Fausto, N., *Biochim. Biophys. Acta*, **279**, 184 (1972).
29. Beck, W. T., Bellantone, R. A., and Canellakis, E. S., *Biochem. Biophys. Res. Comm.*, **48**, 1649 (1972).
30. Wickner, R. B., Tabor, C. W., and Tabor, H., *J. Biol. Chem.*, **245**, 2132 (1970).
31. Coppoc, G. L., Kallio, P., and Williams-Ashman, H. G., *Int. J. Biochem.*, **2**, 673 (1971).
32. Pegg, A. E., and Williams-Ashman, H. G., *Biochem. Biophys. Res. Comm.*, **30**, 76 (1968).
33. Pegg, A. E., and Williams-Ashman, H. G., *J. Biol. Chem.*, **244**, 682 (1969).
34. Williams-Ashman, H. G., and Schenone, A., *Biochem. Biophys. Res. Comm.*, **46**, 288 (1972).
35. Jänne, J., Williams-Ashman, H. G., and Schenone, A., *Biochem. Biophys. Res. Comm.*, **43**, 1362 (1971).
36. Mitchell, J. L. A., and Rusch, H. P., *Federation Proc.*, **31**, 488Abs (Abstract no. 1547) (1972).
37. Feldman, M. J., Levy, C. C., and Russell, D. H., *Biochem. Biophys. Res. Comm.*, **44**, 675 (1971).
38. Feldman, M. J., Levy, C. C., and Russell, D. H., *Biochemistry*, **11**, 671 (1972).
39. Coppoc, G. L., Kallio, P., and Williams-Ashman, H. G., in preparation.
40. Mazelis, M., *J. Biol. Chem.*, **237**, 104 (1962).
41. Israel, M., and Modest, E. J., *J. Med. Chem.*, **14**, 1042 (1971).
42. Pegg, A. E., and Williams-Ashman, H. G., *Arch. Biochem. Biophys.*, **137**, 156 (1970).
43. Jänne, J., and Williams-Ashman, H. G., *Biochem. Biophys. Res. Comm.*, **42**, 222 (1971).
44. Jänne, J., Schenone, A., and Williams-Ashman, H. G., *Biochem. Biophys. Res. Comm.*, **42**, 758 (1971).
45. Raina, A., and Hannonen, P., *FEBS Letters*, **16**, 1 (1971).
46. Hannonen, P., Jänne, J., and Raina, A., *Biochem. Biophys. Res. Comm.*, **46**, 341 (1972).
47. Schenone, A., and Williams-Ashman, H. G., unpublished observations.
48. Hannonen, P., Raina, A., and Jänne, J., *Biochim. Biophys. Acta*, **273**, 84 (1972).
49. Ginsburg, A., and Stadtman, E. R., *Ann. Rev. Biochem.*, **39**, 429 (1970).
50. Russell, D. H., and Potyraj, J. J., *Biochem. J.*, **128**, 1109 (1972).
51. Zappia, V., Zydek-Cwick, C. R., and Schlenk, F., *J. Biol. Chem.*, **244**, 4499 (1969).
52. Kerr, S. J., *J. Biol. Chem.*, **247**, 4248 (1972).
53. Hildesheim, J., Hildesheim, R., and Lederer, E., *Biochemie*, **54**, 431 (1972).
54. Russell, D. H., and Levy, C. C., *Cancer Res.*, **31**, 248 (1971).
55. Andersson, G., and Heby, O., *J. Nat. Cancer Inst.*, **48**, 165 (1972).
56. Russell, D. H., Levy, C. C., Schimpff, S. C., and Hawk, I. A., *Cancer Res.*, **31**, 1555 (1971).
57. Weber, G., Queener, S. F., and Morris, H. P., *Cancer Res.*, **32**, 1933 (1972).
58. Höltä, E., and Jänne, J., *FEBS Letters*, **23**, 117 (1972).
59. Russell, D. H., and Lombardini, J. B., *Biochim. Biophys. Acta*, **240**, 273 (1971).
60. Siimes, M., and Jänne, J., *Acta Chem. Scand.*, **21**, 815 (1967).
61. Pegg, A. E., and Williams-Ashman, H. G., *Biochem. J.*, **108**, 533 (1968).
62. Hayashi, S. I., Yoshiyuki, A., and Noguchi, T., *Biochem. Biophys. Res. Comm.*, **46**, 795 (1972).
63. Katunuma, N., Kominami, E., Kominami, S., and Kito, K., *Adv. Enzyme Regul.*, **10**, 289 (1972).
64. Cavia, E., and Webb, T. E., *Biochem. J.*, **129**, 223 (1972).
65. Weber, G., *Gann Monograph on Cancer Research*, **13**, 47 (1973).

Polyamines in Normal and Neoplastic Growth. Edited by D. H. Russell. Raven Press, New York © 1973.

Polyamine Disposition in the Central Nervous System

Solomon H. Snyder, Edward G. Shaskan,[1] and Sami I. Harik

Departments of Pharmacology and Experimental Therapeutics and Psychiatry and the Behavioral Sciences, Johns Hopkins University School of Medicine, Baltimore, Maryland 21205

The exact function of polyamines in any tissues of animals has not yet been established. Indirect evidence obtained by correlation studies has suggested that polyamines are associated with the disposition of nucleic acids, especially ribosomal RNA. It has not yet been ascertained whether the functions of polyamines in all tissues are the same or whether these compounds subserve one role in one type of tissue and other roles in other tissues. The brain provides an excellent organ for exploring the functions of biological compounds, because of the marked heterogeneity of the brain both in terms of various regions and in terms of different cell types and subcellular organelles.

Some workers have shown that spermidine concentrations are greatest in areas of the brain rich in white matter (1, 2). White matter consists largely of myelin with very few cells, except for the sheath cells. There is unquestionably much less RNA and RNA synthesis in white than in gray matter. Thus these studies do not accord well with the hypothesis that the function of polyamines has to do with RNA formation. By contrast, developmental studies have shown a correlation between polyamine and nucleic acid levels: both undergo a progressive decline with age (2, 3).

In our laboratory we have explored the role of polyamines in the brain utilizing four different approaches: 1) we have evaluated the regional distribution of the polyamines, their precursor putrescine, and the activity of their biosynthetic enzyme S-adenosylmethionine (SAM) decarboxylase; 2) we have begun to assess the influence of drugs upon the disposition of polyamines; 3) we have compared in some detail developmental alterations

[1] Present address: Division of Biological and Medical Sciences, Brown University, Providence, Rhode Island 02912.

of the polyamines, SAM decarboxylase, and nucleic acid content in different brain areas; and 4) we have measured the turnover of polyamines in various brain regions.

I. REGIONAL STUDIES

Earlier investigations (1, 2) had shown a concentration of spermidine in brain regions richest in white matter. To evaluate this issue in detail, we carefully dissected the gray matter in the parietal area of the cerebral cortex from the closely apposing white matter of the internal capsule. Whereas the internal capsule, which is almost exclusively white matter, contained 1.5 times as much spermidine as the gray matter of the cerebral cortex, it was by no means the highest area of the brain in terms of spermidine content (Table 1). Its spermidine levels were only about half those of the medulla pons, which contained the highest spermidine levels of all. Although the medulla pons is rich in white matter, it contains large numbers of neuronal elements, which would be thought of as gray matter, and hence would contain less white matter than the internal capsule. The finding that the medulla pons and midbrain were among the areas of the brain richest in spermidine concurs with the results of previous workers. The cerebellum, which is not particularly rich in white matter, was the second highest in spermidine content.

Highest levels of spermine occurred in the cerebellum, more than twice the lowest levels which were observed in the medulla pons. Since the cere-

TABLE 1. *Regional distribution of spermidine, spermine, and S-adenosyl-L-methionine decarboxylase activity in adult rat brain*

Region	SAM-D activity (pmoles $^{14}CO_2$/hr) / mg protein	Spermidine (nmoles/g)	Spermine (nmoles/g)	Spermidine +spermine	Spermidine / spermine
Cerebral cortex (parietal gray)	652 ± 36	183 ± 10	193 ± 12	376	0.95
Internal capsule	318 ± 25	282 ± 12	131 ± 09	413	2.15
Cerebellum	556 ± 28	406 ± 19	231 ± 18	637	1.76
Medulla pons	407 ± 12	519 ± 17	111 ± 09	630	4.68
Midbrain	286 ± 8	367 ± 18	147 ± 10	514	2.50
Hypothalamus	268 ± 16	311 ± 17	147 ± 05	458	2.12
Corpus striatum	256 ± 11	238 ± 03	173 ± 02	411	1.38

Values represent the mean ±S.E.M. for a minimum of four determinations. Polyamine concentrations are expressed with respect to the wet weight of the tissue sample. Spermidine/spermine represents the molar ratio of spermidine to spermine.

bellum also contained the second highest spermidine concentration, its total polyamine concentration was the greatest of all brain regions studied. By contrast, the second highest levels of spermine occurred in the cerebral cortex, which had the lowest concentration of spermidine. Because of its low spermidine and high spermine concentration, the cerebral cortex displayed the smallest ratio of spermidine:spermine. The medulla pons, on the other hand, which had the highest spermidine and the lowest spermine concentration in the brain, manifested the greatest spermidine:spermine ratio, almost five times that of the cerebral cortex.

In preliminary subcellular studies, we found that most SAM decarboxylase activity was confined to the soluble supernatant fraction of brain homogenates, paralleling the distribution of lactic acid dehydrogenase, a marker for cytoplasm. Accordingly all SAM decarboxylase assays were conducted on soluble supernatant preparations.

There were some striking differences between the regional variations in SAM decarboxylase and those of the polyamines themselves. The cerebral cortex contained the highest SAM decarboxylase activity, almost three times greater than the lowest values, which were observed in the corpus striatum. One might speculate that the high values of SAM decarboxylase in the cerebral cortex are related to the formation of spermine, whose levels are high in the cerebral cortex, rather than spermidine.

Whereas in bacteria the enzymatic activities for the decarboxylation of SAM are readily distinguished from those which transfer the propylamine moiety of decarboxylated SAM to putrescine to form spermidine (4), in mammalian tissues these two enzymatic activities appear to be closely linked (5, 6). Recent evidence shows, however, that even in animal tissues there may be discrete enzymes for SAM decarboxylation, spermidine synthesis, and spermine synthesis (7–9). To assess the significance of SAM decarboxylase activity in the brain, we compared this activity with that of spermidine synthesis in three brain regions (cerebral cortex, cerebellum, and medulla pons) that vary widely in their SAM decarboxylating activity. In these three brain regions the formation of spermidine was accompanied stoichiometrically by the decarboxylation of SAM (Table 2). Thus regional variations in SAM decarboxylase parallel those of spermidine synthesis. We did not assess the activity of spermidine or spermine synthase.

It is unlikely that the regional differences we observed in SAM decarboxylase activity were related to the presence of inhibitors or activators in the different brain regions, because mixing experiments showed that enzyme activity in all brain regions was additive.

Ornithine decarboxylase is thought to be an important rate-limiting enzyme in polyamine biosynthesis (10–12). Unfortunately we have not been

TABLE 2. *Stoichiometry between the formation of* CO_2 *and spermidine from S-adenosylmethionine and putrescine in rat brain regions*

Region	$^{14}CO_2$ Released from SAM-^{14}COOH (pmoles/hr/mg protein)	^{3}H-Spermidine formed from ^{3}H-putrescine (pmoles/hr/mg protein)
Cerebral cortex (parietal gray)	595 ± 40	659 ± 62
Cerebellum	390 ± 30	360 ± 37
Medulla pons	239 ± 35	292 ± 37

The formation of $^{14}CO_2$ from ^{14}C-SAM and the incorporation of ^{3}H-putrescine into ^{3}H-spermidine were assayed in the same experiment. Each value is the mean ± S.E.M. of six determinations. Differences between the mean values for $^{14}CO_2$ released and ^{3}H-spermidine formed for each region studied are not significantly different.

able to detect ornithine decarboxylase in crude brain extracts. One might expect putrescine levels to reflect the activity of ornithine decarboxylase, as has been shown for regenerating rat liver (11–13) and for the liver after treatment with growth hormone (12, 14, 15).

Accordingly, we undertook to develop a sensitive and simple assay for tissue putrescine, because levels of these compounds in most animal tissues are quite low and existing techniques are tedious and relatively nonspecific (4, 16, 17). The enzymatic-isotopic microassay we have developed can measure as little as 20 pmoles of putrescine in tissue samples (18–20). This assay is based on the known ability of putrescine to stimulate the decarboxylation of SAM by baker's yeast SAM decarboxylase (5, 21). Putrescine in tissue samples enhances the evolution of $^{14}CO_2$ from carboxyl-labeled SAM by yeast SAM decarboxylase. The stimulation of yeast SAM decarboxylase is linear with added putrescine over a range of 20–400 pmoles of putrescine. Moreover, of numerous other compounds tested, only agmatine gives a reaction comparable to that of putrescine (presumably due to the presence of arginase in the yeast enzyme preparation which would convert agmatine to putrescine). Cadaverine and 1,3-diaminopropane react only 10 and 5%, respectively, as much as putrescine does. Extensive evaluation of mouse and rat brain and liver by amino acid analyzer procedures both in our laboratory and that of L. Kremzner (personal communication) has failed to indicate the presence of any cadaverine, 1,3-diaminopropane, or agmatine in these tissues. However, 1,3-diaminopropane could be detected in trace amounts in the livers of these animals. Spermidine gives only 0.25% the reaction of putrescine. Accordingly, it is possible to assay putrescine in crude tissue extracts without preliminary purification. Samples are simply homogenized in dilute buffer solution, heated to precipitate protein and destroy endogenous SAM, and then subjected to the enzymatic-isotopic procedure.

The regional distribution of putrescine was fairly similar to that of spermidine. Putrescine content was highest in areas rich in white matter. Maximal values were observed in the spinal cord, where levels were triple those of the hypothalamus, the region with the least putrescine (Table 3).

TABLE 3. *Regional distribution of putrescine in rat brain*

Region	Putrescine content (nmoles/g tissue ± S.E.M.)
Cerebral cortex (white and gray matter)	18.6 ± 0.9
Striatum	17.2 ± 1.3
Hypothalamus	11.3 ± 1.3
Thalamus and mesencephalon	27.9 ± 3.2
Cerebellum	14.4 ± 1.7
Medulla and pons	29.2 ± 4.2
Cervical spinal cord	33.4 ± 1.3

Rats were decapitated and their brains and cervical spinal cords were quickly removed and dissected into the various regions. Putrescine tissue concentrations were determined by the enzymatic-isotopic assay (20). Forty-μl aliquots of tissue extracts, representing 4 mg of tissue, were added to the assay system containing the following ingredients in a final volume of 120 μl: 8 μmoles Tris buffer, pH 7.6; 0.06 μmole of dithiothreitol; 0.82 nmole of S-adenoxyl-L-carboxy ^{14}C-methionine (0.05 μC); and 20 units of baker's yeast S-adenosyl-L-methionine decarboxylase. The radioactivity of $^{14}CO_2$ evolved was converted into moles of putrescine per gram tissue by derivation from a standard curve performed simultaneously with the putrescine assay.

Regional putrescine levels fell into two groups, the medulla pons, midbrain, and spinal cord with levels close to 30 nmoles/g and all of the other brain regions with less than half this value. In studies of the regional distribution of putrescine, cerebral cortex was not dissected into separate white and gray matter areas.

In summary, levels of polyamines and putrescine vary markedly among brain regions. Areas with considerable white matter contain higher levels, although this correlation with white matter is by no means perfect. Except for the cerebral cortex and internal capsule, there is a better correlation between spermidine and SAM decarboxylase than between spermine and SAM decarboxylase. The cerebral cortex, which contained the highest SAM decarboxylase activity in the brain, the lowest spermidine and second highest spermine levels, is a striking exception. Conceivably, in cerebral cortex SAM decarboxylase activity reflects primarily the capacity for the synthesis of spermine, whereas in other areas SAM decarboxylase activity is associated more with spermidine formation. This meshes well with turnover studies, which will be described below, showing that the cerebral cortex is the area of the brain with the greatest conversion of intraventricularly administered ^{3}H-putrescine to ^{3}H-spermine.

II. INFLUENCE OF L-DOPA (L-3,4-DIHYDROXYPHENYLALANINE) ON PUTRESCINE LEVELS IN BRAIN AND LIVER

The availability of a sensitive, specific, and simple assay for putrescine made it possible to study the pharmacology of the polyamines. Initially we screened several drugs. The convulsant drug pentylenetetrazol (Metrazol®), the sedative phenobarbital, and the amino acid histidine failed to influence putrescine levels of brain and liver. By contrast L-DOPA injection increased liver putrescine approximately fivefold, liver ornithine decarboxylase nearly 10-fold, and brain putrescine levels 1.3-fold (Table 4). In experiments in which 400 mg/kg of L-DOPA was administered intraperitoneally in two divided doses over an 8-hr period these increases were even more striking, with a doubling of brain putrescine and a sixfold increase in liver putrescine.

TABLE 4. *Effects of* L-DOPA on putrescine levels in rat brain and liver and on ornithine decarboxylase activity in rat liver

	Control	L-DOPA 200 mg/kg	D-DOPA 200 mg/kg
Brain putrescine (nmoles/g)	33.2 ± 1.7	43.7^{a} ± 3.2	35.6 ± 3.5
Liver putrescine (nmoles/g)	94.0 ± 23.3	460.7^{b} ± 11.6	98.2 ± 22.8
Liver ornithine decarboxylase activity (pmoles CO_2/mg protein/hr)	95 ± 25	937^{a} ± 238	70 ± 11

Experimental values differ from control $^{a}p < 0.05$; $^{b}p < 0.001$

Groups of four 100-g male rats received L-DOPA, D-DOPA, or the vehicle solvent (0.075 N HCl) intravenously 6 hr prior to sacrifice. The brains and livers were assayed for putrescine by the enzymatic-isotopic assay procedure (20). Liver ornithine decarboxylase activity was assayed by estimating the $^{14}CO_2$ evolved upon incubation of 1-^{14}C-ornithine with supernatant preparations of rat liver homogenates by a modification of the method of Jänne and Williams-Ashman (28). The radioactivity of the $^{14}CO_2$ evolved was converted to pmoles CO_2 per mg protein per hour. Data presented are the mean values for each group of rats.

How might DOPA exert this striking effect? The clinical use of L-DOPA in Parkinson's disease is associated with its decarboxylation to dopamine which replaces the depleted dopamine in the brains of parkinsonian patients. DOPA can lower tissue levels of SAM, presumably by acting as a sink for SAM which is consumed in the O-methylation of DOPA by the enzyme catechol-O-methyltransferase (22). Both D- and L-DOPA would be excellent substrates for O-methylation, hence effective depletors of SAM levels, whereas only L-DOPA can be decarboxylated by the enzyme DOPA decarboxylase. Accordingly we compared the influence of D- and L-DOPA upon putrescine and ornithine decarboxylase levels (Table 4). D-DOPA was essentially without effect, indicating that L-DOPA was acting via dopamine.

DOPA decarboxylase occurs in the brain and in many peripheral tissues as well. Apparently via an action upon brain dopamine, L-DOPA is capable of causing a release of growth hormone (23) which is well known to increase liver putrescine (15). Thus DOPA could be influencing liver putrescine via an action on the brain either mediated through growth hormone or through some other mechanism. Alternatively DOPA could be converted to dopamine in the liver itself and thence alter ornithine decarboxylase and putrescine levels. It is possible to distinguish between central and peripheral actions of DOPA. Certain inhibitors of DOPA decarboxylase, such as Ro 4–4602 and MK-486(γ-hydrazino-methyl-DOPA), in relatively small doses are potent inhibitors of peripheral DOPA decarboxylase but do not easily penetrate the blood-brain barrier and hence do not alter brain DOPA decarboxylase.

In larger doses, these compounds are potent inhibitors of peripheral as well as central DOPA decarboxylase. By comparing the influence of peripheral and central inhibition of DOPA decarboxylase, we hope to determine whether the influence of DOPA upon putrescine levels is mediated centrally or peripherally. The possible role of growth hormone in the effects of these drugs will be assessed by examining the action of DOPA in hypophysectomized animals.

III. DEVELOPMENTAL STUDIES

To assess developmental relationships between polyamines and nucleic acids, spermidine, spermine, DNA, RNA, and total acid-insoluble protein were measured simultaneously in the cerebral cortex and cerebellum of rats at five ages between 4 and 45 to 60 days after birth (Tables 5, 6). SAM decarboxylase was also measured at these times (Fig. 1) (24). In the cerebral cortex spermidine concentration fell progressively between 5 and 22 days

TABLE 5. *Developmental distribution of polyamines, nucleic acids, and protein in cerebral cortex (parietal gray)*

Age	DNA[a] (mg/g)	RNA[b] (mg/g)	Protein[c] (mg/g)	Spermidine (nmoles/g)	Spermine (nmoles/g)
5 Days	1.62 ± 0.05	4.20 ± 0.12	43.9 ± 0.8	436 ± 32	291 ± 18
10 Days	0.97 ± 0.05	3.72 ± 0.21	49.0 ± 1.4	319 ± 08	299 ± 12
15 Days	0.86 ± 0.06	3.82 ± 0.11	62.8 ± 1.0	230 ± 14	258 ± 10
22 Days	0.69 ± 0.03	3.08 ± 0.10	65.1 ± 1.2	190 ± 14	217 ± 14
Adult	0.73 ± 0.16	1.86 ± 0.14	67.6 ± 1.8	186 ± 10	201 ± 13

All values are the mean ± S.E.M. for four animals at each age and the concentrations are expressed on a wet weight basis.

[a]Correlation coefficients on DNA: spermidine, $r = 0.93$; spermine, $r = 0.62$.

[b]Correlation coefficients on RNA: spermidine, $r = 0.76$; spermine, $r = 0.75$.

[c]Correlation coefficients on protein: spermidine, $r = -0.48$; spermine, $r = -0.44$.

TABLE 6. *Developmental distribution of polyamines, nucleic acids, and protein in the cerebellum*

Age	DNA[a] (mg/g)	RNA[b] (mg/g)	Protein[c] (mg/g)	Spermidine (nmoles/g)	Spermine (nmoles/g)
5 Days	4.62 ± 0.23	4.10 ± 0.58	41.6 ± 1.6	565 ± 50	278 ± 05
10 Days	7.46 ± 0.25	5.68 ± 0.44	61.7 ± 1.8	748 ± 37	457 ± 10
15 Days	6.26 ± 0.11	4.85 ± 0.42	63.6 ± 1.3	571 ± 14	323 ± 06
22 Days	5.83 ± 0.16	4.45 ± 0.13	71.6 ± 3.1	414 ± 14	309 ± 06
Adult	4.44 ± 0.05	2.26 ± 0.13	75.6 ± 1.8	410 ± 16	248 ± 12

All values represent the mean ± S.E.M. for four animals at each age and the concentrations are expressed on a wet weight basis.

[a]Correlation coefficients on DNA: spermidine, $r = 0.89$; spermine, $r = 0.92$.

[b]Correlation coefficients on RNA: spermidine, $r = 0.82$; spermine, $r = 0.75$.

[c]Correlation coefficients on protein: spermidine, $r = 0.28$; spermine, $r = 0.39$.

when it attained adult levels, less than half the 5-day concentrations. Spermine in the cerebral cortex decreased more gradually and was not significantly reduced until 22 days when its concentration was 70% of the 5-day level. The changes in spermidine correlated more closely with DNA values, which also fell 55% between 5 and 22 days, than with RNA or protein. Spermine alterations in the cerebral cortex paralleled more closely those of RNA than DNA, although these differences were not statistically significant. Since protein concentration increased during the time intervals examined, spermidine and spermine levels correlated negatively with changes in total protein.

Unlike the cerebral cortex in which DNA concentration fell 40% between 5 and 10 days, in the cerebellum DNA increased 60% during this time

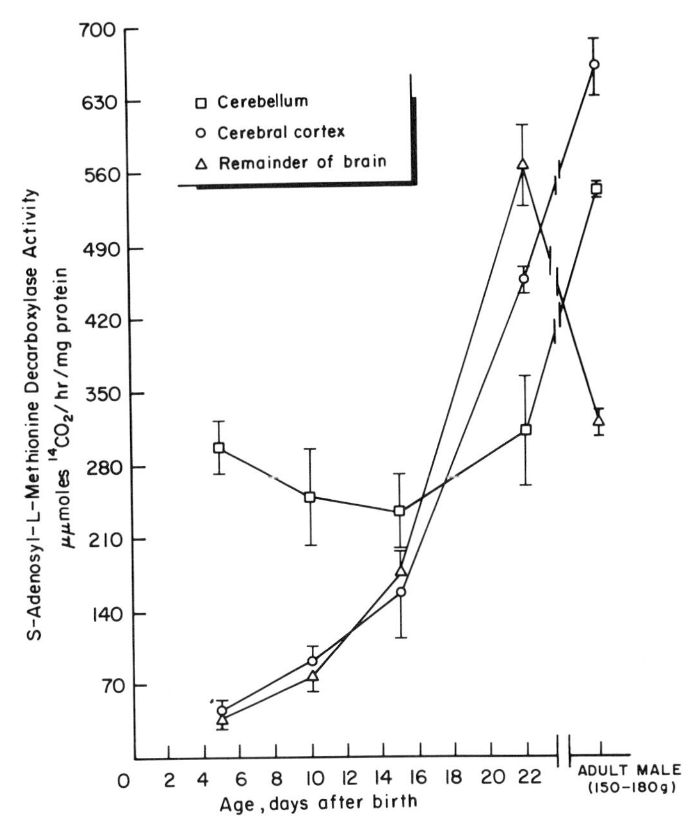

FIG. 1. S-Adenosyl-L-methionine decarboxylase activity as a function of age in the cerebellum, cerebral cortex (parietal gray), and "remainder of brain." Adult (150 to 200 g body wt) rats were between 45 to 60 days of age. Values represent the mean ± S.E.M. of four determinations.

interval and then fell gradually to concentrations in the adult that were the same as in the 5-day-old rat. These changes may relate to the unique rapid growth of the cerebellum between 5 and 10 days. Spermine in the cerebellum changed in a pattern which paralleled the development of DNA, with a 60% increase between 5 and 10 days and a subsequent decline to values in the adult that were the same as in the 5-day-old rat. In the cerebellum, RNA and spermidine showed similar developmental changes, both undergoing small increases of 30% between 5 and 10 days, less than for DNA and spermine. As had been observed in the cerebral cortex, total acid insoluble protein continually increased during development, hence failed to correlate with either spermidine or spermine.

SAM decarboxylase tended to develop in the opposite direction to the changes in polyamines themselves. Enzyme activity in the cerebral cortex increased progressively more than 10-fold between 5 and 45 to 60 days. In the cerebellum, on the other hand, SAM decarboxylase activity was fairly constant between 5 and 22 days, then doubled between 22 and 45 to 60 days. The "remainder of brain" exhibited yet a different pattern of development. Between 5 and 22 days, enzyme activity followed closely the levels in the cerebral cortex with a progressive increase of 10-fold. However, between 22 days and 45 to 60 days, enzyme activity fell 50% in the "remainder of brain."

It is unclear just what the developmental pattern of the polyamines tells us about their function in different brain regions. The tendency of polyamine levels to fall with age would correlate with the gradual decrease in rapid growth of the brain. It is quite unclear why SAM decarboxylase activity should increase in a pattern opposite to that of the polyamines themselves. There are other discrepancies between SAM decarboxylase activity and polyamine levels. SAM decarboxylase in the adult rat brain is about 20% greater than in the liver and higher than any rat tissue except for the pancreas, whereas spermidine and spermine concentrations in the brain are among the lowest in rat tissues, only half the levels observed in the liver.

The correlations we observed between polyamines and nucleic acids also varied between the cerebellum and cerebral cortex. In the cerebellum, both spermidine and spermine correlated better with DNA and RNA, although in the cerebral cortex, spermidine correlated best with DNA and spermine best with RNA. The significance of the correlation with DNA is called into question by the fact that the cerebellum contains four times as much DNA as the cerebral cortex, yet polyamine levels are fairly similar in these two regions.

IV. TURNOVER STUDIES

The rate of turnover of biological compounds can shed considerable light on their functions. For example, with regard to the polyamines, it was found that spermidine turnover in rat liver had a half-life of about 5 days, closely similar to the turnover rate for RNA in the liver (13). In earlier studies the turnover of brain polyamines was assessed after intracisternal administration of radiolabeled putrescine (13). When measuring turnover rates, it is crucial that only a tracer dose of the compound or its precursor be administered. In the previous investigations, the dose of putrescine was more than half the endogenous levels of putrescine in the brain (13). Accordingly, we undertook further studies of polyamine turnover in the brain utilizing lower doses of putrescine, in this case 5 nmoles which represents

15 to 20% of brain putrescine, depending on the region examined. Between 20 min and 2 days after the intraventricular injection of radiolabeled putrescine, virtually all the ^{3}H-putrescine had been converted to radiolabeled spermidine and spermine (Table 7). In some experiments, we observed that about 60 to 70% of the injected dose of putrescine disappeared in the first 20 min, which is similar to results reported previously with tracer doses of other amines such as norepinephrine (25) and histamine (26).

The amount of radioactive label at 20 min varied considerably among different regions. Highest levels occurred in the midbrain, hypothalamus, and cerebellum, whereas the medulla pons, internal capsule, cerebral cortex, and corpus striatum contained 10% or less of the label observed in the midbrain. Presumably these differences are related to factors such as the flow of ventricular fluid and the proximity of various areas to the ventricular system.

Interestingly, the areas with the greatest initial accumulation of radioactivity also subsequently displayed the steepest decline of labeled spermidine. Between 20 min and 19 days after putrescine administration, total radioactivity in the midbrain, hypothalamus, and cerebellum fell approximately 50%, whereas in the other regions total radioactivity in this time interval increased about two- to threefold. Although it is difficult to make an accurate estimate, it seems that the combined fall in midbrain, hypothalamus, and cerebellum during this time was about the same as the increase in the other regions so that the total radioactivity of the brain hardly changed. This suggests that radioactive amines were transferred among brain regions.

In the medulla pons, internal capsule, and corpus striatum, which contained among the lowest initial levels of radioactivity, radiolabeled spermidine levels were approximately constant between 2 and 19 days and in the cerebral cortex even appeared to increase during this time period (Fig. 2). By contrast, after 2 days ^{3}H-spermidine levels in midbrain, hypothalamus, and cerebellum with the half-lives for its specific activity of 16 to 19 days (Fig. 3).

Except for the cerebral cortex no ^{3}H-spermine was detected in any region 20 min after putrescine injection. Between 2 and 19 days in the midbrain and hypothalamus, ^{3}H-spermine levels were fairly constant, although in the cerebellum ^{3}H-spermine doubled between 2 and 14 days but then fell 50% between 14 and 19 days. In the medulla pons, internal capsule, and cerebral cortex, ^{3}H-spermine increased gradually and in the cerebral cortex actually underwent a seven- to eightfold augmentation. Thus the relative amounts of ^{3}H-spermine and ^{3}H-spermidine displayed different patterns in various brain regions. For instance, in the cerebral cortex 19 days after putrescine administration, ^{3}H-spermine levels were as high as those of ^{3}H-spermidine,

TABLE 7. *Regional distribution of radioactive polyamines in adult rat brain*

Region (mean wet weight in g ± S.E.M.)	^{3}H-Amine levels (n-C/g) ± S.E.M.)				
	20 min (N = 7)	2 days (N = 6)	5 days (N = 5)	15 days (N = 4)	19 days (N = 8)
Midbrain (0.12 ± 0.01)					
putrescine	143.0 ± 12.0	1.6 ± 0.6	N.D.[a]	N.D.	N.D.
spermidine	12.2 ± 1.3	113.0 ± 6.4	93.2 ± 3.6	75.9 ± 2.0	54.2 ± 2.4
spermine	N.D.	11.6 ± 1.5	13.0 ± 1.8	13.4 ± 1.3	11.8 ± 1.7
Hypothalamus (0.09 ± 0.01)					
putrescine	50.1 ± 6.8	N.D.	N.D.	N.D.	N.D.
spermidine	4.4 ± 0.8	46.2 ± 3.5	39.3 ± 4.1	28.2 ± 2.2	24.5 ± 3.3
spermine	N.D.	4.6 ± 1.6	3.9 ± 0.4	6.5 ± 0.7	5.9 ± 0.3
Cerebellum (0.27 ± 0.02)					
putrescine	67.2 ± 2.9	1.5 ± 0.7	N.D.	N.D.	N.D.
spermidine	11.9 ± 1.2	70.7 ± 8.1	52.3 ± 3.5	44.3 ± 2.7	33.8 ± 3.7
spermine	N.D.	4.5 ± 0.4	7.7 ± 1.0	10.4 ± 1.3	4.6 ± 0.6
Medulla pons (0.29 ± 0.03)					
putrescine	14.1 ± 1.2	0.6 ± 0.1	N.D.	N.D.	N.D.
spermidine	4.2 ± 0.3	36.0 ± 0.8	36.9 ± 1.4	41.2 ± 1.5	40.1 ± 2.0
spermine	N.D.	2.3 ± 0.4	3.6 ± 0.4	5.1 ± 0.2	5.7 ± 0.3
Internal capsule (0.08 ± 0.02)					
putrescine	15.9 ± 3.1	N.D.	N.D.	N.D.	N.D.
spermidine	1.4 ± 0.3	38.9 ± 4.2	35.5 ± 3.4	29.6 ± 2.5	33.6 ± 1.7
spermine	N.D.	2.4 ± 0.5	2.1 ± 0.8	5.5 ± 1.5	5.6 ± 1.1
Cerebral cortex (0.17 ± 0.02)					
putrescine	7.3 ± 0.8	0.8 ± 0.2	N.D.	N.D.	N.D.
spermidine	1.6 ± 0.4	12.4 ± 1.3	13.9 ± 2.6	14.8 ± 2.2	17.0 ± 2.2
spermine	0.9 ± 0.1	1.9 ± 0.3	5.4 ± 1.0	10.5 ± 1.3	15.3 ± 0.6
Corpus striatum (0.09 ± 0.02)					
putrescine	3.0 ± 0.5	N.D.	N.D.	N.D.	N.D.
spermidine	0.9 ± 0.2	9.2 ± 1.8	9.8 ± 1.0	9.8 ± 1.2	8.9 ± 0.7
spermine	N.D.	N.D.	N.D.	1.4 ± 0.5	0.9 ± 0.2

[a]N.D. = not detectable

^{3}H-Putrescine (5 nmoles; 0.43 μC) was injected intraventricularly. Radioactive polyamines in various brain regions were separated by cation-exchange column chromatography.

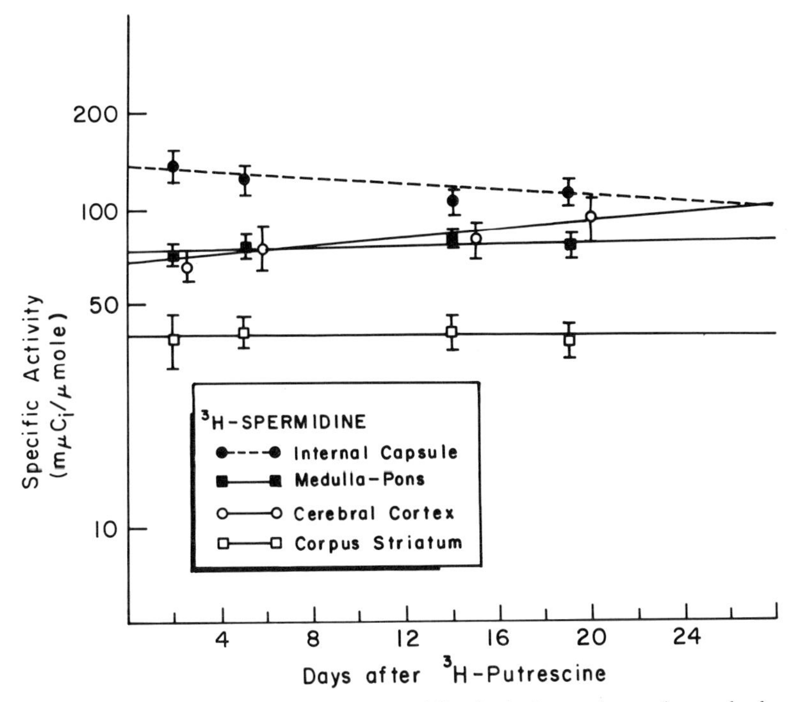

FIG. 2. Changes of the specific activity of spermidine in the internal capsule, cerebral cortex, medulla pons, and corpus striatum of rats following the intraventricular injection of a tracer dose (5 nmoles) of ^{3}H-putrescine. Data are plotted on a semilogarithmic scale. Each point represents the mean ± S.E.M. for a minimum of four rats.

whereas in all of the other regions, ^{3}H-spermine was only 10 to 25% of values of ^{3}H-spermidine. Interestingly, the cerebral cortex, which possesses the highest SAM-decarboxylase activity in the brain, has the lowest endogenous spermidine but the second highest concentration of spermine. The suggestion above that SAM decarboxylase in the cerebral cortex might be concerned primarily with synthesis of spermine is consistent with the observation that the cerebral cortex was the most active brain area in synthesizing spermine *in vivo*.

The failure of total radioactivity of the brain to change during a 3-week period and the apparent transfer of ^{3}H-amines among various brain regions is quite striking. Polyamines bind avidly to negatively charged macromolecules. When they leave one part of the brain, they might rebind in another. They might conceivably recirculate via the cerebrospinal fluid.

The areas in which ^{3}H-amines fell the most contained four to 10 times more ^{3}H-amines initially than did the areas in which there was an increase

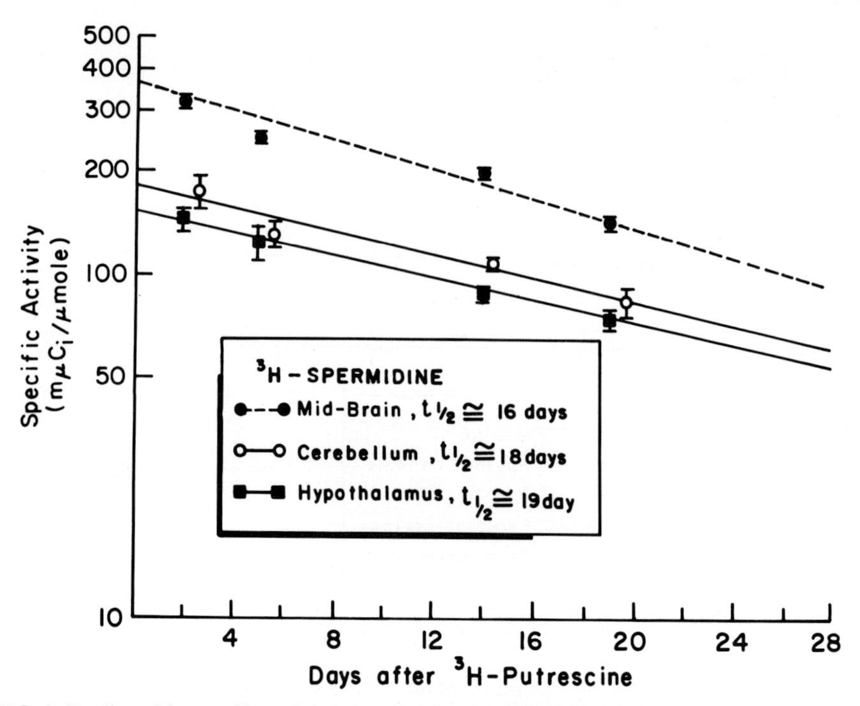

FIG. 3. Decline of the specific activity of spermidine in midbrain, cerebellum, and hypothalamus of rats following an intraventricular tracer dose of ^{3}H-putrescine (5 nmoles). Essentially no ^{3}H-putrescine was detectable by the second day after injection. Specific activity of spermidine is expressed as nC/g wet weight divided by the μmoles/g wet weight of endogenous spermidine determined independently on pooled tissue samples (Shaskan et al., 1972). Data are plotted on a semilogarithmic scale. Each point represents the mean ± S.E.M. for a minimum of four rats.

in ^{3}H-amine with time. Thus a distinguishing characteristic of the areas in which ^{3}H-amines fell the most was that they had the highest initial content, although an increase in ^{3}H-amines was observed in areas with the lowest initial content of radioactivity. These findings might be explained by the following considerations. Let us assume that endogenous polyamines turn over by leaving the cellular constituents in which they were synthesized and subsequently rebind in other tissue constituents. It is possible that all brain areas have a similar polyamine turnover rate and that all participate both in release and rebinding of polyamines. The selective decline of ^{3}H-amines in some brain regions may simply be attributable to their containing the highest initial level of radioactivity. Transfer into these regions from the others might also take place but would not have been detected in our experiments because of the low initial ^{3}H-amine content.

This explanation for the different dynamics of polyamines in various regions suggests that the half-lives we calculated for spermidine in the midbrain, hypothalamus, and cerebellum may be overestimates. Transfer into these regions from other areas would augment the ^{3}H-polyamine content, slowing the apparent turnover. The half-lives we observed in midbrain, hypothalamus, and cerebellum of 16 to 19 days are considerably longer than the 4-day life reported earlier in the liver and brain (13). In the liver the half-life for spermidine turnover was similar to that estimated for ribosomal RNA. Ribosomal RNA in the brain appears to be heterogeneous. Various pools have markedly differing turnover rates, while the overall turnover has a half-life of about 12 days (27). This is fairly comparable to the half-life for spermidine turnover that we observed in the midbrain, hypothalamus, and cerebellum. If the values we calculated are overestimates, then brain RNA and "corrected" spermidine turnover rates would coincide even more closely.

REFERENCES

1. Shimizu, H., Kakimoto, Y., and Sano, I., *J. Pharmacol. Exp. Ther.*, **143,** 199 (1964).
2. Kremzner, L. T., Barrett, R. E., and Terrano, M. J., *Ann. N.Y. Acad. Sci.*, **171,** 735 (1970).
3. Jänne, J., Raina, A., and Siimes, M., *Acta Physiol. Scand.*, **62,** 352 (1964).
4. Tabor, H., and Tabor, C. W., *Pharmacol. Rev.*, **16,** 245 (1964).
5. Pegg, A. E., and Williams-Ashman, H. G., *J. Biol. Chem.*, **244,** 682 (1969).
6. Feldman, M. J., Levy, C. C., and Russell, D. H., *Biochem.*, **11,** 671 (1972).
7. Hannonen, P., Jänne, J., and Raina, A., *Biochem. Biophys. Res. Comm.*, **46,** 341 (1972).
8. Jänne, J., Schenone, A., and Williams-Ashman, H. G., *Biochem. Biophys. Res. Comm.*, **42,** 758 (1971).
9. Raina, A., and Hannonen, P., *FEBS Letters*, **16,** 1 (1971).
10. Pegg, A. E., Lockwood, D. H., and Williams-Ashman, H. G., *Biochem. J.*, **117,** 17 (1970).
11. Russell, D. H., and Snyder, S. H., *Proc. Nat. Acad. Sci. U.S.A.*, **60,** 1420 (1968).
12. Snyder, S. H., Kreuz, D. S., Medina, V. J., and Russell, D. H., *Ann. N.Y. Acad. Sci.*, **171,** 749 (1970).
13. Russell, D. H., Medina, V. J., and Snyder, S. H., *J. Biol. Chem.*, **2445,** 6732 (1970).
14. Russell, D. H., Snyder, S. H., and Medina, V. J., *Endocrinology*, **86,** 1414 (1970).
15. Jänne, J., and Raina, A., *Biochim. Biophys. Acta*, **174,** 769 (1969).
16. Dubin, D. T., *J. Biol. Chem.*, **235,** 783 (1960).
17. Raina, A., *Acta Physiol. Scand.*, **60,** Suppl. 218 (1963).
18. Harik, S., Pasternak, G., and Snyder, S. H., *Proc. Int. Pharmacol. Cong.*, **3,** 95 (1972).
19. Harik, S., Pasternak, G., and Snyder, S. H., *Biochim. Biophys. Acta*, in press.
20. Harik, S., Pasternak, G., and Snyder, S. H., this volume.
21. Jänne, J., Williams-Ashman, H. G., and Schenone, A., *Biochem. Biophys. Res. Comm.*, **42,** 1362 (1971).
22. Wurtman, R. J., Rose, C. M., Matthysse, S., Stephenson, J. and Baldessarini, R., *Science*, **169,** 395 (1970).
23. Boyd, A. E., III, Lebovitz, H. E., and Pfeiffer, J. B., *New Engl. J. Med.*, **283,** 1425 (1970).
24. Shaskan, E. G., Haraszti, J. H., and Snyder, S. H., *J. Neurochem.*, in press.
25. Glowinski, J., Kopin, I. J., and Axelrod, J., *J. Neurochem.*, **12,** 25 (1965).
26. Snyder, S. H., Glowinski, J., and Axelrod, J., *J. Pharmacol. Exp. Ther.*, **153,** 8 (1967).
27. Von Hungen, K., Mahler, H. R., and Moore, W. J., *J. Biol. Chem.*, **243,** 1415 (1968).
28. Jänne, J., and Williams-Ashman, H. G., *J. Biol. Chem.*, **246,** 1725 (1971).

Polyamines in Normal and Neoplastic Growth. Edited by D. H. Russell. Raven Press, New York © 1973.

Specific Increases in Polyamines in Mixed Lymphocyte Reactions

Laurence J. Marton, Kenneth D. Graziano, Michael R. Mardiney, Jr., and Diane H. Russell

Laboratory of Pharmacology, Baltimore Cancer Research Center, National Cancer Institute, Baltimore, Maryland 21211

Several years ago it was found that ornithine decarboxylase, the enzyme which catalyzes the conversion of ornithine to putrescine, was markedly elevated after partial hepatectomy of the rat (1, 2). This elevation was one of the earliest detectable events that occurred after partial hepatectomy as ornithine decarboxylase activity was elevated within the first hour. On the basis of these findings, it was postulated that polyamines may play important roles in the regulation of the growth process. Since that time, enhanced polyamine biosynthesis has been shown to be one of the earliest detectable events in all growth systems studied, i.e., embryonic development (1, 3–5), cardiac hypertrophy (6, 7), and neoplastic growth (8). Further, polyamine biosynthesis and accumulation are enhanced by those hormones which regulate growth processes, i.e., growth hormone in rat liver (9–11), testosterone in rat prostate (12), and estradiol in rat uterus (13, 14).

Since polyamine biosynthesis and accumulation do occur early after a specific stimulus for growth, and since the net increase in polyamines appears to be related to the strength of the stimulus (15–18) (that is, polyamine accumulations parallel in mammalian systems the ability to accumulate higher levels of RNA and thus the ability of a tissue to maintain higher rates of protein synthesis), perhaps the increases in polyamines could be used as a detection system for tissue response.

The stimulation of lymphocytes by allogeneic lymphocytes, antigens, and phytohemagglutinin has been studied extensively because it offers a unique tool in the study of immune mechanisms. It has become of clinical importance as an *in vitro* assay for the study of cell-associated immune functions in both healthy and diseased subjects. In these assays, both the magnitude of the response and the kinetics of the response are used to evaluate cell-associated immune function. With the advent of a rapid, automated method for polyamine determinations (19, 20), it seems feasible at this time that a simple polyamine determination in mixed lymphocyte reactions or in human lymphocytes after PHA or antigen stimulation could serve as an additional reliable index of the proliferative response. With this in mind, we have compared specific increases of polyamine content with ^{3}H-thymidine and ^{3}H-uridine incorporation in mixed leukocyte and mixed lymphocyte reactions.

I. METHODS AND MATERIALS

Methods of culture, the evaluation of thymidine incorporation, and statistical evaluation are as previously described (21, 22). In brief, 3-million leukocytes obtained from sedimented heparinized whole blood were cultured in a final volume of 1.5 ml of Roswell Park Memorial Institute 1640 media (RPMI 1640 media) containing 15 to 20% autologous plasma, 50 units/ml penicillin, 50 μg/ml streptomycin, and 300 μg/ml of glutamine. Mixed leukocyte cultures contained 1.5-million responding cells and 1.5-million allogeneic irradiated cells (1,000 R).

Mixed lymphocyte cultures were performed in an identical manner except that leukocyte populations containing greater than 90% mononuclear cells were obtained by 50 × *g* centrifugation for 15 min as previously described (23). The cell concentration was adjusted to 1 × 10^{6}/ml, and mixed cultures consisted of 0.5 ml of each cell population. All cultures were maintained at an angle of 5° from the horizontal in a 95% air, 5% CO_2 humid atmosphere at 37°C. At appropriate times after cultures were initiated, 0.5 ml of RPMI 1640 media, 0.5 ml of RPMI 1640 media containing 2 μC of ^{3}H-thymidine (1.9 C/mmole, Schwartz BioResearch, Inc.), or 0.5 ml of RPMI 1640 media containing 2 μC of ^{3}H-uridine (4 C/mmole, Schwartz BioResearch, Inc.) was added to each culture tube. Triplicate tubes were incubated for a subsequent 4-hr period and washed with chilled saline and later analyzed for ^{3}H-thymidine incorporation into DNA by methods previously described (21, 22). Incorporation of ^{3}H-uridine into RNA was measured in a similar fashion.

For the determinations of polyamine levels, either six tubes of leukocyte cultures (9 × 10^{6} responding cells) or 18 tubes of purified lymphocyte cultures (9 × 10^{6} responding cells) were pooled, washed twice with saline, and pelleted again for the polyamine assay. To ensure that the cultures used for polyamine assay were treated identically to those used in the isotope incorporation studies, these cultures received 0.5 ml of medium alone at the beginning of the 4-hr pulse period. One ml of 0.1 N HCl was added to each pooled lymphocyte sample, and the cells were disrupted ultrasonically with an E/MC Corporation cell disruptor equipped with a microtip. The samples were deproteinated by adding 1 ml of 6% sulfosalicylic acid, allowed to sit for 10 min in an ice bath, and finally centrifuged at 12,000 × *g* for 15 min in a Sorval RC2-B refrigerated centrifuge. The supernatant was applied directly to the amino acid analyzer column for the polyamine analysis, as described elsewhere in this volume by Marton et al.

Recovery rates for putrescine, spermidine, and spermine were 96 to 99% for all three compounds carried through the preparative steps as outlined above.

One sample divided into four aliquots and then handled independently from the culture stage to the end of the analysis exhibited calculated polyamine values that varied only 3.5% from the mean.

Results are reported as specific increases in polyamine synthesis as well as thymidine and uridine incorporation. The specific increases reported represent the differences of polyamine content and thymidine or uridine incorporation in cultures containing leukocytes or lymphocytes stimulated by their irradiated syngeneic counterpart (AA_x) and in cultures of leukocytes or lymphocytes stimulated by irradiated allogeneic leukocytes or lymphocytes (AB_x).

II. RESULTS AND DISCUSSION

Figure 1 represents the results of an experiment evaluating the specific increase in polyamine content of leukocyte populations in the one-way mixed leukocyte reaction. Specific increases in both spermidine and spermine content are observed on day 5, with peak synthesis occurring on day 7. The general profile of an increase in polyamine levels is temporally analogous to that shown for tritiated thymidine incorporation. A similar profile of polyamine levels was observed in a one-way mixed lymphocyte reaction in which both the responding and stimulatory cell populations consisted of greater than 90% mononuclear cells (Fig. 2). In this case, there was a more dramatic response with maximal specific increases in spermidine and spermine occurring on days 6 and 7 of culture, respectively (Fig. 2, *left*). Both ^{3}H-uridine and ^{3}H-thymidine incorporation were maximal by day 6 (Fig. 2, *right*).

Overall analysis of this data suggests that in both series of experiments the general profile of accumulation of polyamines in cells stimulated by allogeneic leukocytes or lymphocytes was similar to that observed for both

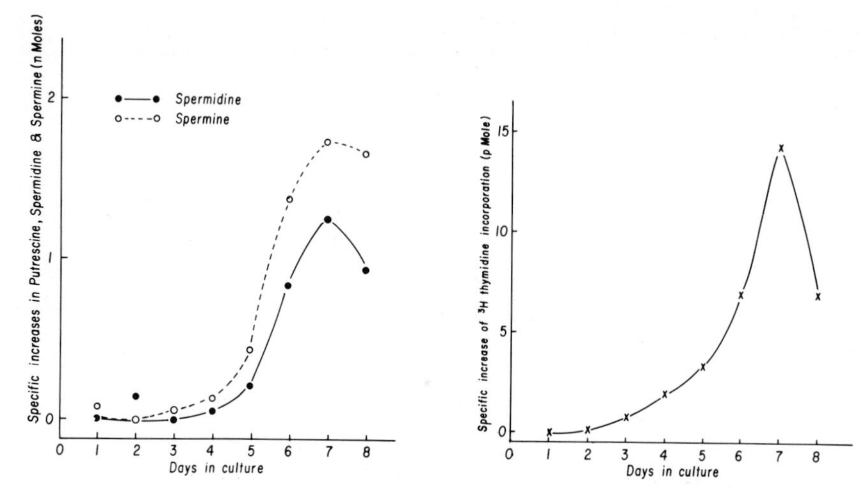

FIG. 1. *Left:* The specific increases in the amounts of spermidine and spermine in a mixed leukocyte culture. Control cultures were analyzed daily together with the appropriate stimulated cultures. The ordinate reflects the amount of each polyamine in the experimental culture minus the amount in the control culture. Each point reflects the specific increase for 9×10^6 responding cells.

Right: The specific increase of ^{3}H-thymidine incorporation into DNA in a mixed leukocyte reaction. The ordinate reflects the amount of ^{3}H-thymidine incorporation in the experimental culture minus the amount in the control culture. Each point represents the ^{3}H-thymidine incorporation of 1.5×10^6 responding cells and reflects the mean of three separate determinations.

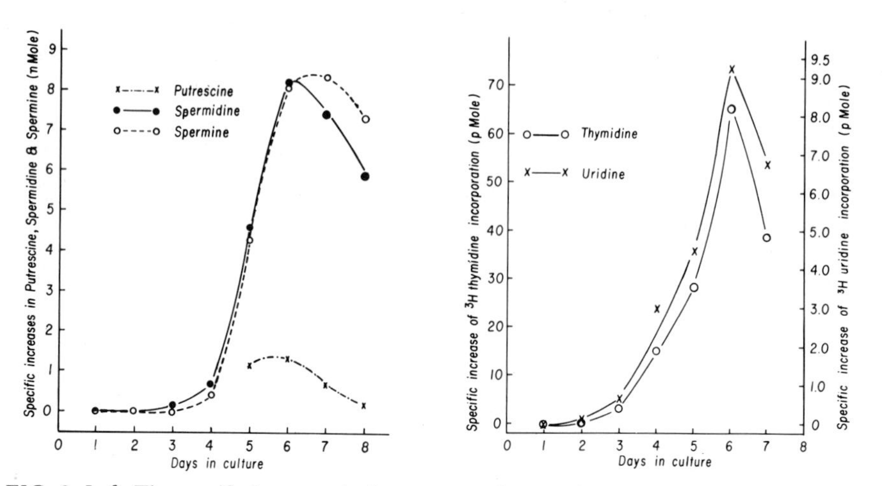

FIG. 2. *Left:* The specific increases in the amounts of putrescine, spermidine, and spermine in a mixed lymphocyte reaction. Control cultures were analyzed daily together with the appropriate stimulated cultures. The ordinate reflects the amount of each polyamine in the experimental

thymidine and uridine incorporation. These data are in accord with previous work, suggesting that polyamine biosynthesis and accumulation parallels the accumulation of RNA after a specific stimulus for growth.

The current ability to automate the amino acid analyzer technique for the study of polyamine determinations with only minimal variability (3.5%) would suggest that it is now feasible to evaluate further the role that polyamines play in the responsiveness of lymphocytes to allogeneic cells, mitogens, or specific antigens. This additional measure of the change within cells undergoing specific immune and mitogen stimulation offers the opportunity of adding another important characteristic to the study of cell-associated immune function in both health and disease.

REFERENCES

1. Russell, D., and Snyder, S. H., *Proc. Nat. Acad. Sci. U.S.A.*, **60,** 1420 (1968).
2. Jänne, J., and Raina, A., *Acta Chem. Scand.*, **22,** 1349 (1968).
3. Russell, D. H., *Ann. N.Y. Acad. Sci.*, **171,** 722 (1970).
4. Russell, D. H., *Proc. Nat. Acad. Sci. U.S.A.*, **68,** 523 (1971).
5. Russell, D. H., and McVicker, R. A., *Biochim. Biophys. Acta,* **259,** 247 (1972).
6. Russell, D. H., Shiverick, K. T., Hamrell, B. B., and Alpert, N. R., *Amer. J. Physiol.*, **221,** 1287 (1971).
7. Feldman, M. J., and Russell, D. H., *Amer. J. Physiol.*, **222,** 1199 (1972).
8. Russell, D. H., and Levy, C. C., *Cancer Res.*, **31,** 248 (1971).
9. Jänne, J., and Raina, A., *Biochim. Biophys. Acta,* **174,** 769 (1969).
10. Russell, D. H., and Snyder, S. H., *Endocrinology,* **84,** 223 (1969).
11. Russell, D. H., Snyder, S. H., and Medina, V. J., *Endocrinology,* **86,** 1414 (1970).
12. Pegg, A. E., and Williams-Ashman, H. G., *Biochem. J.*, **109,** 32P (1968).
13. Cohen, S., O'Malley, B. W., and Stastny, M., *Science,* **170,** 336 (1970).
14. Russell, D. H., and Taylor, R. L., *Endocrinology,* **88,** 1397 (1971).
15. Cohen, S. S., *Introduction to the polyamines,* Prentice-Hall, Englewood Cliffs, N.J., 1971.
16. Dykstra, W. J., Jr., and Herbst, E. J., *Science,* **149,** 428 (1965).
17. Russell, D. H., and Lombardini, J. B., *Biochim. Biophys. Acta,* **240,** 273 (1971).
18. Caldarera, C. M., Barbiroli, B., and Moruzzi, G., *Biochem. J.*, **97,** 84 (1965).
19. Morris, D. R., Koffron, K. L., and Okstein, C. J., *Anal. Biochem.*, **30,** 449 (1969).
20. Marton, L. J., Vaughn, J. G., Hawk, I. A., Levy, C. C., and Russell, D. H., *this volume.*
21. Bredt, A. B., and Mardiney, M. R., Jr., *Transplantation,* **8,** 763 (1969).
22. Mangi, R. J., and Mardiney, M. R., Jr., *J. Exp. Med.*, **132,** 401 (1970).
23. Mangi, R. J., and Mardiney, M. R., Jr., *Europ. J. Clin. Biol. Res.*, **15,** 911 (1970).

←

culture minus the amount in the control culture. Each point reflects the specific increase for 9×10^6 responding cells.

Right: The specific increases of ^{3}H-thymidine incorporation into DNA and RNA, respectively, in a mixed lymphocyte reaction. The ordinate reflects the amount of ^{3}H-thymidine or ^{3}H-uridine incorporation in the experimental culture minus the amount in the control culture. Each point represents the ^{3}H-thymidine or ^{3}H-uridine incorporation of 0.5×10^6 responding cells and reflects the mean of three separate determinations.

Polyamines in Normal and Neoplastic Growth. Edited by D. H. Russell. Raven Press, New York © 1973.

Changes in Polyamine Metabolism in Tumor Cells and Host Tissues During Tumor Growth and After Treatment with Various Anticancer Agents

Olle Heby* and Diane H. Russell

Laboratory of Pharmacology, Baltimore Cancer Research Center, National Cancer Institute, Baltimore, Maryland 21211

There is currently great interest in the relationship between polyamines and growth in both normal and neoplastic systems. Elevated activities of the enzymes in the polyamine biosynthetic pathway and subsequent accumulation of the polyamines are cellular events that occur both *in vitro* and *in vivo* after various growth stimuli (1–41). An apparent priming for growth takes the form in a variety of tissues of dramatically elevated levels of ornithine decarboxylase (4–6, 8, 10, 12, 15–18, 20–23, 25–27, 29, 30, 32–36, 38–41), the enzyme catalyzing the formation of putrescine. In contrast, the ornithine decarboxylase activity in quiescent and nondividing cells is very low. Certainly in mammalian systems, putrescine has been found to exert profound regulatory effects upon the polyamine biosynthetic pathway (5, 17, 38, 40, 42–46). The enzymes in the polyamine biosynthetic pathway, ornithine decarboxylase and putrescine-stimulated and spermidine-stimulated S-adenosyl-L-methionine decarboxylase, have relatively short half-lives (14, 18, 21, 26, 38, 47) and are induced by hormones that are involved in the regulation of growth processes (4, 5, 7–10, 13, 15, 16, 18, 20–23, 30–32, 34–36, 38, 39, 41). In the adult animal there are several populations of cells that neither synthesize DNA nor divide but can be stimulated to do so. The following changes in polyamine metabolism occur generally in cells stimulated to synthesize RNA and DNA and divide from a quiescent state: (a) an early increase in the ornithine decarboxylase activity, followed by an increase in the putrescine concentration; and (b) an increase in the spermi-

* NIH Visiting Fellow

dine/spermine ratio, resulting from a more rapid conversion of putrescine to spermidine than of spermidine to spermine, due to the enhanced pool of putrescine and/or the increased activity of the putrescine-stimulated S-adenosyl-L-methionine decarboxylase. Putrescine is a weak competitive inhibitor of ornithine decarboxylase, an activator of the putrescine-stimulated S-adenosyl-L-methionine decarboxylase, and a competitive inhibitor of the spermidine-stimulated S-adenosyl-L-methionine decarboxylase. In the regenerating rat liver, the spermidine/spermine ratio may be affected also by the conversion of spermine to spermidine (3). To obtain information on which role(s) the polyamines may play in tumor growth, we have studied the polyamine metabolism of transplantable tumors during well-defined growth processes in experimental animals.

In a study of a mouse L1210 lymphoid leukemia (48), it was shown that in the s.c. tumor the ornithine decarboxylase activity was markedly elevated during the earlier phase of tumor growth and declined with increasing size of the tumor (Fig. 1). The S-adenosyl-L-methionine decarboxylase activity was maintained at relatively high levels throughout the tumor growth period. Treatment with some antileukemic agents resulted in decreased concentrations of putrescine and spermidine and decreased activities of ornithine decarboxylase and S-adenosyl-L-methionine decarboxylase in the s.c. tumor.

Andersson and Heby (49) studied the concentrations of polyamines and nucleic acids in an Ehrlich ascites carcinoma (50) at various stages of tumor growth, and found that the concentration of DNA and RNA and that of spermidine and spermine changed in a similar manner during the growth period. They further showed that the growth rate deceleration, which is a consequence of the increasing size of the tumor, was accompanied by a decrease in the concentration of putrescine (Fig. 2). A mathematical model has been formulated by Woo and Simon for the observed relationship between the putrescine concentration and the growth pattern of the Ehrlich ascites tumor, and this model is presented in this volume. Other parameters which reflected the growth rate deceleration of the Ehrlich ascites carcinoma were the ornithine decarboxylase activity (Fig. 1) and the putrescine/spermidine and spermidine/spermine ratios, which all decreased with increasing population doubling time.

In this volume it is reported that hepatomas with different growth rates exhibit dramatically different spermidine/spermine ratios (Russell, this volume). That is, a fast-growing hepatoma, line 3924A, (volume doubling time of 4 days) has a high spermidine/spermine ratio, whereas a slow-growing hepatoma, line 16, (volume doubling time of 24 days) has a very low spermidine/spermine ratio. Characteristically, the slow-growing hepatoma has nondetectable amounts of putrescine, whereas the fast-growing hepa-

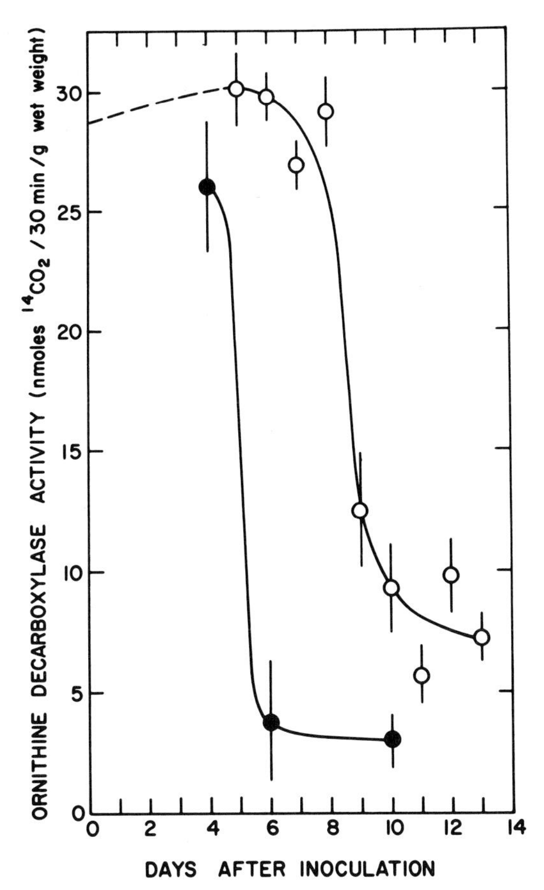

FIG. 1. The levels of ornithine decarboxylase in cells of an L1210 lymphoid leukemia (●—●) and of an Ehrlich ascites carcinoma (○—○) as a function of time after tumor inoculation. Male DBA/2 mice were inoculated s.c. with 10^5 L1210 ascites cells and male Swiss albino mice were inoculated i.p. with 10^7 Ehrlich ascites tumor cells. The tumors were homogenized in 5 vol of 0.05 M sodium-potassium phosphate buffer, pH 7.2, containing 0.1 mM dithiothreitol. The 100,000 × g (90 min) supernatant fraction was assayed for ornithine decarboxylase activity by measuring the release of $^{14}CO_2$ from DL-ornithine-1-^{14}C. The means ± S.E. of five samples for each group are graphed. The day 0 value was determined from the value obtained from the tumor cells 7 days after inoculation, since these were the cells used to transfer the tumor.

toma has considerable amounts of putrescine present. Williams-Ashman et al. (51) have reported similar results studying the polyamine metabolism in a series of hepatomas. They observed that in comparison with normal livers the ornithine decarboxylase activities of fast-growing hepatomas were very high, whereas a number of slow-growing and more differentiated hepatomas exhibited ornithine decarboxylase activities which were considerably

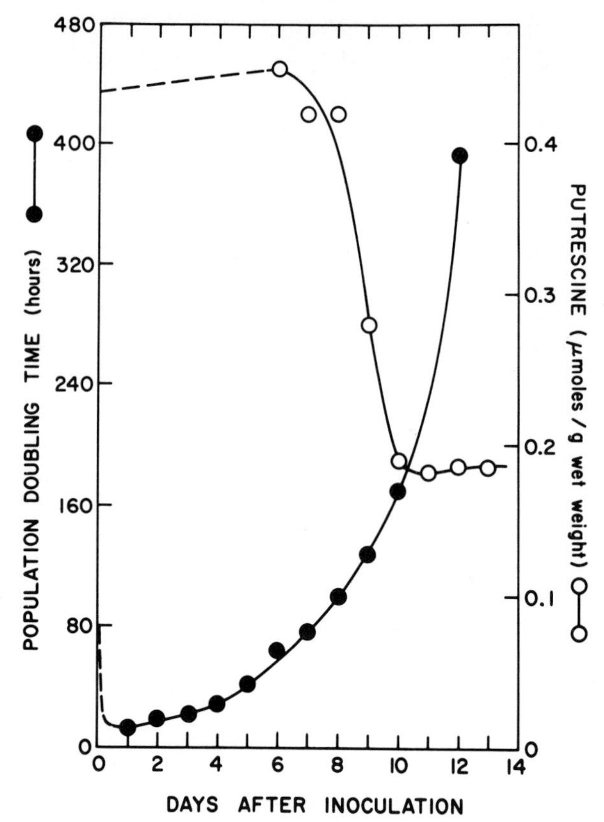

FIG. 2. Changes in the concentration of putrescine in an Ehrlich ascites carcinoma with decreasing growth rate (= increasing population doubling time). Male NMRI albino mice were inoculated i.p. with 4×10^6 Ehrlich ascites tumor cells. Putrescine was assayed spectrophotometrically after alkaline butanol extraction, thin-layer chromatographic separation, and ninhydrin staining. The means of four samples for each group are graphed. Day 0 values were determined from the values obtained from the tumor cells 7 days after inoculation, since these were the cells used to transfer the tumor.

less elevated. The concentration of putrescine in the fast-growing hepatomas was much greater than in the corresponding normal liver, and in the slow-growing hepatomas it was usually also above the normal hepatic range. None of the hepatomas, regardless of their growth rate, had significantly elevated S-adenosyl-L-methionine decarboxylase activities. The observed conservation of the polyamine biosynthetic pathway in hepatomas coupled with the decreases in the activities of the enzymes that utilize ornithine for other pathways (51, 52) adds considerable support to the importance of polyamines in growth systems. The spermidine/spermine ratios reported by these investigators tended to be higher in the hepatomas than in the corresponding normal livers, with a higher spermidine/spermine ratio being detected in fast-growing hepatomas, lines 3924A and 3683F, than in slow-growing hepatoma, line 7800. However, other slow-growing hepatomas had spermidine/spermine ratios which were higher than those of any fast-growing hepatomas studied. Since the polyamine metabolism in tumors is dependent upon the stage of tumor growth (48, 49), and no attempts were made to define the stages of growth at which the hepatomas were examined, it is rather difficult to draw conclusions about the spermidine/spermine ratios which were reported.

The validity of comparisons between various tumors in terms of growth rate and polyamine concentrations is dependent on the factors that determine the growth rate of each particular tumor. The growth rate of a tumor depends on several factors: (a) the length of the cell cycle; (b) the relative size of the growth fraction, i.e., the fraction of tumor cells that participate in the proliferating process; and (c) the rate of cell loss, i.e., the number of tumor cells that are lost through death, metastasis, or exfoliation.

In both the Ehrlich ascites carcinoma and the L1210 lymphoid leukemia, the growth rate deceleration with increasing tumor mass seems to be due to a gradual prolongation of the cell cycle time, a progressive decline in the growth fraction, and a continuous increase in the cell loss factor (53–65). From studies of cell kinetics in various human malignant tumors, it has become increasingly clear that cell loss is a major factor in determining their net rate of enlargement (66, 67). The importance of the cell loss factor (57) is obvious, for example, in basal-cell skin carcinomas, which have a high incidence of mitotic figures, but are classified as slow growing (68). Detailed studies of cell proliferation in these tumors have confirmed that there is indeed a gross discrepancy between their expected and observed rates of enlargement. The most likely explanation for this discrepancy is an extensive and continuous loss of neoplastic cells. Since the growth rate of a tumor is determined on a cell number, weight, or volume basis, tumors of this type will be classified as slow growing in spite of the fact that they may have a very high metabolic activity, including a polyamine pattern that is characteristic of continuously dividing cells. Therefore, a correlation between polyamine metabolism and the growth rate of a tumor is a vague concept unless the population kinetics of the tumor is considered in detail.

I. TUMOR CELL INVASION FROM TRANSPLANTABLE ASCITES TUMORS INTO HOST TISSUES

The degree of infiltration of ascites tumor cells from ascitic fluid into the intra-abdominal organs of mice depends on the invasiveness of the particular tumor inoculated. For example, Ehrlich ascites tumor cells have been

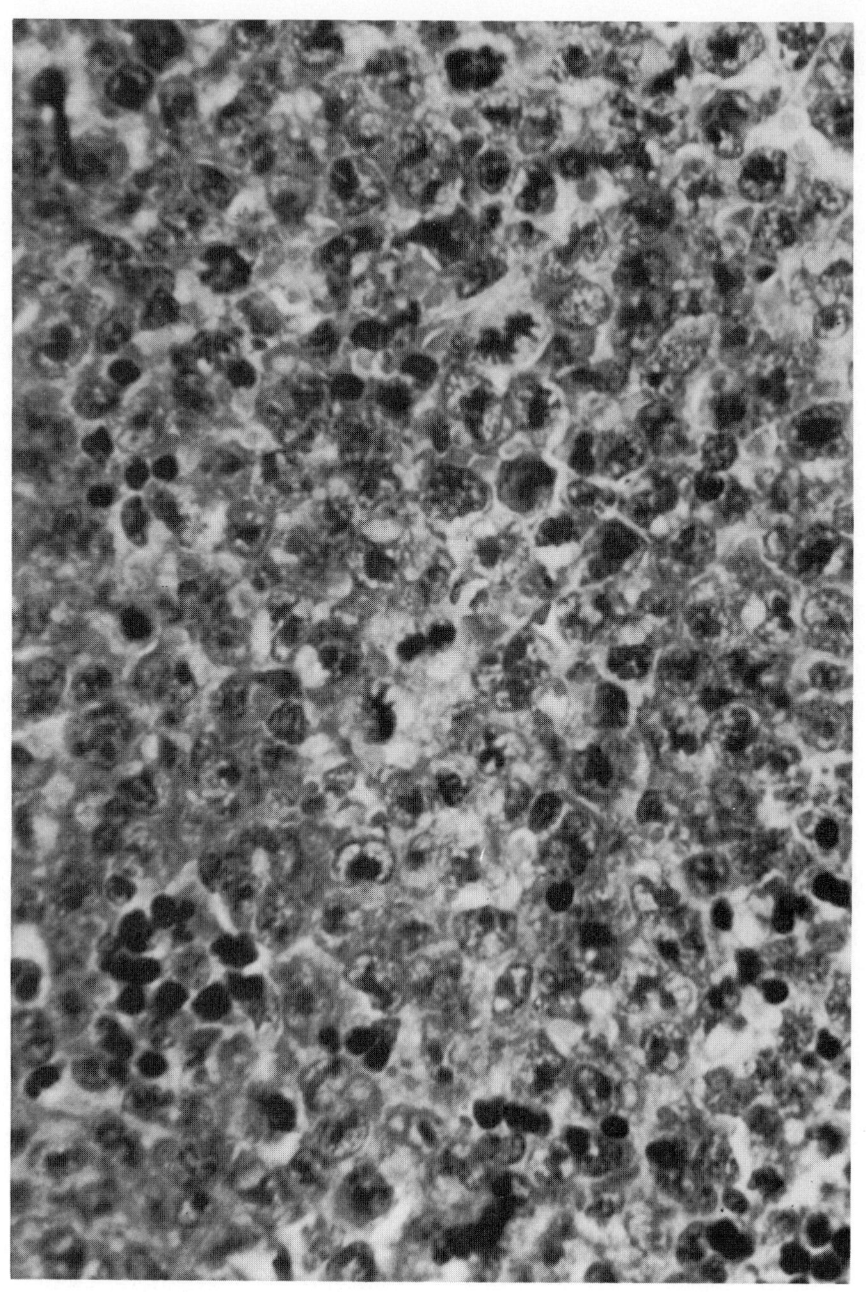

FIG. 3. Section of spleen of a male BDF_1 mouse 7 days after i.p. inoculation of 10^6 L1210 leukemic cells, showing extensive infiltration of leukemic cells. Mitotic figures are frequent. The spleen was distorted due to replacement of its red pulp and the lymphatic nodules of the white pulp by leukemic cells. Hematoxylin and eosin stain.

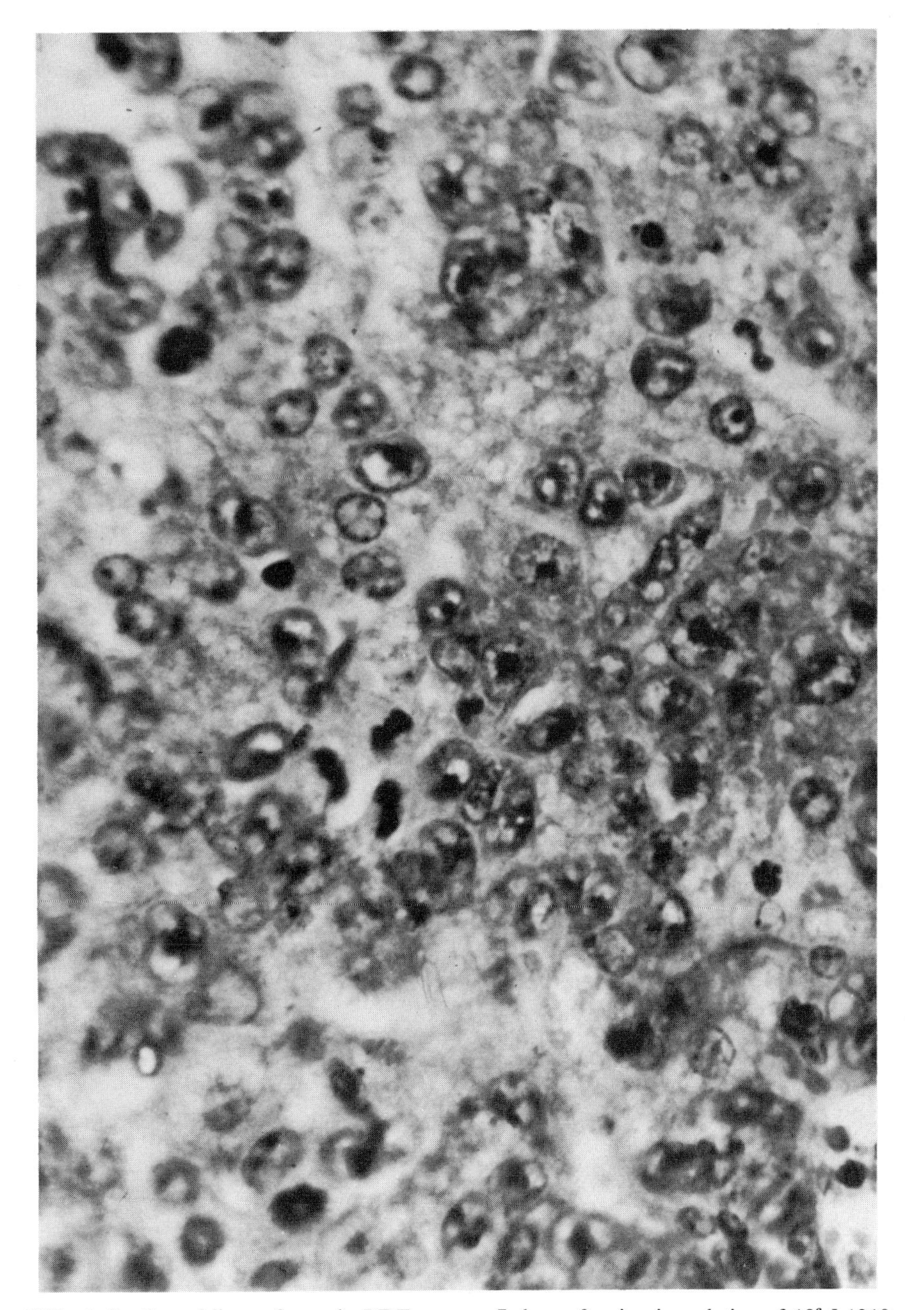

FIG. 4. Section of liver of a male BDF_1 mouse 7 days after i.p. inoculation of 10^6 L1210 leukemic cells, showing intralobular infiltration of leukemic cells. Aggregates of leukemic cells were found in the portal areas, around the central vein, and to a lesser extent in the sinusoids. Hematoxylin and eosin stain.

found to infiltrate certain organs within the abdominal cavity (69–71). The mesenteries and the areolar tissue surrounding the pancreato-splenic lymph node complex become infiltrated at 3 days, the adipose tissue and the pancreas at 6 days, and the parietal peritoneum at 9 days after tumor inoculation of 10^7 cells (71). All other organs studied, including liver and spleen, were found to be resistant to infiltration (71). With a quantitative bioassay for L1210 leukemic cells in mouse tissues, based on the direct relationship between the number of leukemic cells inoculated into appropriate inbred mice and the period elapsing before leukemic death, Skipper at al. (72) found that when 10^5 or 10^6 L1210 ascites cells were inoculated i.p., small numbers of these leukemic cells could be detected in the lungs, thymus, liver, and spleen 3 hr after inoculation. Consistently, iododeoxyuridine-^{125}I- or thymidine-^{3}H-labeled L1210 cells were found to invade body tissues rapidly after i.p. inoculation (62, 73). Increasing numbers of L1210 cells were found in the tissues with subsequent tumor growth (72), and on the median day of death, the bone marrow, liver, and spleen were the most heavily infiltrated tissues (74). In our laboratory a microscopic examination of spleens and livers from mice inoculated 7 days previously with 10^6 L1210 cells revealed a leukemic infiltration that completely obliterated the normal structure of the spleen (Fig. 3) and that was most pronounced in the portal areas and around the central veins of the liver (Fig. 4).

II. EFFECTS OF VARIOUS ANTINEOPLASTIC AGENTS ON POLYAMINE METABOLISM IN SPLEENS INFILTRATED WITH TUMOR CELLS

The marked invasiveness of the L1210 mouse leukemia was the basis for an examination of the possibility of using the activities of polyamine-synthesizing enzymes and endogenous polyamine concentrations in spleens infiltrated with tumor cells as indexes of the efficacy of antineoplastic agents on the process of tumor cell infiltration and proliferation. As discussed earlier in this chapter, it has been shown that the polyamine concentrations were altered markedly in a mouse L1210 s.c. tumor after treatment with certain antineoplastic agents (48).

Polyamine metabolism was studied in spleens of BDF_1 mice after i.p. inoculations of 10^6 L1210 leukemic cells (75, 76). As mentioned above, the spleen is rapidly invaded by the tumor cells and may accurately reflect the tumor activity throughout its time course. Figure 5 illustrates the change in activity of the enzymes in the polyamine biosynthetic pathway as well as endogenous polyamine concentrations in mouse spleen after i.p. inoculation of 10^6 L1210 leukemic cells and drug treatment. The ornithine decarboxylase

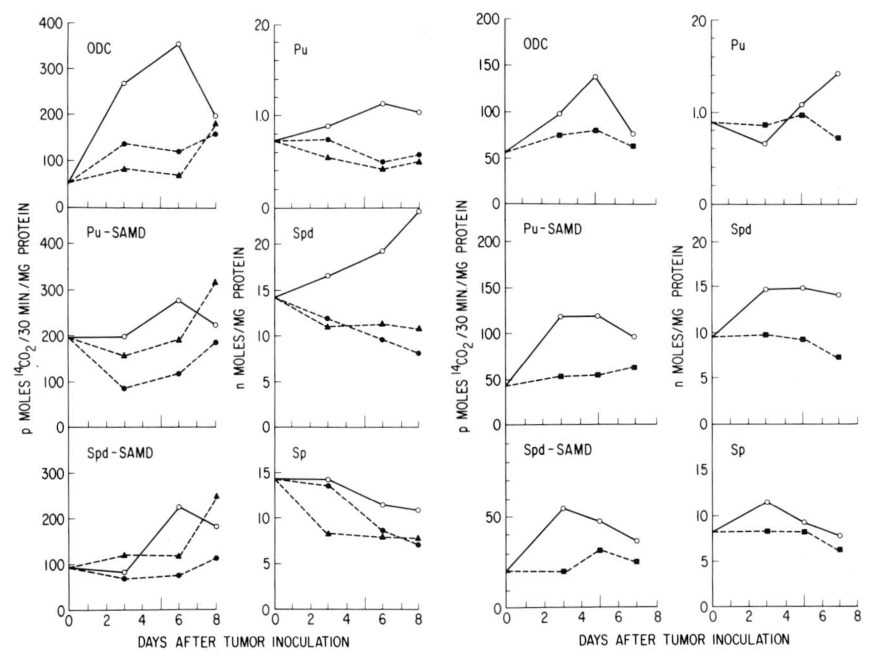

FIG. 5. Effects of methotrexate, cytosine arabinoside, and 5-azacytidine on the polyamine metabolism in spleens of leukemic mice. Lymphoid leukemia L1210 was carried in DBA/2 male mice by weekly i.p. passages. Ascitic fluid was obtained from a DBA/2 donor mouse on the 7th day of tumor growth, and 10^6 L1210 cells were inoculated i.p. into each recipient BDF_1 (C57BL/6 ♀ xDBA/2 ♂) male mouse. From Day 1 until the day of assay methotrexate (●---●) (NSC 740), cytosine arabinoside (▲---▲) (NSC 63 878), and 5-azacytidine (■---■) (NSC 102 816) were administered i.p. in doses of 2, 23, and 3 mg/kg/day, respectively, to three groups of leukemic BDF_1 mice, inoculated on day 0 with tumor cells. There was a control group of untreated leukemic BDF_1 mice (○—○) in each experiment. Every value represents the mean of at least three assays for each enzyme/amine in an extract of five spleens. All values on day 0 correspond to untreated nonleukemic mice and average all separate determinations since no major changes were observed in this group during the experimental period. The activities of the enzymes, L-ornithine decarboxylase (ODC), putrescine-stimulated S-adenosyl-L-methionine decarboxylase (Pu-SAMD), and spermidine-stimulated S-adenosyl-L-methionine decarboxylase (Spd-SAMD) were determined by measuring the release of $^{14}CO_2$ from the carboxyl-labeled substrates. Putrescine (Pu), spermidine (Spd), and spermine (Sp) were assayed spectrophotometrically after alkaline butanol extraction, high-voltage electrophoretic separation, and ninhydrin staining. The protein concentration was quantitatively determined with the Folin phenol reagent after alkaline copper treatment.

activity was markedly elevated within 3 days after tumor inoculation. The activities of the putrescine-stimulated and spermidine-stimulated S-adenosyl-L-methionine decarboxylases were markedly elevated within 3 to 6 days after tumor inoculation. Concomitantly, the concentrations of putrescine and spermidine increased, while that of spermine decreased slightly. When

TABLE 1. *Polyamine metabolism in spleens of leukemic and nonleukemic mice after the 5-azacytidine treatment was stopped*

Day (post-inoculation)	Ornithine decarboxylase activity[a]	S-Adenosyl-L-methionine decarboxylase activities[a]		Putrescine	Spermidine	Spermine	Putrescine/spermidine ratio	Spermidine/spermine ratio
		Putrescine-stimulated	Spermidine-stimulated	(nmoles/mg protein)				
5-Azacytidine-treated leukemic mice								
11	51	77	48	1.03	6.9	6.7	0.15	1.03
14	107	90	66	1.11	8.8	7.0	0.13	1.26
17	69	131	66	0.94	11.1	8.0	0.08	1.39
22	51	85	27	0.85	12.7	9.0	0.07	1.41
5-Azacytidine-treated nonleukemic mice								
11	54	97	45	0.97	10.2	7.0	0.10	1.46
14	97	50	42	2.11	13.0	8.0	0.16	1.63
17	81	94	55	1.31	14.3	9.5	0.09	1.51
22	54	39	30	0.99	12.8	7.7	0.08	1.66
Nonleukemic mice								
11–22	59	42	28	0.87	9.2	8.2	0.09	1.12

The experiment was the continuation of the 5-azacytidine study described in the legend to Fig. 5. The last injection of 5-azacytidine was made on day 9.

[a] pmoles $^{14}CO_2$/30 min/mg protein

methotrexate, cytosine arabinoside, or 5-azacytidine, which are among the most effective antileukemic agents so far tested, was injected i.p. on days 1 through 9 into mice inoculated on day 0 with 10^6 L1210 ascites tumor cells, the mean survival time was increased to 238, 200, and 300% that of untreated leukemic mice, respectively. The weight of the spleens, which more than doubled in leukemic mice by day 7, declined to near normal in the drug-treated leukemic mice. During drug treatment the activities of the enzymes in the polyamine biosynthetic pathway were markedly depressed, and, further, the accumulations of polyamines normally observed in leukemic mice were inhibited (Fig. 5). These drugs did not have a direct inhibitory effect on the enzymes as assayed *in vitro*. After methotrexate and cytosine arabinoside treatments, the enzyme patterns may indicate that tumor cells escape from the drug inhibition by day 8.

The 5-azacytidine study also included samples which were taken after the treatment period. It was found that after the 5-azacytidine injections were terminated on day 9, there was a rapid increase in the spleen weight accompanied by a rapid increase in both the synthesis and accumulation of polyamines in the spleen (Table 1). However, increases in the same parameters also occurred in spleens of nonleukemic mice treated with the same 5-azacytidine regimen, but they were less pronounced. This is probably due to a compensation for the reduction of the lymphatic tissue of the spleen which is the result of the administration of 5-azacytidine to nonleukemic mice (77). The major toxic effect of this agent, when administered to patients with various solid tumors, was hematologic with significant leukopenia and thrombocytopenia, especially at higher dose levels (78).

When the 5-azacytidine treatment was extended beyond day 9, no escape from the drug treatment was observed. However, the mice with the extended treatment died before the group which received the 9-day regimen. Accordingly, nonleukemic mice given daily i.p. 5-azacytidine injections also died when the treatment schedule was prolonged. Thus, it appears that this drug can almost completely suppress the advancement of the disease in the spleen, but the host toxicity that appears when the treatment is prolonged diminishes the therapeutic benefits obtained from the initial 9-day treatment.

III. POLYAMINE CONCENTRATIONS IN LIVERS OF LEUKEMIC MICE TREATED WITH 5-AZACYTIDINE

The concentrations of putrescine, spermidine, and spermine were studied in livers of BDF_1 mice after i.p. inoculations of 10^6 L1210 leukemic cells. As previously mentioned, the liver is rapidly invaded by tumor cells. Putrescine accumulated during the entire growth period and was almost fivefold above normal by day 9 (Table 2). The spermidine and spermine concentrations decreased after tumor inoculation but were above normal by Day 5, and before leukemic death the concentrations were almost twofold above normal. In the drug-treated groups the polyamine concentrations were rather close to normal at all times, even after the treatment was stopped.

TABLE 2. *Polyamine concentrations in mouse liver after i.p. inoculation of 10^6 L1210 leukemic cells and 5-azacytidine treatment*

Day (post-inoculation)	Putrescine	Spermidine	Spermine	Putrescine/spermidine ratio	Spermidine/spermine ratio
	(nmoles/g wet weight)				
Leukemic mice					
3	205	876	830	0.23	1.06
5	340	1,302	1,112	0.26	1.17
7	443	1,933	1,453	0.23	1.33
5-Azacytidine-treated leukemic mice					
3	154	1,129	1,089	0.14	1.04
5	143	1,222	1,160	0.12	1.05
7	116	1,376	1,102	0.08	1.25
14	146	988	796	0.15	1.24
22	171	1,611	825	0.11	1.95
5-Azacytidine-treated nonleukemic mice					
3	88	1,382	1,038	0.06	1.33
5	112	1,344	1,085	0.08	1.24
7	147	1,260	1,081	0.12	1.17
14	115	1,059	657	0.11	1.61
22	124	1,017	984	0.12	1.03
Nonleukemic mice					
3–22	103	1,098	862	0.09	1.27

The experiment was performed as described in the legend to Fig. 5. Leukemic and nonleukemic mice were injected i.p. on days 1 through 9 with 5-azacytidine (3 mg/kg/day).

The fact that the polyamines accumulated only to a minor degree in the liver after the treatment was stopped, in contrast with the marked accumulation observed in the spleen, may indicate that 5-azacytidine provided a greater percentage kill of leukemic cells in the liver than in the spleen. This has actually been shown for methotrexate (74).

The changes occurring in the liver may not be exclusively ascribable to the tumor cell infiltration. Actually, Payne et al. (79) and Lucké and Berwick (80) have substantiated that the presence of a tumor elsewhere in the body markedly alters the metabolism of the liver, including changes in the concentrations and turnover rates of various metabolic components, without secondary involvement of the liver in the neoplastic process. Consistently,

Andersson and Heby (49) demonstrated that the growth of the Ehrlich ascites tumor greatly influenced the spermidine concentration in the liver despite the fact that no tumor infiltration was seen. During the earlier phase of tumor growth, the spermidine concentration dropped precipitously toward a minimum of 72% of the initial, 2-day value at 4 to 5 days after inoculation. Subsequently, there was a progressive increase, reaching 135% of the initial value at 13 days after inoculation. However, the presence of the tumor did not markedly alter the putrescine and spermine concentrations. A comparable accumulation of spermidine has been observed also in the livers of rats bearing solid Rd/3 sarcomas (81) and in the livers of pregnant rats (82). Therefore, in the L1210 system a portion of the changes in the liver polyamine concentrations may be the result of the presence of the tumor alone, and the rest of the changes to the tumor cell infiltration. Actually the concentrations of putrescine and spermidine in the L1210 leukemic cells were higher than in the normal mouse liver, and therefore it would be surprising if the infiltrated liver would not show increased concentrations of these polyamines with the massive infiltration that is occurring.

IV. POLYAMINE CONCENTRATIONS IN SPLEENS OF LEUKEMIC MICE TREATED WITH VARIOUS ANTINEOPLASTIC AGENTS

Our studies on methotrexate (75), cytosine arabinoside (75), and 5-azacytidine (76) suggested that a comparison between polyamine concentration in the spleens of nontreated leukemic mice and leukemic mice treated with various antineoplastic agents would indicate the efficacy of the various agents on the tumor cell infiltration and proliferation. Consequently, polyamine concentrations were determined in spleens of leukemic mice 6 days after the i.p. inoculation of 10^6 L1210 leukemic cells. The antineoplastic agents were administered i.p. on days 1 through 5.

All the antineoplastic agents tested, with the exception of methylglyoxal-*bis*(guanylhydrazone), caused decreases in all polyamine concentrations in spleens of leukemic mice by day 6 (Table 3). Actually, the efficacy of each drug seems to be reflected in the spermidine/spermine ratio at day 6. The difference between the effects of methylglyoxal-*bis*(guanylhydrazone) and of the other drugs tested was most apparent for the putrescine concentration. Instead of a decrease in the concentration of putrescine, there was an almost threefold increase when compared to the concentration observed in nontreated leukemic mice. The spermidine/spermine ratio was also above that for leukemic mice. Actually, this effect was predicted, since methylglyoxal-*bis*(guanylhydrazone) inhibited the activity of S-adenosyl-L-methionine

TABLE 3. *Polyamine concentrations in spleens of leukemic mice treated with various anticancer agents*

Agent	NSC No.	Dose (mg/kg/day)	Survival time (%)	Day (post-inoculation)	Putrescine (nmoles/g wet weight)	Spermidine (nmoles/g wet weight)	Spermine (nmoles/g wet weight)	Putrescine/spermidine	Spermidine/spermine
Leukemic mice									
–	–	–	100	6	230	2,479	1,449	0.093	1.71
Methotrexate	740	2	211	6	101	1,272	1,095	0.079	1.16
"	740	2	232	6	113	1,304	1,159	0.087	1.13
6-Mercaptopurine	755	30	144	6	113	1,256	1,111	0.090	1.13
5-Fluorouracil	19893	20	168	6	122	1,079	966	0.113	1.12
Cytoxan (cyclophosphamide)	26271	45	165	6	109	1,079	950	0.101	1.14
5-Fluorouracil-deoxyriboside	27640	80	156	6	132	1,481	1,047	0.089	1.41
Methylglyoxal-*bis*(guanylhydrazone)	32946	50	170	6	623	2,029	1,143	0.307	1.78
Vinblastine	49842	0.5	135	6	167	1,658	1,079	0.101	1.54
Cytosine arabinoside	63878	25	189	6	105	1,449	1,063	0.072	1.36
5-Azacytidine	102816	3	297	6	119	1,159	1,047	0.103	1.11
L-Asparaginase	109229	15	151	6	203	1,674	1,304	0.121	1.28
1,3-*bis*-(2-chloroetyl)-1-nitrosourea (BCNU)	409962	5	286	6	147	1,626	1,191	0.090	1.37
Nonleukemic mice									
–	–	–	–	6	142	1,530	1,320	0.093	1.16

Male BDF_1 mice were inoculated i.p. on day 0 with 10^6 L1210 leukemic cells and were injected i.p. on days 1 through 5 with various anticancer agents. The polyamines were determined as described in the legend to Fig. 5. Each value represents the mean of at least three determinations of an extract of five spleens from Day 6 postinoculation. In each experiment 20 mice were randomized into a control group to determine the mean survival time after tumor inoculation and drug treatment. The mean survival time after inoculation of 10^6 L1210 leukemic cells was estimated as 8.0 days.

decarboxylase in extracts of spleen from leukemic mice without affecting the ornithine decarboxylase activity. Also, in this volume it is reported by Fillingame and Morris that, during lymphocyte transformation, spermidine and spermine accumulations were inhibited by methylglyoxal-*bis*(guanylhydrazone) whereas putrescine continued to accumulate, at an even higher rate.

V. SUMMARY

It appears that antineoplastic agents which alter the tumor cell kinetics also alter the ability of the tumor to synthesize and/or accumulate polyamines. In the case of the antineoplastic agent methylglyoxal-*bis*(guanylhydrazone), this alteration is due, at least partially, to a direct effect on an enzyme in the polyamine biosynthetic pathway, S-adenosyl-L-methionine decarboxylase. In the cases of the other antineoplastic agents studied, the alterations seen in the polyamine metabolism are more complicated, and appear to be reflections of other altered growth parameters. The continued reliability of the spermidine/spermine ratio as an indicator of the growth rate in a defined growth system lends additional evidence to its usefulness as an index of effectiveness, particularly of antineoplastic agents in mouse L1210 lymphoid leukemia. All the antineoplastic agents that were tested, with the exception of methylglyoxal-*bis*(guanylhydrazone), caused decreased spermidine/spermine ratios. However, all the agents should be tested for possible direct effects on the polyamine synthesizing enzymes.

ACKNOWLEDGMENTS

The authors would like to thank Dr. R. Kirschner for his assistance in preparing the microscopic slides.

REFERENCES

1. Dykstra, W. G., Jr., and Herbst, E. J., *Science,* **149,** 428 (1965).
2. Raina, A., Jänne, J., and Siimes, M., *Biochim. Biophys. Acta,* **123,** 197 (1966).
3. Siimes, M., *Acta Physiol. Scand.,* Suppl. **298,** 1 (1967).
4. Jänne, J., *Acta Physiol. Scand.,* Suppl. **300,** 1 (1967).
5. Jänne, J., Raina, A., and Siimes, M., *Biochim. Biophys. Acta,* **166,** 419 (1968).
6. Russell, D., and Snyder, S. H., *Proc. Nat. Acad. Sci. U.S.A.,* **60,** 1420 (1968).
7. Caldarera, C. M., Moruzzi, M. S., Barbiroli, B., and Moruzzi, G., *Biochem. Biophys. Res. Comm.,* **33,** 266 (1968).
8. Russell, D. H., and Snyder, S. H., *Endocrinology,* **84,** 223 (1969).
9. Moulton, B. C., and Leonard, S. L., *Endocrinology,* **84,** 1461 (1969).
10. Jänne, J., and Raina, A., *Biochim. Biophys. Acta,* **174,** 769 (1969).
11. Seiler, N., Werner, G., Fischer, H. A., Knötgen, B., and Hinz, H., *Hoppe-Seyler's Z. Physiol. Chem.,* **350,** 676 (1969).

12. Fausto, N., *Biochim. Biophys. Acta,* **190,** 193 (1969).
13. Caldarera, C. M., Giorgi, P. P., and Casti, A., *J. Endocrinol.,* **46,** 115 (1970).
14. Russell, D. H., Medina, V. J., and Snyder, S. H., *J. Biol. Chem.,* **245,** 6732 (1970).
15. Pegg, A. E., Lockwood, D. H., and Williams-Ashman, H. G., *Biochem. J.,* **117,** 17 (1970).
16. Stastny, M., and Cohen, S., *Biochim. Biophys. Acta,* **204,** 578 (1970).
17. Schrock, T. R., Oakman, N. J., and Bucher, N. L. R., *Biochim. Biophys. Acta,* **204,** 564 (1970).
18. Russell, D. H., Snyder, S. H., and Medina, V. J., *Endocrinology,* **86,** 1414 (1970).
19. Seiler, N., and Schröder, J. M., *Brain Res.,* **22,** 81 (1970).
20. Cohen, S., O'Malley, B. W., and Stastny, M., *Science,* **170,** 336 (1970).
21. Russell, D. H., and Taylor, R. L., *Endocrinology,* **88,** 1397 (1971).
22. Russell, D. H., and Lombardini, J. B., *Biochim. Biophys. Acta,* **240,** 273 (1971).
23. Kobayashi, Y., Kupelian, J., and Maudsley, D. V., *Science,* **172,** 379 (1971).
24. Heby, O., and Lewan, L., *Virchows Arch. Abt. B. Zellpath.,* **8,** 58 (1971).
25. Russell, D. H., Shiverick, K. T., Hamrell, B. B., and Alpert, N. R., *Am. J. Physiol.,* **221,** 1287 (1971).
26. Russell, D. H., and McVicker, T. A., *Biochim. Biophys. Acta,* **244,** 85 (1971).
27. Bachrach, U., and Ben-Joseph, M., *FEBS Lett.,* **15,** 75 (1971).
28. Caldarera, C. M., Casti, A., Rossoni, C., and Visioli, O., *J. Mol. Cell. Cardiol.,* **3,** 121 (1971).
29. Kay, J. E., and Cooke, A., *FEBS Lett.,* **16,** 9 (1971).
30. Kaye, A. M., Icekson, I., and Lindner, H. R., *Biochim. Biophys. Acta,* **252,** 150 (1971).
31. Moruzzi, G., Barbiroli, B., Corti, A., and Caldarera, C. M., *Ital. J. Biochem.,* **20,** 6 (1971).
32. Richman, R. A., Underwood, L. E., Van Wyk, J. J., and Voina, S. J., *Proc. Soc. Exp. Biol. Med.,* **138,** 880 (1971).
33. Hogan, B. L. M., *Biochem. Biophys. Res. Comm.,* **45,** 301 (1971).
34. Stastny, M., and Cohen, S., *Biochim. Biophys. Acta,* **261,** 177 (1972).
35. Cavia, E., and Webb, T. E., *Biochim. Biophys. Acta,* **262,** 546 (1972).
36. Hayashi, S., Aramaki, Y., and Noguchi, T., *Biochem. Biophys. Res. Comm.,* **46,** 795 (1972).
37. Seiler, N., and Askar, A., *Hoppe-Seyler's Z. Physiol. Chem.,* **353,** 623 (1972).
38. Russell, D. H., and Potyraj, J. J., *Biochem. J.,* **128,** 1109 (1972).
39. Brandt, J. T., Pierce, D. A., and Fausto, N., *Biochim. Biophys. Acta,* **279,** 184 (1972).
40. Hölttä, E., and Jänne, J., *FEBS Lett.,* **23,** 117 (1972).
41. Beck, W. T., Bellantone, R. A., and Canellakis, E. S., *Biochem. Biophys. Res. Comm.,* **48,** 1649 (1972).
42. Pegg, A. E., and Williams-Ashman, H. G., *Biochem. Biophys. Res. Comm.,* **30,** 76 (1968).
43. Pegg, A. E., and Williams-Ashman, H. G., *J. Biol. Chem.,* **244,** 682 (1969).
44. Pegg, A. E., and Williams-Ashman, H. G., *Arch. Biochem. Biophys.,* **137,** 156 (1970).
45. Coppoc, G. L., Kallio, P., and Williams-Ashman, H. G., *Int. J. Biochem.,* **2,** 673 (1971).
46. Ono, M., Inoue, H., Suzuki, F., and Takeda, Y., *Biochim. Biophys. Acta,* **284,** 285 (1972).
47. Russell, D. H., and Snyder, S. H., *Mol. Pharmacol.,* **5,** 253 (1969).
48. Russell, D. H., and Levy, C. C., *Cancer Res.,* **31,** 248 (1971).
49. Andersson, G., and Heby, O., *J. Nat. Cancer Inst.,* **48,** 165 (1972).
50. Andersson, G. K. A., and Agrell, I. P. S., *Virchows Arch. Abt. B Zellpath.,* **11,** 1 (1972).
51. Williams-Ashman, H. G., Coppoc, G. L., and Weber, G., *Cancer Res.,* **32,** 1924 (1972).
52. Weber, G., Queener, S. F., and Morris, H. P., *Cancer Res.,* **32,** 1933 (1972).
53. Baserga, R., *Arch. Path.,* **75,** 156 (1963).
54. Lala, P. K., and Patt, H. M., *Proc. Nat. Acad. Sci. U.S.A.,* **56,** 1735 (1966).
55. Yankee, R. A., DeVita, V. T., and Perry, S., *Cancer Res.,* **27,** 2381 (1967).
56. Lala, P. K., and Patt, H. M., *Cell Tissue Kinet.,* **1,** 137 (1968).
57. Steel, G. G., *Cell Tissue Kinet.,* **1,** 193 (1968).
58. Wiebel, F., and Baserga, R., *Cell Tissue Kinet.,* **1,** 273 (1968).
59. Young, R. C., DeVita, V. T., and Perry, S., *Cancer Res.,* **29,** 1581 (1969).
60. Peel, S., and Fletcher, P. A., *Eur. J. Cancer,* **5,** 581 (1969).

61. Harris, J. W., Meyskens, F., and Patt, H. M., *Cancer Res.*, **30,** 1937 (1970).
62. Hofer, K. G., and Hofer, M., *Cancer Res.*, **31,** 402 (1971).
63. Lala, P. K., *Cancer,* **29,** 261 (1972).
64. Lala, P. K., *Eur. J. Cancer,* **8,** 197 (1972).
65. Dombernowsky, P., and Hartmann, N. R., *Cancer Res.*, **32,** 2452 (1972).
66. Refsum, S. B., and Berdal, P., *Eur. J. Cancer,* **3,** 235 (1967).
67. Iversen, O. H., *Eur. J. Cancer,* **3,** 389 (1967).
68. Kerr, J. F. R., and Searle, J., *J. Path.*, **107,** 41 (1972).
69. Klein, G., *Exptl. Cell Res.*, **2,** 518 (1951).
70. Klein, G., and Révész, L., *J. Nat. Cancer Inst.*, **14,** 229 (1953).
71. Wheatley, D. N., and Ambrose, E. J., *Brit. J. Cancer,* **18,** 730 (1964).
72. Skipper, H. E., Schabel, F. M., Jr., Trader, M. W., and Thomson, J. R., *Cancer Res.*, **21,** 1154 (1961).
73. Hofer, K. G., Prensky, W., and Hughes, W. L., *J. Nat. Cancer Inst.*, **43,** 763 (1969).
74. Skipper, H. E., Schabel, F. M., Jr., Wilcox, W. S., Laster, W. R., Jr., Trader, M. W., and Thompson, S. A., *Cancer Chemother. Rep.*, **47,** 41 (1965).
75. Russell, D. H., *Cancer Res.*, **32,** 2459 (1972).
76. Heby, O., and Russell D. H., *Cancer Res.*, **33,** 159 (1973).
77. Šorm, F., and Vesely, J., *Neoplasma,* **11,** 123 (1964).
78. Weiss, A. J., Stambaugh, J. E., Mastrangelo, M. J., Laucius, J. F., and Bellet, R. E., *Cancer Chemother. Rep.*, **56,** 413 (1972).
79. Payne, A. H., Kelly, L. S., and White, M. R., *Cancer Res.*, **12,** 65 (1952).
80. Lucké, B., and Berwick, M., *J. Nat. Cancer Inst.*, **15,** 99 (1954).
81. Neish, W. J., and Key, L., *Int. J. Cancer,* **2,** 69 (1967).
82. Neish, W. J., and Key, L., *Biochem. Pharmacol.*, **17,** 497 (1968).

Polyamines in Normal and Neoplastic Growth. Edited by D. H. Russell. Raven Press, New York © 1973.

The Effect of Growth Conditions on the Synthesis and Degradation of Ornithine Decarboxylase in Cultured Hepatoma Cells

Brigid L. M. Hogan, Susan Murden, and Anne Blackledge

Biochemistry Group, School of Biological Sciences, University of Sussex, Falmer, Brighton BN1 9QG, Sussex, England

I. INTRODUCTION

When nondividing quiescent cells or tissues are exposed to appropriate growth-promoting agents [e.g., steroid or polypeptide hormones, phytohemagglutinin (PHA), or serum], a number of biochemical changes are elicited, leading eventually to cell growth, DNA synthesis, and mitosis. These early events include an increase in membrane permeability to amino acids, to nucleosides (other than adenosine), and to glucose, an increase in RNA polymerase activity, a stimulation of protein synthesis and the intracellular content of polyribosomes, and a decrease in the overall rate of intracellular protein degradation (for numerous references, see ref. 1). Another change which seems to be associated invariably with the transition of cells from the nongrowing to the growing state is an early increase in the activity of ornithine decarboxylase (ODC), the first enzyme on the pathway of polyamine synthesis. For example, such an increase has been reported in rat liver after various hormone treatments (2) or partial hepatectomy (3), in rat uterus after estradiol administration (4), and in human lymphocytes after PHA stimulation (5). A study of the factors controlling the amount of ODC in mammalian cells might therefore be expected to throw light on the mechanism by which these cells switch from the nongrowing to the growing state, a problem central to an understanding of cancer. To facilitate such a study, we have used mammalian cells in tissue culture where growth conditions can be manipulated easily.

II. MATERIALS AND METHODS

Rat hepatoma (HTC) cells were grown in suspension in Tricine-buffered Swim's S77 medium (6) containing 10% calf serum. Calf serum and a mixture of the nonessential amino acids normally omitted from Eagle's minimal essential medium were obtained from Flow Laboratories, Irvine, Scotland. ODC activity in the $10,000 \times g$ supernatant fraction of cell extracts was measured by the release of $^{14}CO_2$ from L-ornithine-1-^{14}C, as described previously (7). Incorporation of amino acids into protein was determined by incubation of approximately 5×10^5 cells with 0.5 μC ^{14}C-leucine or ^{14}C-Phe for 30 min, followed by washing, incubation with 0.1 N NaOH for 15 min at 37°C, and precipitation with 10% trichloracetic acid.

III. RESULTS

It had previously been shown (7) that ODC activity increases rapidly after diluting high-density, stationary-phase hepatoma cells into fresh medium with serum, reaching a maximum several hundredfold higher than the original level approximately 4 hr after dilution. About 60% of this increase is still observed when RNA synthesis is inhibited by more than 95% with actinomycin D. This observation, and the fact that no simple inhibitor of the enzyme appeared to be present in high-density cells, suggested that the step-up conditions might result in an increased rate of synthesis of ODC from preexisting messenger RNA or a decrease in the rate of degradation of the enzyme, or both. Experiments were therefore carried out to measure the apparent turnover of the enzyme under different growth conditions.

When protein synthesis in high-density cells was inhibited with 50 μg/ml cycloheximide, ODC activity was lost very rapidly, with a half-life of between 5 and 13 min. By contrast, once the cell extracts had been prepared, the enzyme activity was completely stable during incubation at 37°C, and the addition of 50 μg/ml cycloheximide did not inhibit the enzyme reaction. In order to see if dilution of high-density cells changed the apparent rate of turnover of ODC, cells were stepped up; 2 and 3.5 hr later the cells were treated with cycloheximide and the loss of ODC activity measured. As shown in Fig. 1, dilution of the cells resulted in approximately a 250-fold increase in ODC activity at 3.5 hr. Before dilution, the apparent half-life was 5 min; after 2 hr, this had increased to about 90 min, and at 3.5 hr it was 31 min. In another experiment with the same design, the apparent half-life before dilution was 12 min, and had risen to 27 min by 3 hr and 38 min by 5 hr.

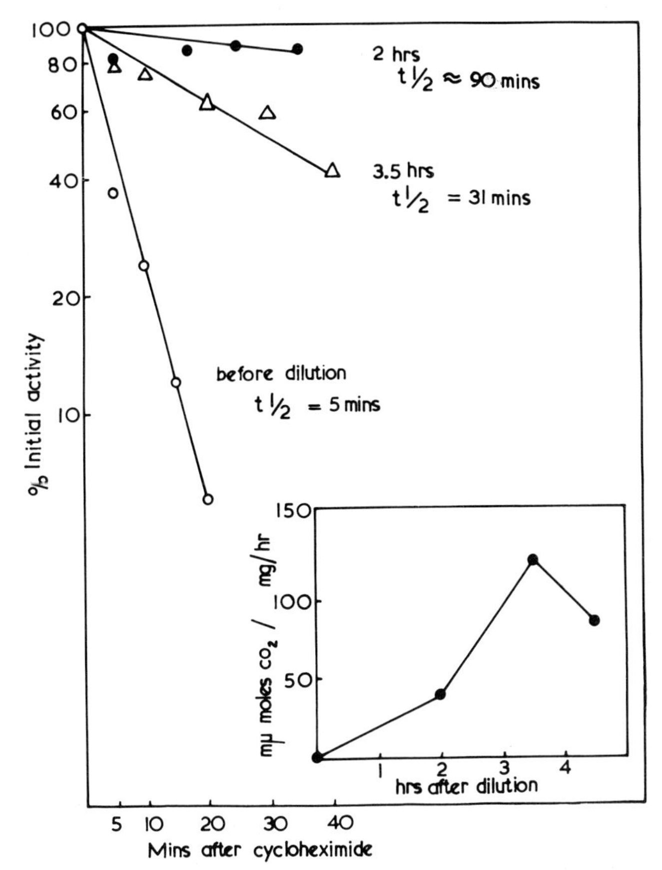

FIG. 1. Effect of dilution of high-density, stationary-phase cells on the apparent turnover of ODC. Cells at 8×10^5/ml were diluted to 2.5×10^5/ml with fresh medium plus serum. Cycloheximide was added to a final concentration of 50 μg/ml, and the subsequent loss of ODC activity as measured in cell extracts was followed before (○—○), and 2 hr (●—●), and 3.5 hr (△—△) after dilution. The inset shows the change in the absolute level of ODC activity with time after dilution.

Growth of high-density cells might be expected to be limited by several factors, including (a) a lack of one or more amino acids, (b) a lack of glucose, (c) depletion of one or more of the high molecular weight components of serum required for cell growth, or (d) low pH and accumulation of toxic by-products, all of which would be corrected upon dilution into fresh medium with serum. Therefore, in order to see which factors were most important in bringing about the increase in the level of ODC activity and the decrease in its apparent turnover after dilution, cells were shifted into medium containing salts, vitamins, and glucose but no amino acids or serum, into normal Swim's S77 medium without serum, and into S77 containing 10% calf serum. As shown in Fig. 2, restoring the initial level of glucose and correcting the pH were insufficient to stimulate ODC activity, but the additional presence of amino acids did elicit such an effect, which could be further enhanced by the presence of serum.

In the next set of experiments, the effect of supplementing high-density cells with a variety of amino acids was tested. As shown in Fig. 3, glutamine added to five times the normal level in S77 (5 × 2 mM = 10 mM) brought

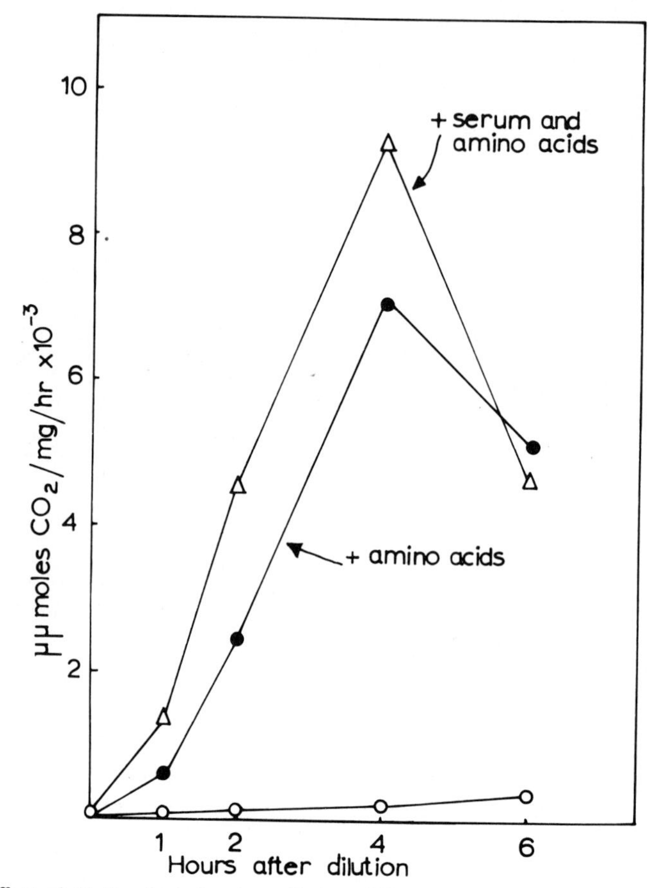

FIG. 2. Effect of diluting high-density cells into different media on the intracellular level of ODC. Cells at 7.5×10^5/ml were diluted to 2×10^5/ml into Tricine-buffered medium containing the same salts, vitamins, and glucose as Swim's S77 (○—○), or into Swim's S77 medium without serum (●—●), or with 10% calf serum (△—△).

about an eightfold increase in ODC activity after 4 hr, and five times the level of nonessential amino acids (Ala, Asn, Asp, Glu, Gly, Pro, and Ser) normally added to Eagle's tissue culture medium (5 × 0.1 mM = 0.5 mM) produced a fourfold stimulation. In other experiments, 10 mM glutamine stimulated ODC activity between two and 14 times after about 3 hr. A mixture of essential amino acids produced either a small increase or slight inhibition when added in high concentration. L-Ornithine added to a final concentration of 2 mM to high-density or serum-starved cells (see below) did not produce any increase in ODC activity, suggesting that the amino acids were not acting through conversion to ornithine, which, as the sub-

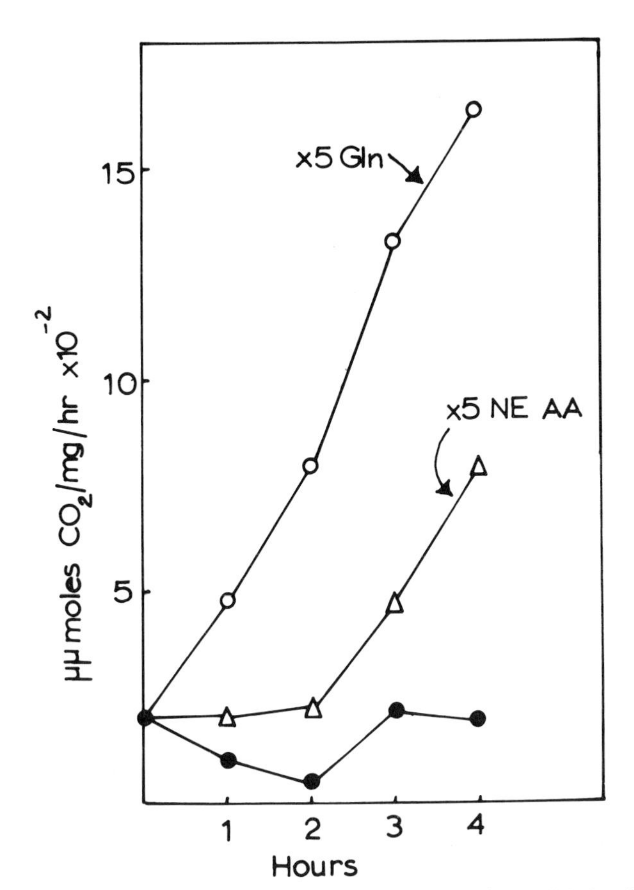

FIG. 3. Effect of amino acid supplementation on the ODC activity of high-density cells. Cells at 8.5 × 10^5/ml were either incubated alone (●—●), or after addition of glutamine (Gln) to a final concentration of 10 mM (○—○), or a mixture of nonessential amino acids (NE AA) to 0.5 mM (△—△).

strate of the enzyme, might be expected to protect ODC from degradation. Figure 4 shows that 4 hr after supplementing the medium with 10 mM glutamine, the apparent half-life of ODC in high-density cells had increased from 10 to 17 min. During the incubation period, ODC activity had increased 2.6 times, from 316 to 807 pmoles CO_2/mg protein/hour, and the overall rate of protein synthesis as measured by the incorporation of ^{14}C-leucine into protein had increased twofold.

Whereas the experiments described above show that supplementing high-density cells with amino acids can increase the intracellular level of ODC up to about 10 times, the following experiments show that addition of serum to serum-depleted cells also results in an increase in ODC activity, usually on the order of 30-fold. If cells are grown to a high density and then centri-

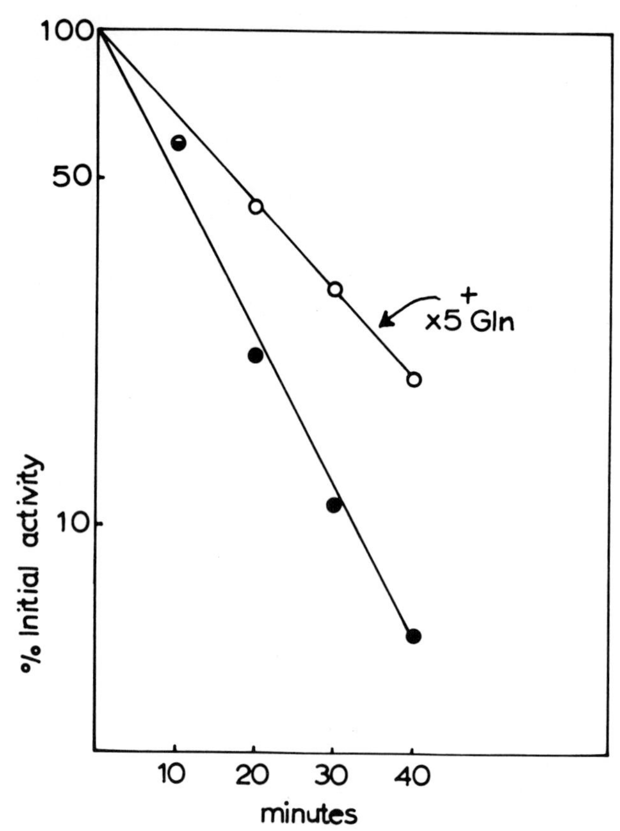

FIG. 4. Effect of glutamine on the apparent turnover of ODC. Fifty μg/ml cycloheximide was added to cells at 4.5 × 10^5/ml either before (●—●) or 4 hr after (○—○) supplementation with 10 mM glutamine, and the decline in ODC activity followed.

fuged down and resuspended in S77 medium without serum, there is an initial increase in ODC activity, peaking at about 4 to 6 hr (see Fig. 2), followed by a slow decline to a very low basal level by 20 hr. The cells remain viable for up to 36 hr under these conditions, but the cell number only increases by about 10% (the normal doubling time being 24 hr). If serum is added back to a final concentration of 10% to cells starved for serum for 20 hr, there is an increase in ODC activity within an hour, reaching a maximum about 36 times the basal level 5 to 6 hr later (see Fig. 5). If serum is added after only 15 hr of incubation without serum (when the ODC activity is still above the basal level), the increase in ODC activity is more rapid, but

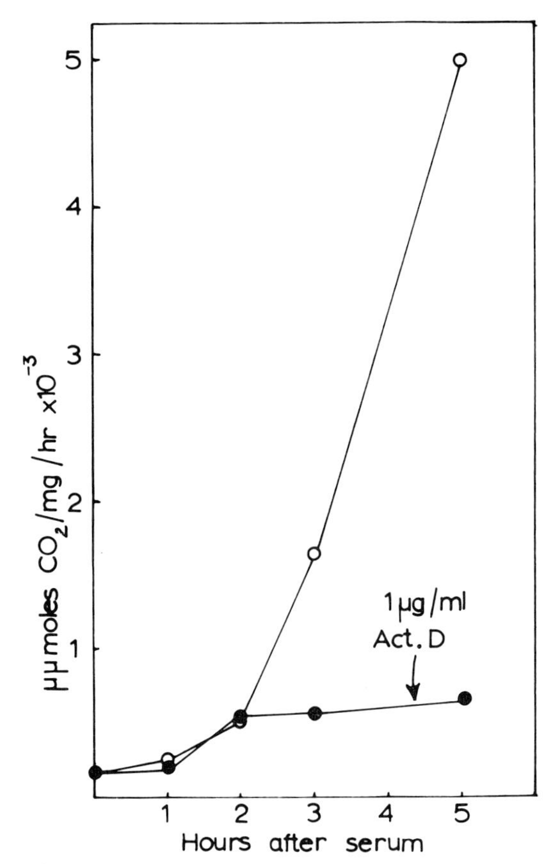

FIG. 5. Effect of serum on ODC activity of serum-starved cells. Cells were resuspended at a concentration of about 6×10^5/ml in S77 medium without serum and incubated for 20 hr. Calf serum then was added to a final concentration of 10% with (●—●) or without (○—○) 1 μg/ml actinomycin D.

if it is not added until 36 hr, then there is a lag of about 2 hr before an increase in ODC can be detected. The increase in ODC activity on the addition of serum is obtained even if the serum is extensively dialyzed against medium lacking amino acids, and most of the increase over the first 2 hr is observed when RNA synthesis is inhibited by 95% with actinomycin D (see Fig. 5).

Figure 6 shows the effect of serum addition on the apparent turnover of ODC. Cells maintained without serum for 14 hr were either incubated for a further 1.5 hr without serum or were supplemented with 10% serum. Cycloheximide was then added and the decline in ODC activity followed. Serum addition resulted in a sixfold increase in ODC activity (from 231 to 1480 pmoles CO_2/mg protein/hr) and a 50% increase in the apparent half-life, from 8 to 13 min. Experiments in which the rate of ^{14}C-Phe incorporation into total protein was measured showed that the addition of serum resulted

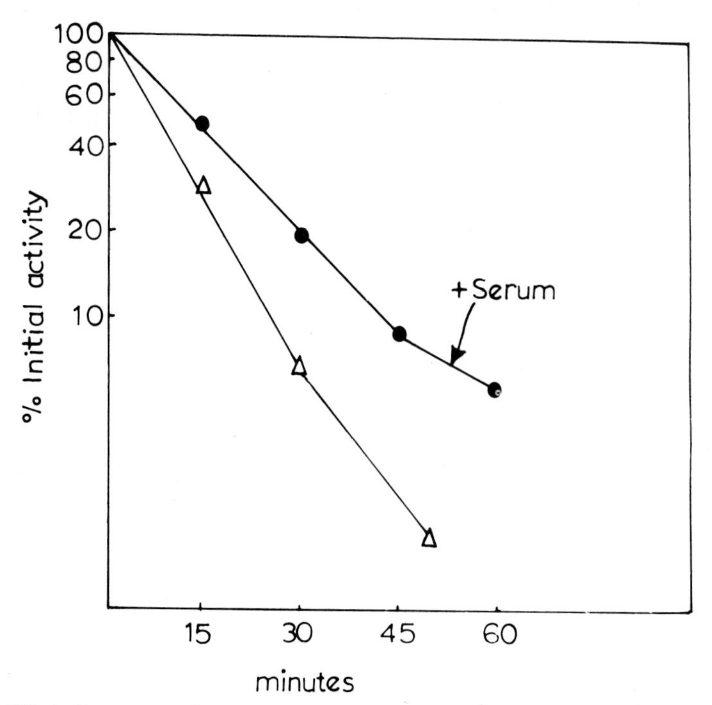

FIG. 6. Effect of serum on the apparent turnover of ODC. Cells were resuspended at a concentration of 7.4×10^5/ml in S77 without serum. After 14 hr, serum was added to a concentration of 10% to one half of the cells. Ninety min later, 50 μg/ml cycloheximide was added to both control (△—△) and serum-supplemented (●—●) cells, and the decline in ODC activity followed.

in a 50 to 70% stimulation of protein synthesis, which reached a plateau within 2 hr (see also ref. 8).

IV. DISCUSSION

ODC in rat liver has been reported to have a very short half-life of about 10 min (9). We have shown here that when hepatoma cells in tissue culture reach the high-density, stationary phase of the growth cycle, the enzyme also appears to have a very rapid turnover, with a half-life of between 5 and 13 min. However, until the loss of enzyme activity following inhibition of protein synthesis can be shown to be associated with a loss of intact enzyme molecules, this conclusion must, of course, remain tentative. After dilution of high-density cells into fresh medium, there is a rapid increase in ODC activity, reaching a maximum several hundred times higher than the initial level. In part, this increase appears to result from a decrease in the rate of degradation of the enzyme. For example, as shown in Fig. 1, there is an increase in the half-life of ODC from 5 min before dilution to approximately 90 min 2 hr after dilution, and 31 min at 3.5 hr. However, during this time there was a 250-fold increase in the level of ODC activity within the cells, suggesting that there had also been an increase in the rate of synthesis of the enzyme, either from preexisting or newly synthesized messenger RNA.

The decrease in the apparent turnover of ODC after dilution of high-density cells may be a result of the increased availability both of certain amino acids which were previously in limited supply and of high molecular weight components of serum which had also been depleted by cell growth. We have shown that addition of amino acids, in particular glutamine, to high-density cells can increase the half-life of ODC in HTC cells. A similar effect of amino acid supplementation on the turnover of ODC in PHA-stimulated human lymphocytes has also been reported (10). Tyrosine amino-transferase (TAT) and general cell protein turnover in HTC cells are both increased when the cells are incubated in medium lacking amino acids (11, 12). However, the mechanism by which the amino acid supply affects protein turnover in animal cells is not yet known (for discussion, see ref. 1).

Addition of serum to serum-depleted HTC cells results in an approximately 30-fold increase in ODC activity. This is accompanied by a decrease in the apparent turnover of the enzyme (see Fig. 6), but it probably also involves an increase in its rate of synthesis since the overall change in activity observed cannot be entirely accounted for in terms of the decrease in degradation. The turnover of both TAT and general cell protein in HTC cells is enhanced when cells are incubated in medium lacking serum (11, 12), but, as in the case of amino acid starvation, the mechanism of this effect is not yet understood. Components of serum which stimulate thymidine incorporation and phosphate uptake in serum-depleted cells have been partially purified and tentatively identified as proteins (13, 14). It is possible, but by no means certain, that these serum components do not enter cells but interact only with the plasma membrane, either to elicit some signal which is transmitted to the interior of the cell, or to change the permeability of the membrane so that other growth regulatory compounds enter or leave in greater amounts (15). In either case, the net effect is a change in several metabolic processes, of which the synthesis and degradation of ODC is one example.

ACKNOWLEDGMENTS

We thank the Science Research Council for financial support of this work and the Medical Research Council for a studentship for A.B.

REFERENCES

1. Hershko, A., Mammont, P., Shields, R., and Tomkins, G. M., *Nature (New Biol.)*, **232**, 206 (1971).
2. Panko, W. B., and Kenney, F. T., *Biochem. Biophys. Res. Comm.*, **43**, 346 (1971).
3. Russell, D. H., and Snyder, S. H., *Proc. Nat. Acad. Sci. U.S.A.*, **60**, 1420 (1968).
4. Kaye, A. M., Icekson, I., and Lindner, H. R., *Biochim. Biophys. Acta*, **252**, 150 (1971).
5. Kay, J. E., and Cooke, A., *FEBS Letters*, **16**, 9 (1971).
6. Granner, D. K., Thompson, E. B., and Tomkins, G. M., *J. Biol. Chem.*, **245**, 1472 (1970).
7. Hogan, B. L. M., *Biochem. Biophys. Res. Comm.*, **45**, 301 (1971).
8. Gelehrter, T. D., and Tomkins, G. M., *Proc. Nat. Acad. Sci. U.S.A.*, **64**, 723 (1969).
9. Russell, D. H., and Snyder, S. H., *Mol. Pharmacol.*, **5**, 253 (1969).
10. Kay, J. E., Lindsay, V. J., and Cooke, A., *FEBS Letters*, **21**, 123 (1972).
11. Auricchio, F., Martin, D. W., and Tomkins, G. M., *Nature*, **224**, 806 (1969).
12. Hershko, A., and Tomkins, G. M., *J. Biol. Chem.*, **246**, 710 (1971).
13. Holley, R. W., and Kiernan, J. A., in G. E. W. Wolstenholme and J. Knight (Editors), *Growth control in cell cultures*, Churchill, Livingstone, 1971, p. 3.
14. Cunningham, D. D., and Pardee, A. B., *Proc. Nat. Acad. Sci. U.S.A.*, **64**, 1049 (1969).
15. Sheppard, J. R., *Nature (New Biol.)*, **236**, 14 (1972).

Polyamines in Normal and Neoplastic Growth. Edited by D. H. Russell. Raven Press, New York © 1973.

Accumulation of Polyamines and Its Inhibition by Methyl Glyoxal Bis-(Guanylhydrazone) During Lymphocyte Transformation

Robert H. Fillingame and David R. Morris

Department of Biochemistry, University of Washington, Seattle, Washington 98195

High levels of the aliphatic polyamines have been repeatedly observed in tissues undergoing growth and division (see ref. 1 for review). For example, increases in polyamine levels have been observed in regenerating liver (2–4) and in the prostate of orchiectomized animals during testosterone-induced growth (5). Increases in RNA content accompany polyamine accumulation in these systems. This correlation and other evidence (6, 7) has led to the suggestion that polyamines may serve a regulatory role in RNA metabolism (4, 6).

To elucidate the role of polyamines in the growth of mammalian cells, we have sought to develop a system that would permit greater experimental manipulation than is possible in whole-animal studies. Our initial experiments indicate that the transformation of small lymphocytes in culture is a suitable system for this purpose. Small lymphocytes are normally not capable of division. However, when cultured in the presence of Concanavalin A (ConA), they undergo a sequential series of events which transforms them into cells capable of division. During transformation we have found that there are large and easily quantitatable changes in the cellular levels of DNA, RNA, and protein. These changes are accompanied by large increases in the cellular levels of putrescine, spermidine, and spermine (8). A careful analysis of the initial changes leading to transformation indicates that the increase in cellular RNA precedes polyamine accumulation. In addition, we will show here that spermidine and spermine accumulation can be inhibited by methyl glyoxal *bis*-(guanylhydrazone) (MGBG) and that this inhibition has no effect on net cellular RNA accumulation.

I. EXPERIMENTAL PROCEDURES

Lymphocytes were obtained from bovine suprapharyngeal lymph nodes and purified and cultured as previously described (8). The methods used in determination of the DNA, RNA, protein, and polyamine levels have also been described (8). Methyl glyoxal *bis*-(guanylhydrazone) dihydrochloride monohydrate (Aldrich Chemical Co.) was dissolved in phosphate-buffered saline and neutralized with sodium hydroxide. The relative rate of RNA synthesis was measured by pulsing 3-ml aliquots obtained from mother cultures with ^{3}H-uridine (New England Nuclear, 25.4 C/mmole). The radioactivity in the trichloroacetic acid-precipitable, KOH-hydrolyzable fraction was counted.

II. CHARACTERISTICS OF ConA-INDUCED LYMPHOCYTE TRANSFORMATION

We routinely obtain 1 to 2×10^{10} purified lymphocytes from two bovine lymph glands. This is enough material to carry out an experiment such as is illustrated in Figs. 1–3. Because there is some cellular debris and nuclei from broken cells in the cellular suspension we inoculate into culture, we routinely preincubate the cultures for 24 hr prior to the addition of ConA. After this time the debris has disappeared, leaving greater than 98% of the components in the culture as cells. More than 95% of these cells are small or medium lymphocytes. A 15 to 20% decline in cellular concentration normally occurs during the preincubation period. After preincubation the rate of cell death is approximately 10% per day in the absence of ConA. During the 3 days following the addition of ConA, approximately 90% of the cells transform into either large lymphoblasts (60%) or cells which are intermediate in size between the small lymphocyte and the lymphoblasts (20 to 30%). Cells begin dividing about 36 hr after the addition of ConA, and division is fairly asynchronous between 48 and 84 hr (1 to 1.5% of the cells entering mitosis per hour).

The decline in DNA per culture (Fig. 1) parallels the previously described decline in cell number during preincubation and following 3 days of culture in the absence of ConA. In the presence of ConA, the decline in DNA parallels the untreated control for the first 36 hr but then increases until at least 72 hr, the net increment being approximately 1.5-fold. DNA synthesis is initiated at about 24 hr as shown by the increase in ^{3}H-thymidine incorporation into DNA (Fig. 1). In the following sections, we have chosen to

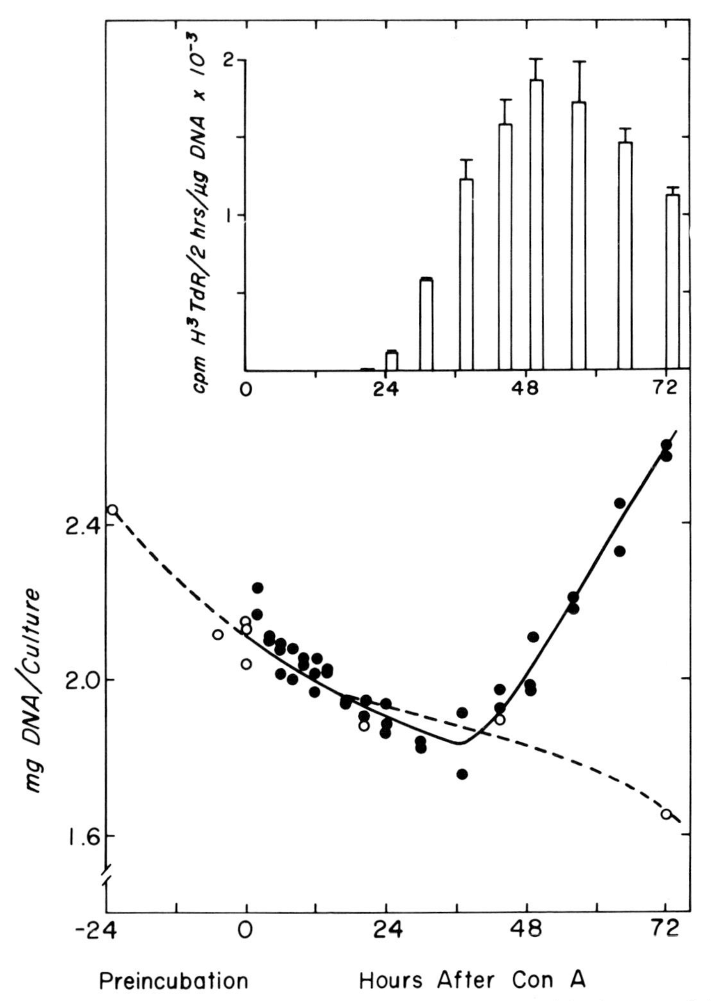

FIG. 1. Changes in the DNA content of lymphocyte cultures and in the rate of thymidine incorporation into DNA during transformation by ConA. ConA was added to most of the cultures after a 24-hr preincubation (●) and others were incubated without ConA (○). The rate of thymidine incorporation was measured with aliquots taken from mother culture (*insert*). This figure is taken from ref. (8).

normalize RNA, protein, and polyamine values to DNA since there are slight variations in the number of cells per culture. DNA is proportional to cell number for at least 24 hr, and this normalization allows a more precise temporal comparison of the various parameters measured.

The accumulation of RNA during transformation is shown in Fig. 2A.

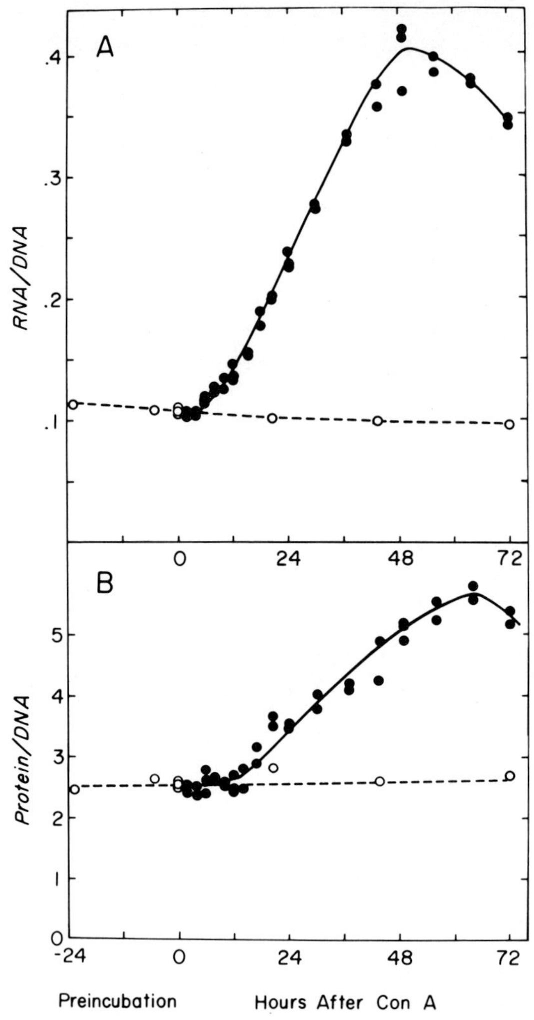

FIG. 2. Changes in the cellular levels of RNA and protein during transformation. The cellular RNA and protein contents (in mg) of the cultures were normalized to the DNA content (in mg). ConA was added to most of the cultures after preincubation (●). The other cultures were incubated without ConA (○). This figure is taken from ref. (8).

RNA (normalized to DNA) remained constant for the first 4 hr and then increased at a constant rate until 49 hr. The net increase was approximately fourfold in this experiment. The RNA/DNA ratio declines after 49 hr (Fig. 2A) because of the increase in DNA at this time. The amount of RNA per culture actually continued to increase until 64 hr and then remained constant.

The increase in cellular protein during transformation is shown in Fig. 2B. Protein (normalized to DNA) remained constant for approximately 12 hr and then increased at a fairly constant rate until 56 hr. The protein content per culture continued to increase until 72 hr, although the protein normalized to DNA decreases in Fig. 2B because of large increases in DNA. The increase in protein per culture or per DNA is about 2.5-fold.

III. POLYAMINES DURING LYMPHOCYTE TRANSFORMATION

Nonstimulated small lymphocytes contain three to four times more spermine than spermidine. In most other tissues, e.g., liver, the amount of spermidine exceeds or equals the amount of spermine. The actual levels of spermidine, spermine, RNA, and protein per cell (DNA) are also much lower in small lymphocytes than in a tissue such as liver (see ref. 8). This may not be surprising since the small lymphocyte is quiescent not only with respect to DNA synthesis and division but also in terms of RNA and protein synthesis.

The changes in polyamine content during a typical lymphocyte-transformation experiment are shown in Fig. 3. The first significant increase in the putrescine level was seen 10 hr after the addition of ConA. The putrescine pool increased rapidly until 36 hr (a 15-fold increase) and then remained elevated until 72 hr. Increases in the spermidine level were not seen until 12 hr after the addition of ConA. Between 12 and 48 hr spermidine accumulated very rapidly, the net increase being sixfold. Spermidine did not reach its maximum per culture or per DNA until 64 hr. The decline in spermidine between 64 and 72 hr (only 5 to 10% per culture) is magnified in Fig. 3 because of the rapid increase in DNA over this interval. Spermine, which is at a relatively high level in unstimulated lymphocytes, increased gradually between 12 and 24 hr. After 24 hr, there was a more rapid increase in spermine/DNA, and it reached a maximum by 49 hr. However, spermine per culture continued to accumulate to at least 72 hr.

To compare unequivocally the kinetics of RNA and polyamine accumulation, it was necessary to show that the increases in polyamine levels were not delayed because of leakage of polyamines to the culture media during harvesting. If transforming cells were more subject to leakage, analysis of only the cellular level could introduce an artifactual delay in the kinetics. To do these medium analyses, it was necessary to harvest the cultures in the presence of an amine oxidase inhibitor, because our medium is supplemented with calf serum which contains an amine oxidase capable of oxidizing spermidine and spermine. As shown in Table 1, the putrescine content of the medium was very low after harvesting cells which had been inoculated

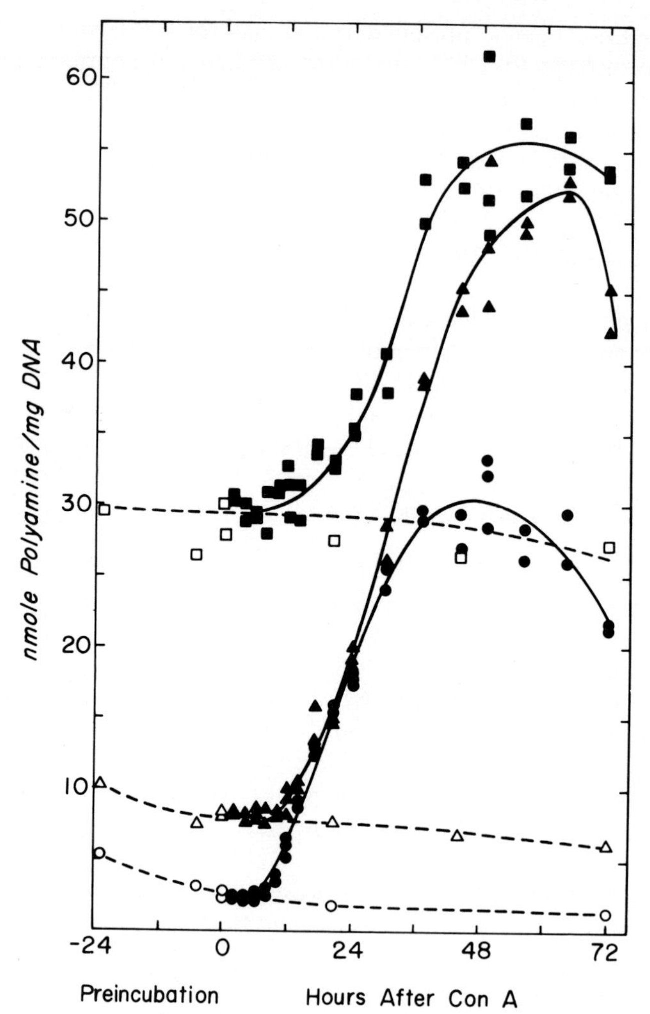

FIG. 3. Polyamine levels during transformation. The cellular levels of putrescine (○), spermidine (△), and spermine (□) were determined in the same cultures analyzed in Figs. 1 and 2 and were normalized to the DNA content. Polyamine contents were determined in the presence (closed symbols) or absence of ConA (open symbols). This figure is taken from ref. (8).

into culture 1 hr earlier (23 hr prior to the addition of ConA). However, by the end of the preincubation the putrescine level in the medium increased and actually exceeded the level found in the cells. The amount of putrescine found in the medium at this time can easily be accounted for by oxidation of the spermidine and spermine which are released from dying cells during this period (one of the secondary end products of the oxidation is putrescine).

TABLE 1. *Polyamine levels in cells and medium*

Hours after ConA	nmoles putrescine/100 ml culture		nmoles spermidine/100 ml culture	
	cells	medium	cells	medium
−23	12.8	1.5	24.7	0.4
0	5.5	20.9	15.6	0.7
6	5.0	22.7	18.4	0.7
12	9.9	22.4	18.4	0
24	32.5	24.6	35.5	1.2
39	61.5	20.6	78.9	1.1
49	63.5	15.0	107.3	1.2

Cultures were preincubated with 5 mM Iproniazid phosphate for 15 min and harvested as previously described (8).

Because of the high background level of putrescine found in the medium after preincubation, it is impossible to determine if there is significant leakage during harvest in the hours immediately following the addition of ConA. Thus, if susceptibility to leakage does increase between 0 and 8 hr after the addition of ConA, it is possible that putrescine increases at an earlier time than shown in Fig. 3. Analysis of the medium for spermidine indicated that at no time was more than 5% of the cellular spermidine released during harvesting (Table 1). Nor does it appear that the relative amount of leakage varied at different times. We can thus state unequivocally that leakage from the cells during harvesting does not contribute to the lag in spermidine accumulation shown in Fig. 3. Spermine was never detected in medium samples and apparently does not pose a leakage problem.

Since the lags in spermidine and spermine, and probably putrescine, accumulation are not artificially induced by leakage, we are in a position to compare the initial time course of RNA and polyamine accumulation. In Fig. 4 the data from six experiments, including those presented in Fig. 3, are averaged and plotted. These data have also been analyzed statistically. It is clear that the initial increase in RNA, which is statistically significant by 6 hr, occurs prior to increases in any of the polyamines. The increase in putrescine is significant by 8 hr and clearly follows RNA. Spermidine and spermine do not increase significantly until 10 hr. The simplest interpretation of these data is that polyamines cannot be stimulating RNA accumulation. If they are involved in RNA metabolism, they would appear to be increasing in response to RNA rather than vice versa. As we have stated earlier, it is possible that putrescine is increasing at an earlier time. However, it seems unlikely that a minor increase in putrescine could have such a profound stimulatory effect, especially since putrescine is generally considered to play only a precursor role in polyamine biosynthesis and is not required for optimal growth of *Escherichia coli* mutants (9). Another pos-

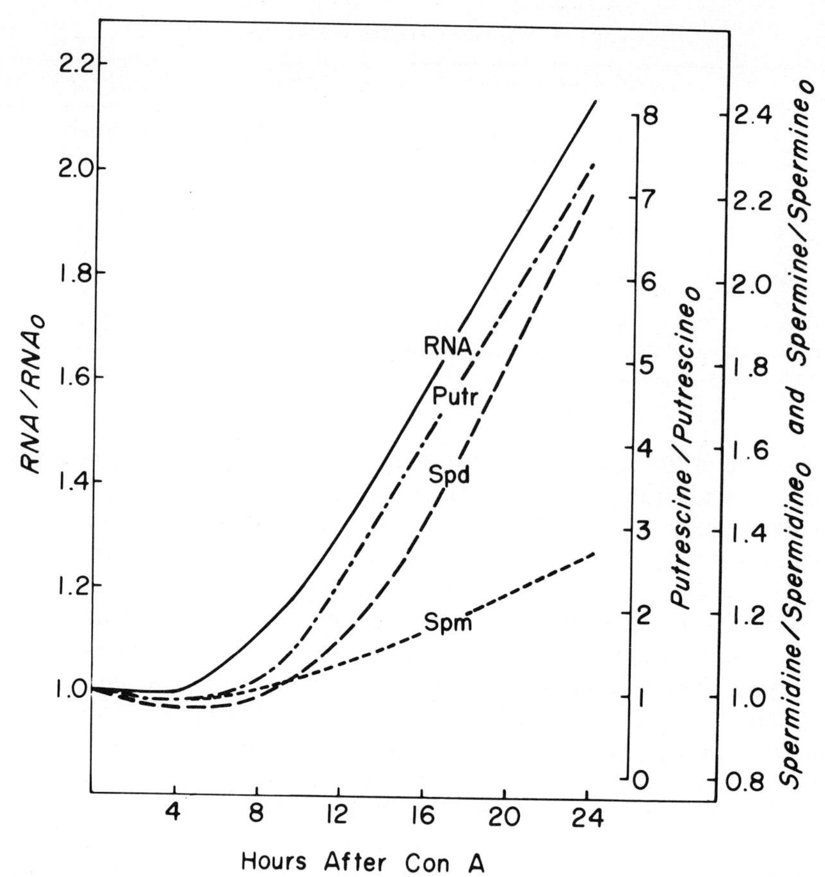

FIG. 4. Relative changes in the RNA, putrescine, spermidine, and spermine levels during the first 24 hr of culture after addition of ConA. The values from six experiments were normalized by dividing the value at a given time by the value at 0 hr. The fitted lines drawn through the averages are shown. This figure is taken from ref. (8).

sibility we cannot exclude is that spermidine and/or spermine redistribute themselves intracellularly during the first 6 hr after ConA, and thus have a stimulatory effect on the RNA accumulation in this way.

IV. EFFECT OF INHIBITION OF POLYAMINE SYNTHESIS ON RNA ACCUMULATION

Early in 1972, Williams-Ashman and Schenone reported that MGBG was a potent inhibitor of the rat prostate and yeast S-adenosylmethionine decarboxylases (10). We have confirmed this report with the S-adenosylmethionine decarboxylase activity obtained from crude extracts of transformed lymphocytes. Activity is inhibited 50% at 0.3 μM MGBG, and completely inhibited at 10 μM MGBG. The degree of inhibition does not appear to vary with the concentration of putrescine used as activator of the enzyme. As shown in Fig. 5A and B, addition of MGBG to transforming

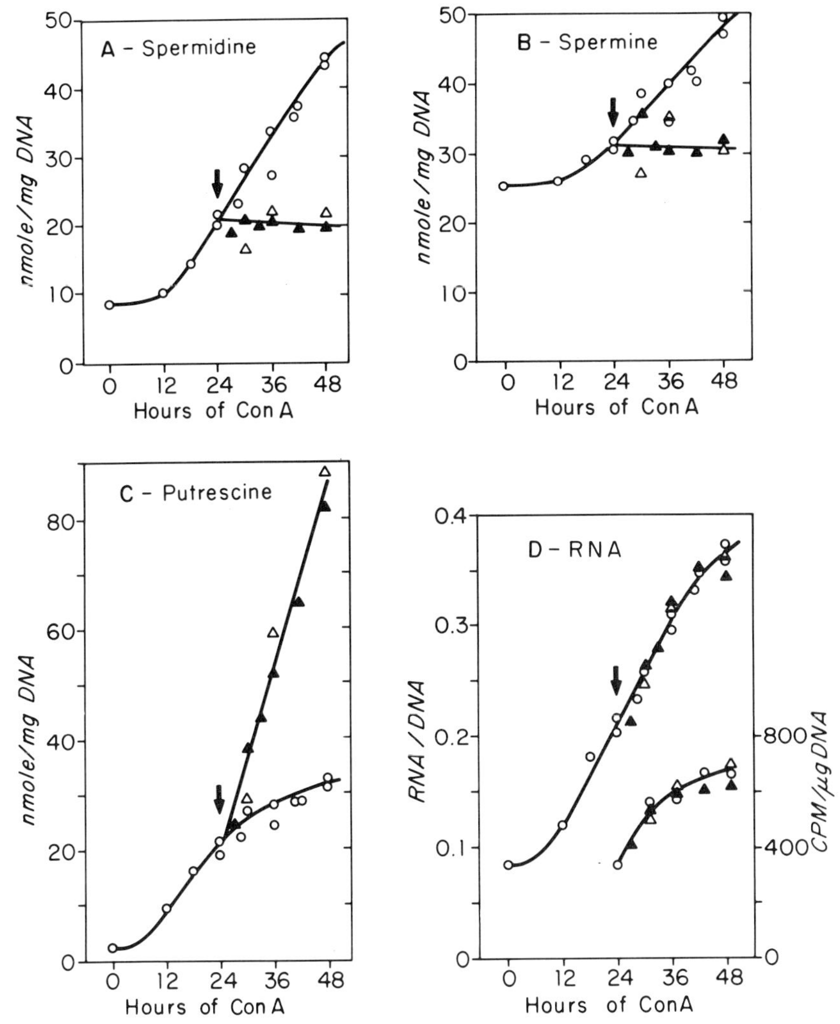

FIG. 5. Effects of MGBG on polyamine and RNA accumulation during lymphocyte transformation. ConA-stimulated cultures to which no inhibitor was added are indicated by (○). MGBG was added to the other cultures at 24 hr at a final concentration of 8 μM (△) or 40 μM (▲). A, Spermidine accumulation; B, spermine accumulation; C, putrescine accumulation; D, RNA accumulation and insert showing ^{3}H-uridine incorporation (3.3 μC/ml) into RNA during 2-hr pulses of 3-ml aliquots removed from the respective mother cultures.

lymphocyte cultures at 24 hr after stimulation results in an abrupt cessation of spermidine and spermine accumulation. The concentration of MGBG used in these experiments was 8 or 40 μM. More recent experiments indicate that concentrations as low as 0.5 μM MGBG may be as effective in inhibiting spermidine and spermine accumulation. As shown in Fig. 5C, putrescine accumulation continues in the presence of MGBG, as would be anticipated from a block in the pathway between putrescine and spermidine (spermine) formation. In fact, putrescine accumulates faster than would be expected if this block were the only factor involved. We questioned if this accelerated rate of putrescine formation was modulated by an increase in ornithine decarboxylase activity. Although ornithine decarboxylase is close to its maximal activity at 24 hr (8), addition of MGBG at this time results in a twofold increase in ornithine decarboxylase over the next 10 hr, as shown in Table 2. When we assayed the same dialyzed extracts for adenosylmethionine decarboxylase, we were surprised to find that its activity increased over 100-fold between 24 and 42 hr in the presence of MGBG (Table 2). We normally observe a biphasic increase in adenosylmethionine decarboxylase during lymphocyte transformation. The level is 10-fold higher at 24 hr than in nonstimulated lymphocytes. Between 24 and 36 hr a second increase in adenosylmethionine decarboxylase takes place, which accounts for the difference in control values in Table 2. The increased levels of ornithine decarboxylase and adenosylmethionine decarboxylase in MGBG-treated cultures may indicate that the cells are trying to compensate for the polyamine deficiency.

Despite the fact that spermidine and spermine accumulation are completely inhibited in the presence of MGBG, RNA accumulation proceeds at an unabated rate, as shown in Fig. 5D. The rate of uridine incorporation

TABLE 2. *Effect of MGBG on ornithine and adenosylmethionine decarboxylase activities*

Hours after ConA	Ornithine decarboxylase[a]		S-Adenosylmethionine decarboxylase[a]	
	Control	+MGBG	Control	+MGBG
24	9.47[b]	–	0.72[b]	–
28	8.69	15.3	0.94	18.5
34	5.95	18.5	1.92	39.0
42	4.06	14.7	0.77	77.5

Cultures were made 40 μM in MGBG 24 hr after the addition of ConA and then harvested at the times indicated.

[a] Activity was measured after dialysis of the supernatant fraction of the homogenate.

[b] Specific activity (nmole/60 min/mg)

into RNA during 2-hr pulses is also shown in Fig. 5D. There is no effect on uridine incorporation despite the fact that spermidine and spermine accumulation has ceased; nor is there an effect on the rate of uridine incorporation during 15-min pulses, which should be more reflective of the rate of synthesis of unstable as well as stable RNA. We have thus inhibited spermidine and spermine accumulation over the 24-hr interval when both polyamines and RNA are normally increasing at a maximal rate. This inhibition does not result in any readily discernible changes in either the rate of cellular RNA accumulation or the rate of uridine incorporation into RNA. Interpretation of these results is complicated by the fact that putrescine continues to increase in the presence of MGBG. It is possible that putrescine alone is capable of serving any role that spermidine and spermine normally serve in RNA synthesis and accumulation. Secondly, we do not know whether the RNA that is synthesized during polyamine limitation is normal RNA or is of an aberrant type. Resolution of this question will require detailed analysis of the types of RNA made and their processing.

Although gross RNA accumulation and synthesis appear to be normal during polyamine limitation, we have preliminary indications that other events in the transformation response may be affected. The rate of leucine incorporation into cellular protein normally increases 1.5-fold between 24 and 48 hr after the addition of ConA. In one experiment where MGBG was added at 24 hr (8 or 40 μM) the rate of incorporation did not increase significantly over this interval. In a similar experiment, where the rate of thymidine incorporation into DNA was measured, the normal increase between 24 and 48 hr was reduced by about 50% in the presence of MGBG. In addition, the rate of division at 48 hr, as measured by the accumulation of metaphase cells in the presence of colcemid, was reduced by about 50% when MGBG was present between 24 and 48 hr. However, we do not know whether these effects are the direct result of polyamine limitation or are due to secondary effects of MGBG. If these effects are due to polyamine limitation, they should be observable at the minimum concentration of MGBG required to inhibit spermidine and spermine synthesis and should be reversed by addition of spermidine and/or spermine to the culture medium.

V. CONCLUSIONS

The lymphocyte transformation system seems well suited to studies of the function of polyamines in mammalian cell growth. Large quantities of relatively homogeneous lymphocytes can be obtained and grown in culture under defined conditions. A sequential order of changes occurs during lymphocyte transformation, and the magnitude of these changes is large enough to be easily quantitated. Inhibitor studies can be performed without many of the complications inherent in whole animal studies.

A careful kinetic analysis of cellular RNA accumulation and polyamine accumulation indicates that polyamines increase after RNA. The simplest interpretation of these results is that polyamines cannot be stimulating RNA accumulation. This interpretation is supported by the results with MGBG, which indicate that spermidine and spermine accumulation can be completely inhibited and not have any gross effects on either RNA accumulation or the rate of uridine incorporation into RNA.

ACKNOWLEDGMENTS

This work was supported by a research grant from the National Institute of General Medical Sciences (GM 13957). R. H. F. was supported by training grant GM 00052 from the National Institutes of Health.

REFERENCES

1. Cohen, S. S., in *Introduction to the polyamines,* Prentice-Hall, Inc., Englewood Cliffs, N.J., 1971.
2. Dykstra, W. J., Jr., and Herbst, E. J., *Science,* **149,** 428 (1965).
3. Raina, A., Jänne, J., and Siimes, M., *Biochim. Biophys. Acta,* **123,** 197 (1966).
4. Russell, D. H., and Lombardini, J. B., *Biochim. Biophys. Acta,* **240,** 273 (1971).
5. Pegg, A. E., Lockwood, D. H., and Williams-Ashman, H. G., *Biochem. J.,* **117,** 17 (1970).
6. Caldarera, C. M., Barbiroli, B., and Moruzzi, G., *Biochem. J.,* **97,** 84 (1965).
7. Raina, A., and Cohen, S. S., *Proc. Nat. Acad. Sci. U.S.A.,* **55,** 1587 (1966).
8. Fillingame, R. H., and Morris, D. R., in preparation.
9. Morris, D. R., and Jorstad, C. M., *J. Bacteriol.,* **101,** 731 (1970).
10. Williams-Ashman, H. G., and Schenone, A., *Biochem. Biophys. Res. Comm.,* **46,** 288 (1972).

Polyamines in Normal and Neoplastic Growth. Edited by D. H. Russell. Raven Press, New York © 1973.

The *In Vivo* Chemical Stimulation of Hepatic Ornithine Decarboxylase Activity: Modifications of Activity at the Transcriptional and Post-Transcriptional Levels of Protein Synthesis

William T. Beck and E. S. Canellakis

Department of Pharmacology, Yale University School of Medicine, New Haven, Connecticut 06510

I. INTRODUCTION

During the course of our work on the purification and appearance in the cell cycle of ornithine decarboxylase (ODC) (1, 2), we became interested in various conditions which modify hepatic levels of this important enzyme in polyamine synthesis. It has been shown by a number of investigators that the administration of hormones to intact, hypophysectomized, or adrenalectomized rats increases hepatic ODC activity (3–8).

Because cyclic AMP is known to be a mediator of many hormonal responses and has been shown to increase the activity of a number of hepatic enzymes *in vivo* (9–15), we thought it worthwhile in this series of experiments to investigate the effect of exogenously administered dibutyryl cyclic AMP, as well as drugs which may increase endogenous levels of the cyclic nucleotide, on the hepatic ODC activity of intact rats. In addition, we have studied the effects of treatment with inhibitors of RNA and protein synthesis on ODC activity under conditions of exposure to the other drugs.

We found that dibutyryl cyclic AMP, as well as theophylline and dexamethasone, stimulated hepatic ODC activity. When actinomycin D, which had no independent effect on the basal level of the enzyme, was given with dibutyryl cyclic AMP, ODC activity was inhibited. Moreover, this same dose of actinomycin D only partially inhibited the increase in ODC activity produced by theophylline and dexamethasone. However, when the animals were dosed with either puromycin alone or with puromycin plus dibutyryl cyclic AMP at levels of puromycin which inhibited protein synthesis, the administration of the antibiotic alone resulted in the stimulation of ODC activity and it potentiated the activity due to dibutyryl cyclic AMP.

II. MATERIALS AND METHODS

Male Sprague-Dawley rats, weighing 140 to 160 g, were purchased from either Perfection Breeders (Douglasville, Pa.) or Carworth Laboratories, Inc. (New City, N.Y.). The enzyme responses were the same, regardless of the source of the animals.

The following compounds were purchased: (1-^{14}C)-DL-ornithine·HCl, 10 to 12 mC/mmole, (1-^{14}C)-L-leucine, 29 mC/mmole, (4, 5-^{3}H)-L-leucine, 30 to 50 C/mmole, and Liquifluor from New England Nuclear (Boston, Mass.); N^{6}-2′-O-dibutyryl cyclic adenosine monophosphate (crystalline) from Boehringer-Mannheim (New York, N.Y.); theophylline from Calbiochem (San Diego, Calif.); Decadron (dexamethasone phosphate) and actinomycin D from Merck, Sharp and Dohme (Rahway, N.J.); puromycin dihydrochloride from Nutritional Biochemicals Corp. (Cleveland, Ohio); 2-mercaptoethanol from Eastman Chemicals (Rochester, N.Y.); puromycin aminonucleoside, 6-dimethylaminopurine, Trizma Base, pyridoxal-5′-phosphate, and bovine albumin, fraction V, from Sigma Chemicals (St. Louis, Mo.). All other chemicals were of reagent grade and were obtained from Fischer Scientific Co. (Springfield, N.J.).

The following drugs were prepared for i.p. injection in 0.2 ml of 0.9% NaCl/100 g body weight: dibutyryl cyclic AMP, 2 mg; theophylline, 10 mg; puromycin, 12 or 40 mg, neutralized; puromycin aminonucleoside, 7.5 mg; 6-dimethylaminopurine, 4.2 mg; actinomycin D, 80 μg dissolved in 50% ethanol and brought to volume with NaCl. Decadron was used as supplied, and 1 mg/100 g body weight/0.25 ml was injected. The specific activities of ^{14}C- and ^{3}H-leucine were adjusted to approximately 30 mC/mmole, and 15 μC ^{14}C- or 33 μC ^{3}H/0.2 ml/100 g body weight was injected.

The rats were fasted overnight, with a minimum of 15 hr of food deprivation before sacrifice. The animals were decapitated and bled at intervals after dosing, and the livers were removed and immediately homogenized in 3 volumes of 0.25 M sucrose. The homogenates were centrifuged for 1 hr at 100,000 $\times$ *g*, after which the postmitochondrial supernatant fluids were transferred to separate iced tubes.

ODC activity was assayed no later than 1 day after preparation of the 100,000 $\times$ *g* supernatant fractions. No loss of enzyme activity occurred during this interval when the supernatants were stored at -20°C. The ODC enzyme activity was determined by measuring the release of $^{14}CO_2$ from (1-^{14}C)-DL-ornithine as previously described (1). The enzyme fractions analyzed contained 20 to 30 mg protein in a final assay volume of 2 ml. The reaction was linear with time and enzyme concentration for the 30-min incubation conditions. One unit of ODC activity is defined as 1 pmole $^{14}CO_2$ released/mg protein/30 min. Tyrosine aminotransferase (TAT) activity was assayed according to the method of Granner and Tomkins (16). Units of TAT enzyme activity are expressed as μmoles *p*-hydroxyphenylpyruvate formed/mg protein/15 min at 37°C. Serine dehydratase (SDH) activity was assayed according to the method of Suda and Nakagawa (17). Units of SDH activity are expressed as μmoles pyruvate formed/mg protein/5 min at 37°C. Statistical analysis of the data was determined using the *t* test. Protein was measured by the biuret reaction (18) or by the method of Lowry et al. (19), using bovine serum albumin as a standard.

To determine the amount of ^{14}C or ^{3}H incorporated into liver protein, aliquots of the whole homogenate were precipitated with 5% trichloroacetic acid (TCA), heated at 80 to 90°C for 15 min, cooled, and centrifuged. The samples were washed two times in 5% TCA, extracted with hot ethanol (70°C for 15 min), centrifuged, resuspended in TCA, and centrifuged again. The pellets were dissolved in 0.1 N NaOH, and aliquots were counted for radioactivity. Activity is expressed as cpm ^{14}C or ^{3}H incorporated/mg total protein. It was ascertained that the ^{14}C-leucine administered to the animals did not interfere with the determination of ODC activity.

III. RESULTS

In Table 1 it is shown that dibutyryl cyclic AMP administration stimulated ODC activity to 33.9 units 3 hr after dosing, approximately six times greater than the control activity. Simultaneous injections of actinomycin D along with the dibutyryl cyclic AMP significantly inhibited the dibutyryl cyclic AMP-stimulated activity by 91% to an average value of 3.16 units. Table 1 also shows that 3 hr after the i.p. administration of theophylline to normal rats, the mean hepatic ODC activity was 50.5 units, a 10-fold increase over the corresponding saline control average activity of 5.08 units. However, simultaneous administration of actinomycin D along with the theophylline only inhibited theophylline-stimulated activity by 62%, bringing it to an average value of 19.0 units. It is also shown in Table 1 that the average 3-hr dexamethasone-stimulated hepatic ODC activity was 108 units, a more than 20-fold increase above control levels.

Moreover, simultaneous administration of actinomycin D with dexamethasone also inhibited dexamethasone-stimulated activity by 58% to an average value of 45.4 units which was, nevertheless, greater than the maximum stimulation of 33.9 units achieved with dibutyryl cyclic AMP alone. Thus, the same dose of actinomycin D reduced both theophylline- and dexamethasone-stimulated activity by 60%, even though the dexamethasone-stimulated activity was twice that due to theophylline. It is further seen in Table 1 that actinomycin D had no effect on the 3-hr endogenous, nonstimulated ODC activity. Three hr after treatment with actinomycin D alone, the activity of 6.6 units was not significantly different than that of the saline controls. Finally, at the doses of drugs and actinomycin D used, the inhibitor was much more effective in blocking the stimulatory effects of dibutyryl cyclic AMP than those of theophylline and dexamethasone.

Since the results in Table 1 suggested that cyclic AMP may be involved in the stimulation of hepatic ODC activity, the dibutyryl cyclicAMP effect was investigated in greater detail. Figure 1 shows that a single i.p. injection of dibutyryl cyclic AMP caused an approximate sevenfold increase in hepatic ODC activity to 33.9 units 3 hr after dosing, when compared with 0 time control values of 4.9 units. The activity reached a maximum at 3 hr and then declined to 12.1 units by 5 hr, a value approximately 30% of the 3-hr peak. Moreover, it is also seen in Fig. 1 that rats given simultaneous injections of dibutyryl cyclic AMP and actinomycin D have only a twofold increase in ODC activity to 10.6 units 2 hr after dosing. The activity then returned to below control levels, ranging from 2.0 to 3.3 units for the next 3 hr; thus, the concurrent administration of actinomycin D abolished the 3-hr maximum dibutyryl cyclic AMP-stimulated activity. Finally, Fig. 1 also shows that when actinomycin D was administered 1 hr after dibutyryl cyclic AMP, there was no significant inhibition of ODC activity at any time

TABLE 1. *Effect of drug treatment on rat liver ODC activity 3 hr after dosing*[a]

Treatment	ODC activity[b] Drug only	ODC activity[b] Drug + actinomycin D	p	Percent inhibition
Saline	5.08 ± 1.15	6.22 ± 1.85	N.S.	–
Dibutyryl cyclic AMP	33.9 ± 6.47	3.16 ± 0.85	<0.05	91
Theophylline	50.5 ± 8.16	19.0 ± 5.85	<0.02	62
Dexamethasone	108 ± 12.8	45.4 ± 7.85	<0.01	58

[a] Drugs were administered i.p. at the following doses per 100 g body weight: dibutyryl cyclic AMP, 2 mg; dexamethasone, 1 mg; theophylline, 10 mg; actinomycin D, 80 μg.

[b] ODC activity is expressed as pmoles $^{14}CO_2$ released/mg protein/30 min (units). The values represent the mean ± S.E. of seven to 18 animals. The statistical significance of the results was determined by the *t* test. Data taken from Beck et al. (36).

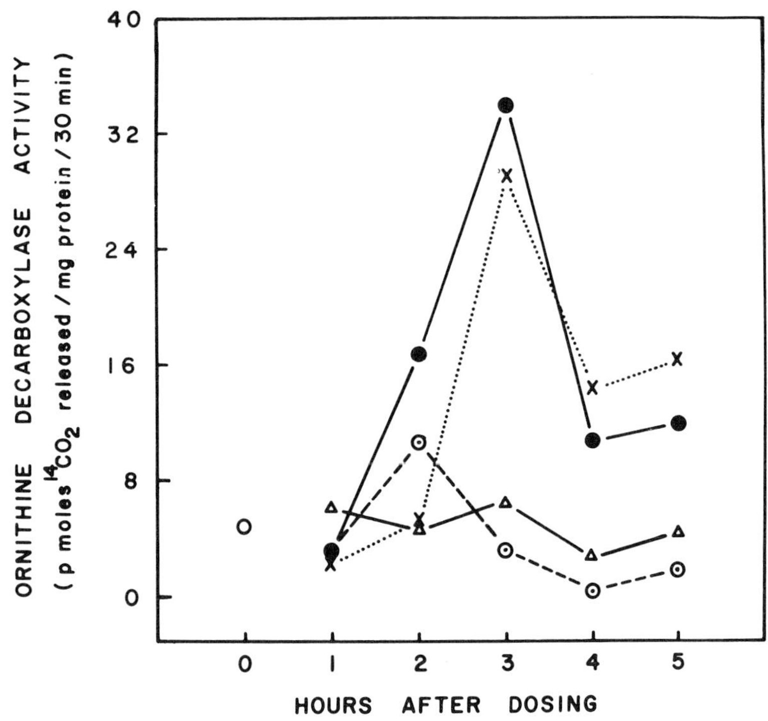

FIG. 1. Effect of dibutyryl cyclic AMP and actinomycin D on rat liver ODC activity. Animals were injected i.p. at 0 hr with either dibutyryl cyclic AMP, 2 mg/100 g body weight (●), actinomycin D, 80 μg/100 g body weight (△) or dibutyryl cyclic AMP + actinomycin D (⊙). Another group of animals received dibutyryl cyclic AMP at 0 hr, followed by actinomycin D 1 hr later (×). The ODC activity of control rats (○) was 4.9 ± 1.0 units. ODC activity was assayed by $^{14}CO_2$ entrapment as described in Methods. Each point represents the mean of from three to eight rats; each assay was done in triplicate. The S.E. ranged from 15 to 25% of the mean values. Data taken from Beck et al. (36).

point studied, the maximum 3-hr activity of 28.9 units comparing favorably with the activity from the livers of those animals treated with dibutyryl cyclic AMP alone. Furthermore, when administered alone, actinomycin D was seen to have no effect on the baseline ODC activity.

Our results with actinomycin D would suggest, therefore, that the increased ODC activity in response to chemical stimuli may be dependent upon new messenger RNA formation that is completed within 1 hr after administration of the stimulus.

Thus, these previous data suggested that cyclic AMP may stimulate ODC activity at the transcriptional stage. To determine if there was any involvement at the post-transcriptional level of protein synthesis, rats were given puromycin simultaneously with dibutyryl cyclic AMP; the results are shown in Fig. 2A. Again, a single i.p. administration of dibutyryl cyclic AMP to normal rats stimulated hepatic ODC activity; however, peak activity was broader than previously noted, occurring 2 to 3 hr after dosing.

A synergistic effect of dibutyryl cyclic AMP and puromycin was seen when these two compounds were administered concurrently. Whereas at 2 hr a preliminary decrease in ODC activity was seen (10.9 units, a 73% decrease compared to animals given dibutyryl cyclic AMP alone), by 3 hr the ODC activity was 167 units. The peak puromycin/dibutyryl cyclic AMP-stimulated activity (175 units at 3.5 hr) occurred when the dibutyryl cyclic AMP-stimulated activity had returned to 13.2 units. The puromycin/dibutyryl cyclic AMP-stimulated activity then decreased to 48.9 units by 5 hr, a value equivalent to the maximum activity obtained with dibutyryl cyclic-AMP alone (44.6 units).

The result of administering puromycin alone to normal rats is also shown in Fig. 2A. In further contrast to actinomycin D, which was shown in Fig. 1 to have no effect of its own on ODC activity, Fig. 2A demonstrates that puromycin alone clearly stimulated hepatic ODC activity, reaching a peak of 71.7 units by 3.5 hr after dosing, a more than 14-fold increase over 0 time controls. The activity then decreased to a value of 15.2 units, or slightly more than 20% of the 3.5-hr peak activity. In addition, it is seen that saline had no effect of its own. Finally, the greater than 35-fold stimulation over the zero time value was more than additive, since the combined activity of the dibutyryl cyclic AMP and puromycin alone was 79.3 units as compared to 167 units for the concurrent administration of these compounds. It appeared, therefore, that the enzyme was in some way sensitive to a kind of translational interference.

An attempt was made to correlate the time sequence of inhibition of protein synthesis by puromycin with the stimulation of ODC activity. In Fig. 2B, it is seen that ^{14}C incorporation into liver protein of saline-treated animals was relatively constant, ranging between means of 1,621 and 2,046/cpm/mg protein over the 5-hr period of study. Furthermore, the pattern of ^{14}C incorporation into liver protein of rats treated with dibutyryl cyclic AMP alone was essentially the same as that obtained for the controls, in spite of the increased ODC activity (Fig. 2A) with mean values of 1,447 to 1,936 cpm incorporated/mg protein. It should be noted that these values were all (except for the 5-hr point) slightly lower than controls, the 1-hr point being 1,447 cpm/mg protein, or about 20% less than the zero time value of 1,803 cpm/mg protein, but none of these differences proved to be statistically significant.

The i.p. administration of either puromycin or puromycin plus dibutyryl

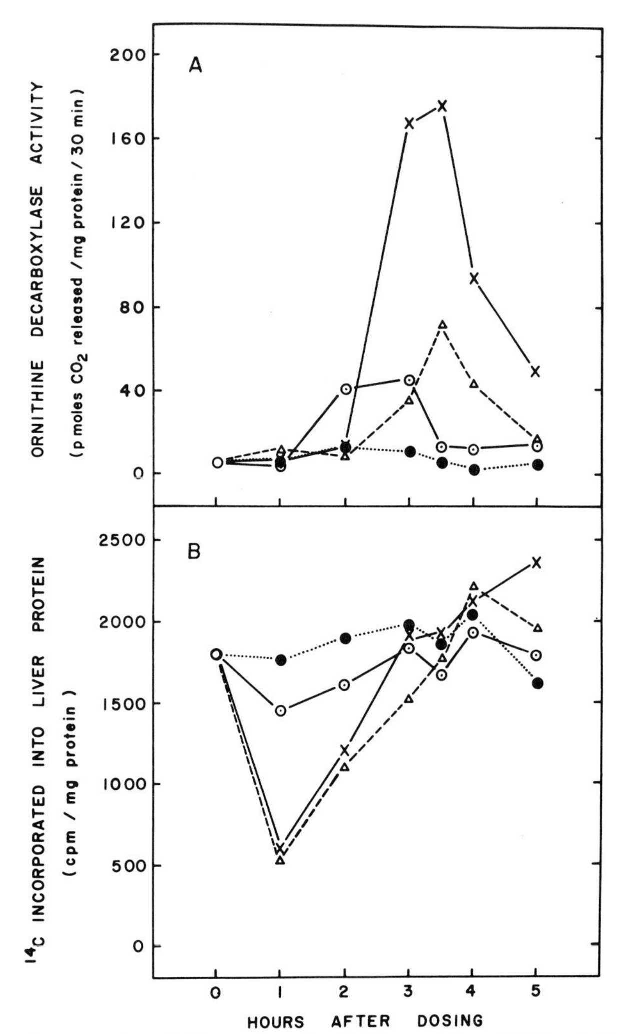

FIG. 2. Effect of puromycin and dibutyryl cyclic AMP on rat liver ODC activity (A) and total protein synthesis (B). Rats were injected i.p. at 0 hr with either puromycin (△), 12 mg in 0.2 ml isotonic NaCl, neutralized/100 g body weight, dibutyryl cyclic AMP (⊙), 2 mg in 0.2 ml isotonic NaCl/100 g body weight, puromycin plus dibutyryl cyclic AMP (×), or isotonic NaCl (●), 0.2 ml/100 g body weight.

A: ODC activity was assayed by the method of $^{14}CO_2$ entrapment as described in Methods. B: All animals received an i.p. injection of 15 μC/100 g body weight (1-^{14}C)-L-leucine (30 mC/mmole) 20 min before killing. It was ascertained experimentally that the ^{14}C-leucine administered to the rats did not interfere with the assay of ODC activity. The determination of incorporation of radioactivity into liver protein is described in Methods. The results presented for each point represent the mean of three to eight animals. The S.E. range from 15 to 20% (A) or 5 to 20% (B) in all cases. Data taken from Beck et al. (38).

cyclic AMP depressed the incorporation of ^{14}C into liver protein by more than 60% after 1 hr, at a time when no differences in ODC activity were discernible (Fig. 2A). However, ODC activity increased and reached a maximum when the puromycin inhibition of protein synthesis was terminated (3 hr).

There are reports in the literature demonstrating that puromycin has metabolic effects independent of its action on protein synthesis (20–23). Two hydrolysate analogues of puromycin, its aminonucleoside and 6-dimethylaminopurine, share these other metabolic effects with puromycin, but do not inhibit protein synthesis. To determine whether the stimulation of ODC activity by puromycin was related to its effect on protein synthesis or to its other metabolic effects, rats were injected i.p. with equimolar doses of the non-protein synthesis inhibitory analogues. The results are shown in Table 2. Clearly, by 3.5 hr after dosing, only puromycin caused a stimulation of hepatic ODC activity (66.6 units), whereas neither puromycin aminonucleoside nor dimethylaminopurine had any effect on ODC activity (3.23 and 5.85 units, respectively) when compared with controls at the same time period.

The data in Table 2 and Fig. 2 suggested, therefore, that the inhibition and/or return of protein synthesis is necessary for the stimulation of rat liver ODC activity by puromycin. To substantiate this point, normal rats were given a larger dose of puromycin (40 mg/100 g body weight). The effects on ODC activity and protein synthesis are shown in Fig. 3. Data for the lower dose of puromycin, 12 mg/100 g body weight, taken from Fig. 2, are shown for comparison. It is seen in the top half of Fig. 3 that, following a single i.p. injection of the larger dose of puromycin, ODC activity reached a peak of 61.7 units 5 hr after dosing, when compared with the 3.5-hr peak activity of 66.6 units following administration of the lower dose of the

TABLE 2. *Effect of puromycin and puromycin analogue treatment on rat liver ODC activity 3-½ hr after dosing*[a]

Treatment	ODC activity[b]	*p*
Saline	5.08 ± 1.15	–
Puromycin dihydrochloride	66.6 ± 7.4	<0.01
Puromycin aminonucleoside	3.23 ± 0.54	N.S.
Dimethylaminopurine	5.85 ± 2.23	N.S.

[a] Compounds were administered i.p. at equimolar doses/100 g body weight: puromycin dihydrochloride, 12 mg; puromycin aminonucleoside, 7.5 mg; dimethylaminopurine, 4.2 mg.

[b] ODC activity is expressed as pmoles $^{14}CO_2$ released/mg protein/30 min (units). The values represent the mean ± S.E. of five to eight animals. Statistical significance of the results was determined by the t test.

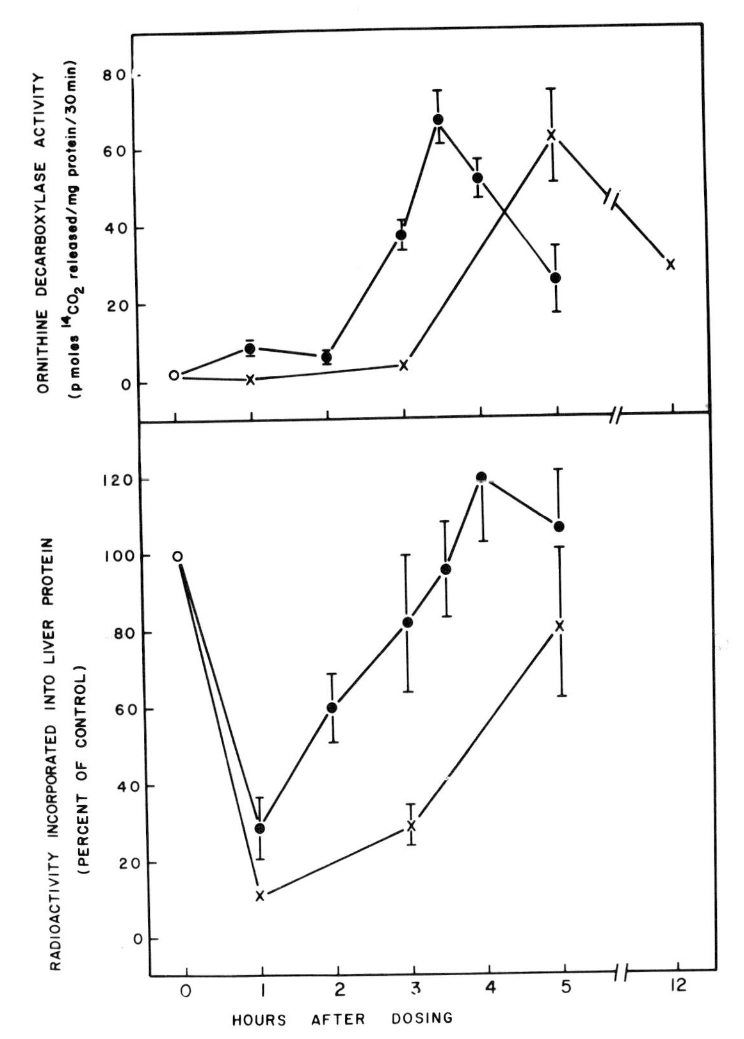

FIG. 3. Effect of different doses of puromycin on rat liver ODC activity (top) and total protein synthesis (bottom). Rats were injected i.p. at 0 hr with the following doses of neutralized puromycin in 0.2 ml isotonic NaCl/100 g body weight: 12 mg (●); 40 mg (×). *Top:* ODC activity was assayed by the method of $^{14}CO_2$ entrapment as described in the Methods. *Bottom:* Twenty min before killing, all animals received an i.p. injection of either ^{14}C-leucine (15 μC/100 g body weight) or ^{3}H-leucine (33 μC/100 g body weight). The specific activities of both ^{14}C- and ^{3}H-leucine were approximately 30 mC/mmole. It was ascertained experimentally that the ^{14}C-leucine administered to the rats did not interfere with the assay of ODC activity. The determination of incorporation of radioactivity into liver protein is described in the Methods. The results presented for each point represent the mean of three to eight animals and the vertical bars represent the S.E.M.

antibiotic. Thus, rather than increasing the peak 3.5-hr ODC activity, puromycin at the larger dose merely delays the same peak activity to 5 hr. Concomitantly, the return of protein synthesis as measured by incorporation of radioactivity into total liver protein (lower panel) was also delayed by the large dose of puromycin. Approximately 30% of control isotope incorporation was reached at 3 hr following administration of the large dose of puromycin, whereas the same value was reached by only 1 hr after dosing with the smaller amount of the antibiotic. Similarly, the return to near control levels was also delayed by the 40 mg/100 g dose of puromycin, 80% of control levels being achieved by 5 hr, rather than 3 hr as observed following the administration of the 12 mg/100 g amount. It would appear, therefore,

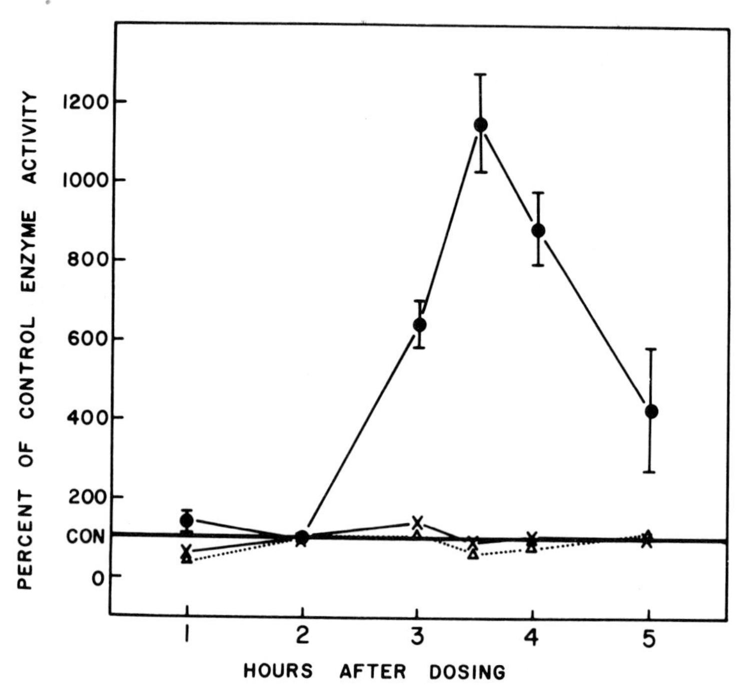

FIG. 4. Rat hepatic ODC, TAT, and SDH activities following puromycin treatment. Rats were injected i.p. at 0 hr with puromycin, 12 mg/100 g body weight/0.2 ml NaCl, neutralized. ODC activity was assayed by the method of $^{14}CO_2$ entrapment as described in the Methods. Units of ODC enzyme activity are expressed as pmoles $^{14}CO_2$ released/mg protein/30 min at 37°C. TAT activity was assayed according to the method of Granner and Tomkins (16). Units of TAT enzyme activity are expressed as μmoles *p*-hydroxyphenylpyruvate formed/mg protein/15 min at 37°C. SDH activity was assayed according to the method of Suda and Nakagawa (17). Units of SDH activity are expressed as μmoles pyruvate formed/mg protein/5 min at 37°C. The results presented for each point represent the mean of three to eight animals and the vertical bars represent the S.E.M. Symbols: ODC (●); TAT (×); SDH (△).

that the stimulation of hepatic ODC activity following puromycin administration is related to the return of protein synthesis.

In addition to puromycin, dibutyryl cyclic AMP and dexamethasone have been shown in Table 1 to stimulate hepatic ODC activity. These latter two compounds have also been shown to stimulate the *in vivo* activity of tyrosine aminotransferase and serine dehydratase (11–14). However, Fig. 4, which compares rat hepatic ODC, tyrosine aminotransferase, and serine dehydratase activities following puromycin treatment, clearly shows that puromycin stimulated only the activity of ODC to 1,148% of control levels by 3.5 hr. In marked contrast, puromycin had no effect on the activities of tyrosine aminotransferase or serine dehydratase over the same time period. These results would suggest, therefore, a unique role for ODC in response to a post-transcriptional modifier.

IV. DISCUSSION

A. Dibutyryl cyclic AMP and actinomycin D effects on ODC activity

It has been shown that the i.p. administration of dibutyryl cyclic AMP to normal rats increased hepatic ODC activity. This is another instance whereby exogenously administered dibutyryl cyclic AMP increased the synthesis or activity of an hepatic enzyme, since, for example, dibutyryl cyclic AMP has been shown to increase the activities *in vivo* of tyrosine aminotransferase (11, 13, 14), serine dehydratase (11–13), phosphoenol pyruvate carboxykinase (13, 14), glucose-6-phosphatase (10), phosphopyruvate carboxylase (9), and L-alanine-glyoxylate aminotransferase (15). Furthermore, a variety of other compounds have also been shown to stimulate hepatic ODC activity (3–5, 7, 8, 24–27). It has been shown in this chapter that theophylline and dexamethasone increased the activity of this enzyme. Our results with dibutyryl cyclic AMP and theophylline suggest that cAMP may be involved in the stimulation of hepatic ODC activity.

Moreover, the data demonstrate that, whereas actinomycin D almost completely inhibited the dibutyryl cyclic AMP stimulation of hepatic ODC activity, it had less of an effect on the ODC activity stimulated by either theophylline or dexamethasone. The results also show that when actinomycin D was administered 1 hr after dosing with dibutyryl cyclic AMP, the stimulation of ODC activity was not blocked. This finding is in agreement with those of others (4, 5, 28) where delayed administration of actinomycin D did not inhibit the stimulation of hepatic ODC activity. These results with actinomycin D would suggest, therefore, that the increased ODC activity in response to these chemical stimuli may be dependent upon the formation of new messenger RNA within 1 hr after administration of the stimulus.

B. Puromycin effect on ODC activity

In contrast to the effects of actinomycin D, puromycin, when administered simultaneously with dibutyryl cyclic AMP, produced a more than 35-fold stimulation of hepatic ODC activity. Furthermore, puromycin alone also stimulated ODC activity, and these results appear to be related to the effect of the antibiotic on protein synthesis. In apparent confirmation of this point, it was seen in Table 2 that neither puromycin aminonucleoside nor dimethylaminopurine, analogues of puromycin which do not inhibit protein synthesis, had any effect on hepatic ODC activity. Moreover, it was observed in Fig. 3 that a larger dose of puromycin delayed the return of protein synthesis and subsequently delayed the increase of ODC activity. These data suggest, therefore, that the effect of puromycin on hepatic ODC activity is related to the inhibition and/or return of protein synthesis. Finally, the puromycin effect appeared to be specific for ODC, for it was shown in Fig. 4 that neither tyrosine aminotransferase nor serine dehydratase activities were stimulated by the antibiotic. These results suggest that the puromycin stimulus for increasing ODC activity is ineffective in increasing the activities of the other two enzymes.

On the basis of these results, we may consider some explanations for the puromycin stimulation of hepatic ODC activity. If one assumes that a protein inhibitor of ODC exists which is more sensitive than ODC to the actions of puromycin, then following the administration of this drug, at a time of maximal depression of protein synthesis, the production of both the inhibitor protein and the ODC should cease. However, as puromycin levels fall, protein synthesis would resume, but the inhibition of ODC would be relieved first, thus causing a rise in ODC activity. Alternatively, it may be possible that because puromycin causes an extensive inhibition of protein synthesis, this may produce an accumulation of the ODC messenger RNA, resulting in a "rebound" phenomenon of greatly enhanced ODC activity when the drug effect was terminated. Both of these hypotheses are consistent with the kinetics of inhibition presented in Figs. 2 and 3 and are amenable to experimental verification.

However, such facile explanations should be tempered by the complexities inherent in the metabolism of eukaryotic cells, and the fact that other explanations for the puromycin effect are possible. For example, partial hepatectomy results in an increased amino acid content of hepatic cells (29–31); at the same time, ODC activity increases (4, 24, 26, 28, 32–34). In addition, cyclic AMP increases the hepatic content of free amino acids (35), and administration of dibutyryl cyclic AMP (36; Figs. 1 and 2) or a mixture of amino acids (4) increases ODC activity. It is thus conceivable that the inhibition of protein synthesis by puromycin also increases hepatic amino acid pools which then act as a causal stimulus for increased ODC activity.

Other possibilities should be considered. (a) Puromycin may increase endogenous cyclic AMP levels with consequent increases in ODC activity; such an explanation has been proposed for changes in the activities of glycolytic enzymes by puromycin (21, 37). However, the same dose of actinomycin D which completely inhibits the dibutyryl cyclic AMP stimulation of ODC activity (36; Fig. 1) only partially inhibits the puromycin effect (38). (b) The drug may release nascent polysomal ODC molecules. It is known that both cyclic AMP and puromycin are capable of releasing nascent protein and active enzyme from polysomes *in vitro* (39, 40), but these effects were not found *in vivo* (41). Furthermore, this explanation might require that the peak in ODC activity occur at the time of the maximal effect of puromycin on protein synthesis; this, as has been shown in Figs. 2 and 3, is not the case. (c) It has been shown that steroids increase ODC activity (7, 8, 36; Table 1). Thus, the possibility exists that puromycin administration may stress the animals severely enough to cause the release of adrenal cortical hormones, although the results in Fig. 4 show that the activities of TAT and SDH, which can be stimulated by adrenal cortical hormones, were not stimulated by puromycin. Nevertheless, all of the above considerations are suitable for further experimental investigation.

Finally, the enhancement of ODC activity in normal rats should not be confused with the "paradoxical" effect of puromycin in augmenting the already high enzymatic activities of fetal tissues or of hydrocortisone-stimulated preparations (41–45). Whereas a low dose of actinomycin D inhibits the puromycin-stimulated ODC activity (38), the drug has been shown to "superinduce" TAT in tumor cells (46, 47).

C. Conclusion

In any consideration of our data, it should be appreciated that these experiments have been done on whole, intact animals; it is difficult to determine whether we are observing direct or indirect effects of the administration of these compounds. Nevertheless, it would be of great interest to know whether the puromycin-stimulated ODC activity following the inhibition and/or return of protein synthesis has any relation to or bearing upon rapidly growing systems. If so, would it be possible to exploit this relationship in such a way as to slow or prevent this growth? To summarize, therefore, (a) the administration of dibutyryl cyclic AMP, theophylline, or dexamethasone to intact rats increased hepatic ODC activity, and this increase was blocked by simultaneous administration of actinomycin D; (b) puromycin administration to rats increased ODC activity; (c) simultaneous administration of puromycin and dibutyryl cyclic AMP caused a synergistic stimulation of hepatic ODC activity; (d) the effect of puromycin on the enhanced ODC activity appeared to be related to the inhibition and/or return of protein synthesis; (e) the effect of puromycin was apparently specific for ODC, and not other enzymes. We are currently investigating a number of possible explanations for the puromycin effect on ODC.

V. SUMMARY

The hepatic ODC activity of normal rats was stimulated more than sevenfold 3 hr after a single i.p. injection of dibutyryl cyclic AMP. The 3-hr ODC activity was also stimulated by single injections of either theophylline or dexamethasone (10- and 21-fold, respectively). The simultaneous administration of actinomycin D with either dibutyryl cyclic AMP, theophylline, or dexamethasone reduced the 3-hr ODC activity by 91, 62, and 58%, respectively. However, simultaneous administration of puromycin plus dibutyryl cyclic AMP led to a 35-fold stimulation of ODC activity by 3-½ hr, and, in contrast to actinomycin D, puromycin alone increased ODC activity 14-fold. The stimulation of ODC activity following puromycin administration was shown to be related to the inhibition and/or return of protein synthesis. Moreover, the effect of puromycin appeared to be specific for ODC, since tyrosine aminotransferase and serine dehydratase activities were not affected by administration of the antibiotic.

ACKNOWLEDGMENTS

The authors wish to thank Ms. Rill Ann Bellantone for her excellent technical assistance in these endeavors.

This work was supported by U.S. Public Health Service grant CA-10748, U.S. Public Health Service grant CA-04823, and a U.S. Public Health Service Research Career Award 5-K06-GM 03070 to E.S.C.

REFERENCES

1. Friedman, S. J., Halpern, K. V., and Canellakis, E. S., *Biochim. Biophys. Acta,* **261,** 181 (1972).
2. Friedman, S. J., Bellantone, R. A., and Canellakis, E. S., *Biochim. Biophys. Acta,* **261,** 188 (1972).
3. Jänne, J., and Raina, A., *Biochim. Biophys. Acta,* **174,** 769 (1969).

4. Fausto, N., *Biochim. Biophys. Acta,* **238,** 116 (1971).
5. Russell, D. H., Snyder, S. H., and Medina, V. J., *Endocrinology,* **86,** 1414 (1970).
6. Kostyo, J. L., *Biochem. Biophys. Res. Comm.,* **23,** 150 (1966).
7. Panko, W. B., and Kenney, F. T., *Biochem. Biophys. Res. Comm.,* **43,** 346 (1971).
8. Cavia, E., and Webb, T. E., *Biochim. Biophys. Acta,* **262,** 546 (1972).
9. Yeung, D., and Oliver, I. T., *Biochemistry,* **7,** 3231 (1968).
10. Greengard, O., *Biochem. J.,* **115,** 19 (1969).
11. Jost, J.-P., Hsie, A. W., and Rickenberg, H. V., *Biochem. Biophys. Res. Comm.,* **34,** 748 (1969).
12. Jost, J.-P., Hsie, A., Hughes, S. D., and Ryan, L., *J. Biol. Chem.,* **245,** 351 (1970).
13. Wicks, W. D., Kenney, F. T., and Lee, K., *J. Biol. Chem.,* **244,** 6008 (1969).
14. Wicks, W. D., Lewis, W., and McKibbin, J., *Biochim. Biophys. Acta,* **264,** 177 (1972).
15. Snell, K., *Biochem. J.,* **123,** 657 (1971).
16. Granner, D. K., and Tomkins, G. M., *Methods Enzymol.,* **17A,** 633 (1970).
17. Suda, M., and Nakagawa, H., *Methods Enzymol.,* **17B,** 346 (1971).
18. Gornall, A. G., Bardawill, C. J., and David, M. M., *J. Biol. Chem.,* **177,** 751 (1949).
19. Lowry, O. H., Rosebrough, N. J., Farr, A. L., and Randall, R. J., *J. Biol. Chem.,* **193,** 265 (1951).
20. Hofert, J. F., and Boutwell, R. K., *Arch. Biochem. Biophys.,* **103,** 338 (1963).
21. Appleman, M. M., and Kemp, R. G., *Biochem. Biophys. Res. Comm.,* **24,** 564 (1966).
22. Søvik O., *Acta Physiol. Scand.,* **66,** 307 (1966).
23. Garren, L. D., Davis, W. W., Crocco, R. M., and Ney, R. L., *Science,* **152,** 1386 (1966).
24. Fausto, N., *Biochim. Biophys. Acta,* **190,** 193 (1969).
25. Fausto, N., *Cancer Res.,* **30,** 1947 (1970).
26. Schrock, T. R., Oakman, N. J., and Bucher, N. L. R., *Biochim. Biophys. Acta,* **204,** 564 (1970).
27. Russell, D. H., *Biochem. Pharmacol.,* **20,** 3481 (1971).
28. Russell, D. H., and Snyder, S. H., *Mol. Pharmacol.,* **5,** 253 (1969).
29. Christensen, H. N., Rothwell, J. T., Sears, R. A., and Streicher, J. A., *J. Biol. Chem.,* **175,** 101 (1949).
30. Braun, G. A., Marsh, J. B., and Drabkin, D. L., *Metabolism,* **11,** 957 (1962).
31. Ferris, G. M., and Clark, J. B., *Biochim. Biophys. Acta,* **273,** 73 (1972).
32. Raina, A., Jänne, J., and Siimes, M., *Biochim. Biophys. Acta,* **123,** 197 (1966).
33. Russell, D., and Snyder, S. H., *Proc. Nat. Acad. Sci. U.S.A.,* **60,** 1420 (1968).
34. Russell, D. H., and Snyder, S. H., *Endocrinology,* **84,** 223 (1969).
35. Mallette, L. E., Exton, J. H., and Park, C. R., *J. Biol. Chem.,* **244,** 5713 (1969).
36. Beck, W. T., Bellantone, R. A., and Canellakis, E. S., *Biochem. Biophys. Res. Comm.,* **48,** 1649 (1972).
37. Blatt, L. M., Scamahorn, J. O., and Kim, K. H., *Biochim. Biophys. Acta,* **177,** 553 (1969).
38. Beck, W. T., Bellantone, R. A., and Canellakis, E. S., *Nature,* **241,** 275 (1973).
39. Khairallah, E. A., and Pitot, H. C., *Biochem. Biophys. Res. Comm.,* **29,** 269 (1967).
40. Chuah, C.-C., and Oliver, I. T., *Biochemistry,* **10,** 2990 (1971).
41. Grossman, A., and Boctor, A., *Proc. Nat. Acad. Sci. U.S.A.,* **69,** 1161 (1972).
42. Chuah, C.-C., Holt, P. G., and Oliver, I. T., *Int. J. Biochem.,* **2,** 193 (1971).
43. Grossman, A., and Mavrides, C., *J. Biol. Chem.,* **242,** 1398 (1967).
44. Kenney, F. T., *Science,* **156,** 525 (1967).
45. Schimke, R. T., *Nat. Cancer Inst. Monogr.,* **27,** 301 (1967).
46. Reel, J. R., and Kenney, F. T., *Proc. Nat. Acad. Sci. U.S.A.,* **61,** 200 (1968).
47. Thompson, E. B., Granner, D. K., and Tomkins, G. M., *J. Mol. Biol.,* **54,** 159 (1970).

Polyamines in Normal and Neoplastic Growth. Edited by D. H. Russell. Raven Press, New York © 1973.

Polyamines in Marine Invertebrates

Carol-Ann Manen and Diane H. Russell

Department of Zoology, University of Maine, Orono, Maine 04473 and Laboratory of Pharmacology, Baltimore Cancer Research Center, National Cancer Institute, Baltimore, Maryland 21211

Polyamines are known to be present ubiquitously in microorganisms, plants, and vertebrates. There are two reports of the occurrence of spermine in sea urchins and tunicates but no data concerning quantitation (1, 2). No study has been conducted, to our knowledge, to confirm the general presence of polyamines in marine invertebrates and also to determine which amines are present and to what extent. Invertebrates with extensive skeletal material were not included in this study to ensure that the quantitation reflects the amine concentrations per gram of cellular material. The naturally occurring polyamines, putrescine, spermidine, and spermine, are present in the eight phyla and in the several species within each phyla that were surveyed (Table 1). In those crustaceans, gastropods, and echinoids surveyed, there tended to be very high concentrations of spermine, along with

TABLE 1. *Polyamine content of certain marine invertebrates*

Animal	Putrescine	Spermidine	Spermine
	(nmoles/g wet weight)		
Phylum: Porifera			
Class: Demospongiae			
Polymastia mammillaris	nd	<40	110
Trichostemma hemisphericum	50	130	nd
Phylum: Cnidaria			
Class: Hydrozoa			
Tubularia couthouyi	460	520	710
Tubularia crocea	230	390	300
Obelia commissuralis	550	400	<40
Sertularia pumila	530	860	70
Class: Anthozoa			
Metridium senile	80	120	200

TABLE 1. (*continued*)

Animal	Putrescine	Spermidine	Spermine
	(nmoles/g wet weight)		
Phylum: Nemertea			
Class: Anopla			
Lineus bicolor	<40	230	1320
Cerebratulus sp.[a]	90	40	490
Phylum: Annelida			
Class: Polychaeta			
Nereis pelagica	330	800	960
Lepidonotus sp.[a]	40	60	400
Chaetopterus variopedatus[a]			
Head	<40	40	630
Gut	210	130	530
Gonad	<40	100	660
Class: Oligochaeta			
Clitellio arenarius	390	840	880
Phylum: Arthropoda			
Class: Crustacea			
Mysis stenolepis	nd	180	920
Gammarus sp.	150	360	1530
Phylum: Mollusca			
Class: Gastropoda			
Acmaea testudinalis	120	410	740
Dendronotus frondosus	nd	580	490
Coryphella rufibranchialis	nd	400	1140
Phylum: Echinodermata			
Class: Echinoidea			
Strongylocentrotus purpuratus[a]			
Ovary	<40	40	1140
Testis	<40	58	1585
Gut	<40	<40	648
Lytechinus pictus[a]			
Ovary	60	62	1203
Testis	<40	110	1540
Gut	82	100	693
Phylum: Chordata			
Class: Ascidiacea			
Didemnum albidum	270	150	140
Botryllus schlosseri	280	140	160
Boltenia echinata	780	100	120

nd = nondetectable.
Unless otherwise stated, organisms were collected in the Gulf of Maine.
[a] Obtained from Pacific Bio-Marine, Venice, California.

small amounts of putrescine and spermidine. The hydrozoans contained considerable amounts of putrescine, spermidine, and spermine, and the ascidiaceans have more putrescine than either spermidine or spermine. In the phyla Nemertea and Annelida, there was considerable variation in the three amines in different species of the same class, which may be a reflection of their ecological diversity. The concentrations of polyamines in a marine organism appear to be related to the ecology of that organism. For example, more active animals such as *Nereis pelagica* and *Lineus bicolor* contained much higher amounts of the polyamines than did less active animals such as *Botryllus schlosseri* and *Metridium senile* (Table 1). These data corroborate the theory that polyamine content is related to the metabolic demand on the organism.

I. POLYAMINE BIOSYNTHESIS IN SEA URCHINS

Polyamine biosynthesis was studied in several species of developing sea urchins in order to understand better the relationship of polyamine synthesis to the development of a marine invertebrate (3, 4). Most striking, in relation to the pattern found in other organisms, was the presence in sea urchin gametes of large amounts of spermine coupled with low amounts of putrescine and spermidine. This contrasts with the pattern of polyamines present in both microorganisms and mammals. In general, microorganisms contain mainly putrescine and spermidine, and adult mammalian tissues contain rather similar concentrations of spermidine and spermine, with putrescine present in lower amounts (5). However, the putrescine concentration increases rapidly when tissues undergo growth processes, i.e., in regenerating rat liver (6, 7), in cardiac hypertrophy (8, 9), and in appropriate mammalian tissues after hormonal stimulation (10–15).

Since the polyamine biosynthetic pathway in sea urchins had not been explored previously, we wanted to ascertain whether it was similar to the pathway present in other organisms. We found that labeled ornithine was incorporated into labeled putrescine in sea urchin embryos, and, further, that ^{14}C-putrescine added just after fertilization, followed by a pulse of cold putrescine, resulted in rapid labeling of both the spermidine and spermine pools. The specific activity of the polyamines after such labeling is shown in Fig. 1. The low specific activity of spermine is a reflection of the very large pool of spermine present in these organisms.

Evidence accumulated to date indicates that the enzymes involved in spermidine and spermine synthesis in sea urchins are similar to those reported for amphibians (17) and mammals (18, 19) and differ from the bacterial system (5). In bacteria, the conversion of putrescine to spermidine involves two separate enzymatic reactions (5). The first is the enzymatic decarboxylation of S-adenosyl-L-methionine (SAM) to form carbon dioxide and 5′deoxy-5′-S-(3-methylthiopropylamine) sulfonium adenosine (decarboxylated SAM). This SAM decarboxylase requires Mg^{++} and contains

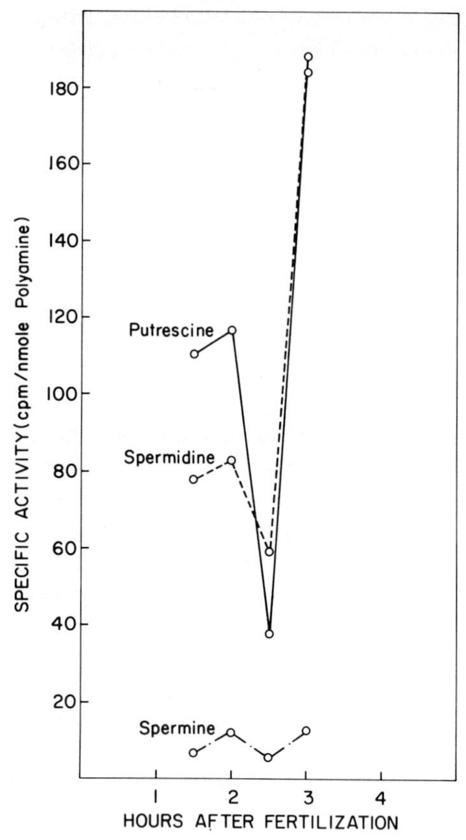

FIG. 1. Specific activities of ^{14}C-putrescine, ^{14}C-spermidine, and ^{14}C-spermine after incubating fertilized eggs with ^{14}C-putrescine (2mC) for 30 min, followed by a cold putrescine wash (10-fold greater concentration). Both the concentrations of amines and the radioactivity of each pool were determined by extraction of the amines into alkaline 1-butanol and separation by high-voltage electrophoresis (16). Each point represents the mean for three determinations.

pyruvate as a prosthetic group. A propylamine transferase then catalyzes the formation of spermidine from putrescine and a propylamine molecule which derives from decarboxylated SAM. In sea urchins as well as in mammals, at least in crude homogenates, there appears to be coupling of the decarboxylase and transferase. Utilizing an amino acid analyzer technique for the separation of polyamines (which is discussed elsewhere in this volume by Marton et al.), we have assayed SAM decarboxylase activity utilizing ^{14}C-putrescine and cold SAM as the substrates. We have found that SAM decarboxylase from sea urchins which has been purified through (a) G-100 Sephadex and (b) DEAE cellulose still exhibits coupling of the decarboxylation of SAM with the transfer of the propylamine moiety to putrescine to form spermidine (unpublished observations). Whether or not these two functions can be uncoupled after extensive purification, the important issue is the coupling of crude homogenates both in mammalian systems and in sea urchins. The rate-limiting step in spermidine or spermine synthesis appears to be the activity of SAM decarboxylase. Therefore, the most accurate estimates of spermidine and spermine synthesis can be obtained from the measurement of putrescine-stimulated SAM decarboxylase and spermidine-stimulated SAM decarboxylase, respectively. Under physiological conditions, changes in these enzyme activities appear to correlate well with changes in the pools of spermidine and spermine.

Through the use of the double-reciprocal plot, the apparent K_m for putrescine of SAM decarboxylase of sea urchins was determined. Saturating levels of SAM were used, and partially purified SAM decarboxylase preparations from *Lytechinus pictus* were utilized as the enzyme source in these assays. The K_m for putrescine under these conditions, 2.5×10^{-5} M, is an order of magnitude lower than the K_m for putrescine of this enzyme in rat liver. The calculated K_m for spermidine obtained from enzyme preparations of *L. pictus* was 6.9×10^{-4} M. This is similar to the K_m calculated for the partially purified enzyme from *Strongylocentrotus purpuratus* (5×10^{-4} M). In both cases, the values are lower than those obtained from enzyme preparations of rat liver. The lower amounts of putrescine and spermidine necessary to optimize their conversion to spermidine and spermine, respectively, may account for both the lower amounts of putrescine present in sea urchins and the higher levels of spermine that accumulate.

With these preliminary experiments, which indicate that the polyamine biosynthetic pathway in sea urchins is similar to that in other organisms, we have studied the pattern of enzymatic synthesis as well as polyamine levels in the developing sea urchin. Although there are some species-specific differences in temporal expression, there are not diverse differences, and therefore we will present the pattern in one sea urchin, *S. purpuratus*. Data on other species will be published elsewhere (3, 4).

II. ORNITHINE DECARBOXYLASE ACTIVITY IN DEVELOPING EMBRYOS OF *S. PURPURATUS*

It is of interest to study the early changes which occur after fertilization of sea urchin eggs, since synchrony of cell division is exhibited at least through the first two cell cycles. Interestingly, ornithine decarboxylase (ODC) activity exhibits cyclical alteration in the first 3 hr after fertilization (Fig. 2). There is a maxima of activity at ½ hr after fertilization with ODC

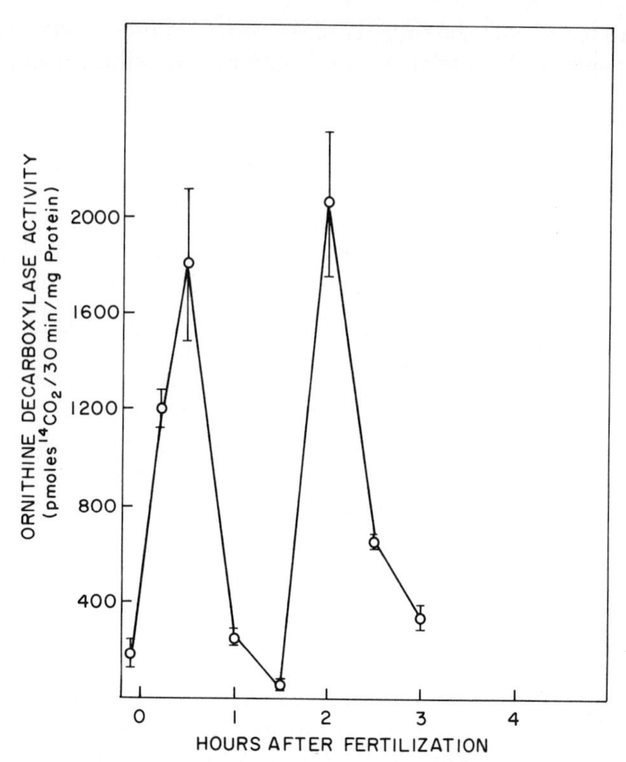

FIG. 2. Ornithine decarboxylase activity in early cleavage stages of *S. purpuratus*. Activity was determined by measuring the evolution of $^{14}CO_2$ from ornithine-1-^{14}C (11). Each point represents the mean ± S.E. for separate determinations on four pools of embryo.

activity ninefold greater than that present in the unfertilized eggs. Thereafter, the activity declines and exhibits a minima at 1-½ hr after fertilization. This is approximately the time of the first cell division. There is another maxima of ODC activity 2 hr after fertilization, at which time ODC activity was 14-fold above that detectable in unfertilized eggs. This maxima is then followed by another rapid decline in activity with low activity detectable 3 hr after fertilization, the time of the second cell division (Fig. 2). The relationship of these enzyme cycles to the phases of the cell cycle is hard to assess. The early cleavage stages of sea urchins, at least in *S. purpuratus,* reportedly do not have a G_1 phase, and exhibit a very short S phase (20). There is general agreement that the period of DNA synthesis is short and occurs early in interphase (21, 22). ODC activity appears to be enhanced prior to the onset of DNA synthesis.

The overall pattern of ODC activity at various developmental stages in

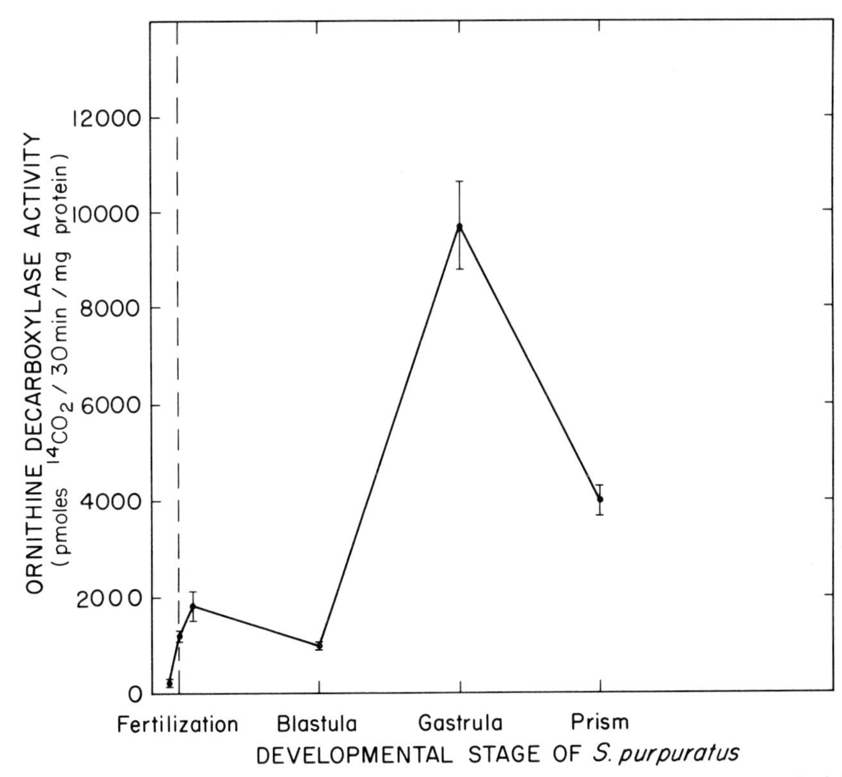

FIG. 3. Ornithine decarboxylase activity in developing embryos of *S. purpuratus*. Activity was determined by measuring the evolution of $^{14}CO_2$ from ornithine-1-^{14}C (11). Each point represents the mean ± S.E. of at least five determinations on at least five separate pools of embryos.

S. purpuratus indicates that its activity at blastulation is approximately fivefold that detectable in unfertilized eggs and, by gastrulation, ODC activity is about 50-fold that found in unfertilized eggs. In the early prism stage, ODC activity has declined from the very high level present at gastrulation, but it is still markedly elevated (Fig. 3).

III. PUTRESCINE- AND SPERMIDINE-STIMULATED S-ADENOSYL-L-METHIONINE DECARBOXYLASE ACTIVITIES IN *S. PURPURATUS* EMBRYOS

Early cyclical patterns were detected for both putrescine- and spermidine-stimulated SAM decarboxylase in *S. purpuratus* (Fig. 4). However, the cyclical nature of spermidine-stimulated SAM decarboxylase is much less

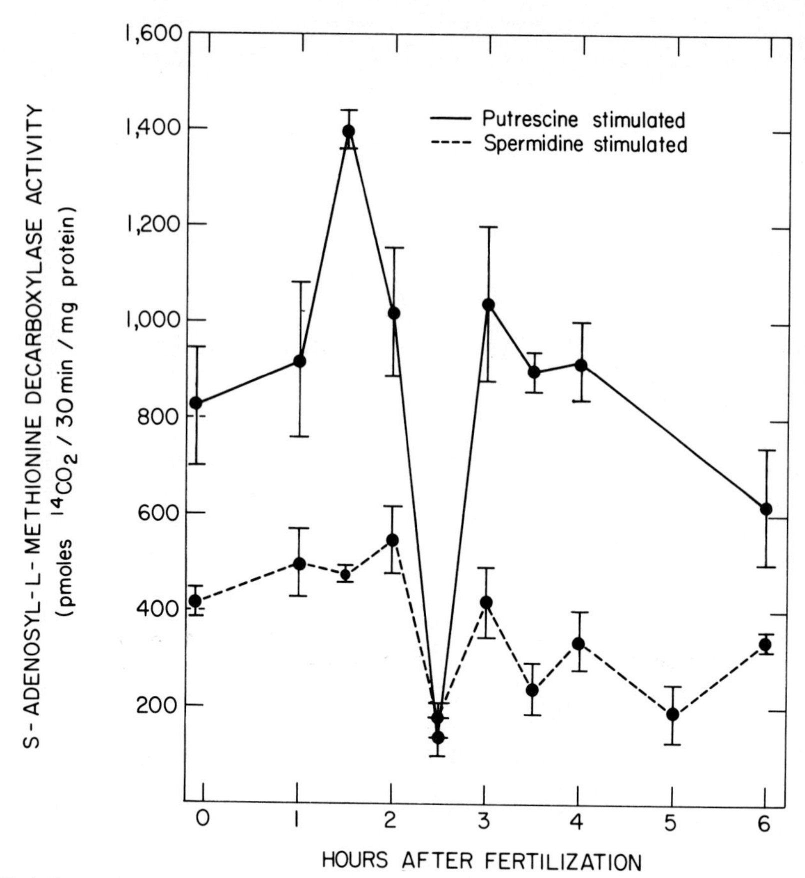

FIG. 4. Putrescine- and spermidine-stimulated SAM decarboxylase activities in early cleavage stages of *S. purpuratus*. Activity was assayed by measuring $^{14}CO_2$ released from ^{14}COOH-SAM in the presence of the appropriate amine (18). Each point represents the mean ± S.E. for four separate determinations.

pronounced than for putrescine-stimulated SAM decarboxylase. The first maxima of activity occurred 1-½ hr after fertilization, again the time of the first cell division. Thereafter there is a rapid decline in both enzymatic activities and at 2-½ hr after fertilization, there is no difference in the activity of putrescine-stimulated versus spermidine-stimulated SAM decarboxylase. Another less dramatic peak of activity occurs at 3 hr after fertilization for both these enzymes. Note that these enzymes exhibit maxima after the maximum for ODC activity. Therefore, ODC activity is minimal at the time of cell division, and putrescine- and spermidine-stimulated SAM decarboxylase activities are maximal at the time of cell division.

TABLE 2. *S-Adenosyl-L-methionine decarboxylase activity of developing embryos of* S. purpuratus

Stage	Putrescine -stimulated	Spermidine-stimulated
	(pmoles $^{14}CO_2$/30 min/mg protein)	
Unfertilized egg	830 ± 129	466 ± 80
Fertilized (1 hr)	920 ± 166	537 ± 82
Blastula	2,740 ± 210	646 ± 85
Gastrula	3,780 ± 446	910 ± 105
Prism	2,030 ± 145	456 ± 38

Activities were assayed by measuring $^{14}CO_2$ evolution from ^{14}COOH-SAM in the presence of the appropriate amine (18). Each point represents the mean ± S.E. for four separate determinations.

Enzymatic expression for both these enzymes at various developmental stages of *S. purpuratus* is rather similar. That is, there is maximal enzymatic activity present at the time of gastrulation for both putrescine- and spermidine-stimulated SAM decarboxylase and a decline thereafter (Table 2). There is a difference in the ratio of the activities at blastulation; for putrescine-stimulated SAM decarboxylase there is nearly a threefold increase in activity by blastulation, whereas there is only a slight increase in spermidine-stimulated SAM decarboxylase by the blastula state. These data correlate well with the enhanced levels of spermidine that occur after fertilization (Table 3). At the two to four cell stage, there is approximately 10 times more spermine than spermidine, whereas by gastrulation the concentration of spermidine doubles while the concentration of spermine remains essentially constant.

TABLE 3. *Polyamine content of developing embryos of* S. purpuratus

Stage	Putrescine	Spermidine	Spermine
	(nmoles/mg protein)		
Sperm (5)	0.61 ± 0.05	0.38 ± 0.12	4.84 ± 1.06
Eggs (5)	0.15 ± 0.03	0.16 ± 0.02	5.20 ± 0.30
2–4 cell (1)	0.77	0.63	7.03
Blastula (3)	0.58 ± 0.02	0.67 ± 0.15	6.47 ± 0.82
Gastrula (3)	0.68 ± 0.25	1.15 ± 0.15	5.63 ± 1.33

(n) = Number of separate pools assayed.
Pools of embryos were assayed for endogenous levels of putrescine, spermidine, and spermine, by extraction of amines into alkaline 1-butanol and separation by high-voltage electrophoresis as described previously (16). Data are presented as the mean ± S.E. for duplicate determinations of the number of separate pools shown in parentheses.

IV. VARIATIONS IN THE POLYAMINE POOLS DURING EARLY CLEAVAGE STAGES

Preliminary results with *S. purpuratus* embryos indicate marked fluctuations in the levels of polyamines during early cleavage stages. The most striking event appears to be the marked increases in putrescine, spermidine,

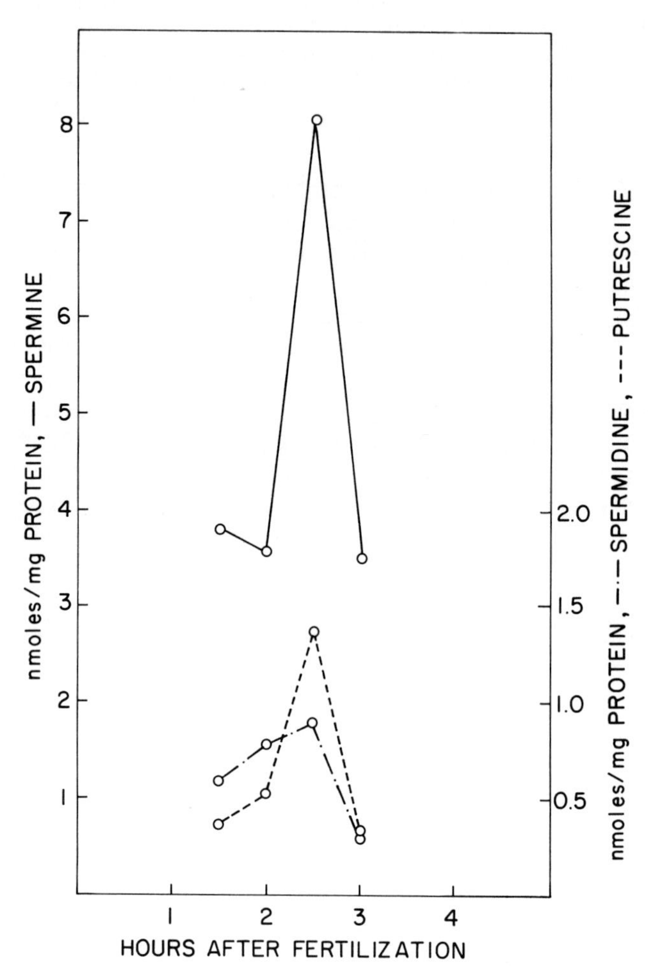

FIG. 5. Fluctuations in the levels of putrescine, spermidine, and spermine in early cleavage stages of *L. pictus*. Pools of embryos were assayed for endogenous levels of putrescine, spermidine, and spermine by extraction of these amines into alkaline 1-butanol and separation by high-voltage electrophoresis (16). Each point represents the mean for two or more determinations.

and spermine levels at 2-½-hr postfertilization, followed by a rapid decline in the concentrations of all three amines. The maxima are just prior to second cleavage, and the minima occur at cleavage.

This same cyclical pattern in polyamine levels is exhibited by *L. pictus* embryos. Figure 5 illustrates a typical pattern, and the experiment has been repeated with similar results. Again, there is a marked increase just prior to second cleavage followed by a rapid decline. These rapid fluctuations in polyamine levels may mean that there is either active secretion of polyamines at discrete times or active degradation.

In summary, then, both polyamine biosynthesis and the levels of polyamines in early cleavage stages of sea urchins exhibit cyclical variations. There are numerous reports of cyclical variations of metabolic parameters during the early cleavage stages of sea urchins (23–25) and these variations occur at a definite time in relation to cell division. For instance, protein synthesis in sea urchins in early cleavage is elevated during prometaphase-metaphase and is depressed during anaphase-telophase of the next mitotic division (25). Studies conducted on the cell cycle in a variety of cells capable of being synchronized in some manner indicate that the above-mentioned cell cycle stage specificity of synthesis is a general phenomenon. The interesting physiological questions that this chapter raises have to do with the elucidation of the precise role(s) played by the polyamines. What physiological significance can be found for the early synthesis of polyamines and then the equally rapid decrease in the levels just at the time of cell division? Further studies are in progress to ascertain whether these dramatic fluctuations in polyamine levels are caused by secretion processes or by metabolic enzymes that degrade polyamines. Since little is known about the metabolic fate of polyamines in mammalian systems, an understanding of metabolism in sea urchins might serve as a clue to possible mechanisms in mammalian systems.

REFERENCES

1. Ogata, A., and Komada, T., *J. Pharm. Soc. Jap.*, **63,** 653 (1943).
2. Ackermann, D., and Janka, R., *Hoppe-Seyler's Z. Physiol. Chem.*, **296,** 279 (1954).
3. Manen, C. A., and Russell, D. H., *J. Embryol. Exp. Morphol.*, in press.
4. Manen, C. A., and Russell, D. H., in preparation.
5. Tabor, H., and Tabor, C. W., *Pharmacol. Rev.*, **16,** 245 (1964).
6. Dykstra, W. G., Jr., and Herbst, E. J., *Science*, **149,** 428 (1965).
7. Jänne, J., and Raina, A., *Acta Chem. Scand.*, **22,** 1349 (1968).
8. Russell, D. H., Shiverick, K. T., Hamrell, B. B., and Alpert, N. R., *Amer. J. Physiol.*, **221,** 1287 (1971).
9. Feldman, M. J., and Russell, D. H., *Amer. J. Physiol.*, **222,** 1199 (1972).
10. Pegg, A. E., and Williams-Ashman, H. G., *Biochem. J.*, **109,** 32P (1968).
11. Russell, D. H., and Snyder, S. H., *Endocrinology*, **84,** 223 (1969).
12. Jänne, J., and Raina, A., *Biochim. Biophys. Acta*, **174,** 766 (1969).

13. Russell, D. H., Snyder, S. H., and Medina, V. J., *Endocrinology,* **86,** 1414 (1970).
14. Russell, D. H., and Taylor, R. L., *Endocrinology,* **88,** 1397 (1971).
15. Russell, D. H., and Potyraj, J. J., *Biochem. J.,* **128,** 1109 (1972).
16. Russell, D. H., Medina, V. J., and Snyder, S. H., *J. Biol. Chem.,* **245,** 6732 (1970).
17. Russell, D. H., *Proc. Nat. Acad. Sci. U.S.A.,* **68,** 523 (1971).
18. Pegg, A. E., and Williams-Ashman, H. G., *J. Biol. Chem.,* **244,** 682 (1969).
19. Feldman, M. J., Levy, C. C., and Russell, D. H., *Biochemistry,* **11,** 671 (1972).
20. Hinegardner, R. T., Rao, B., and Feldman, D. E., *Exp. Cell Res.,* **36,** 53 (1964).
21. Nemer, M., *J. Biol. Chem.,* **237,** 143 (1962).
22. Ficq, A., Aiello, F., and Scarano, E., *Exp. Cell Res.,* **29,** 128 (1963).
23. Nagano, H., and Mano, Y., *Biochim. Biophys. Acta,* **157,** 546 (1968).
24. Løvtrup, S., and Iversen, R. M., *Exp. Cell Res.,* **55,** 25 (1969).
25. Mano, Y., *Devel. Biol.,* **22,** 433 (1970).

Polyamines in Normal and Neoplastic Growth. Edited by D. H. Russell. Raven Press, New York © 1973.

The Stimulation of RNA Synthesis in Mature Amphibian Oocytes by the Microinjection of Putrescine

C. C. Wylie and Diane H. Russell

Department of Anatomy, University College London, Gower Street, London WC1E 6BT, England, and Laboratory of Pharmacology, Baltimore Cancer Research Center, National Cancer Institute, Baltimore, Maryland 21211

I. INTRODUCTION

Polyamines have been known for some time to be implicated in rapid tissue growth: they are used as growth factors in tissue culture (1), and their concentration rises in a variety of conditions of rapid growth, e.g., regenerating rat liver (2, 3), early embryogenesis of the chick (4) and rat (5).

More recently, several lines of evidence suggest that polyamines are involved in ribosomal RNA (rRNA) metabolism. The anucleolate mutant of *Xenopus laevis,* unlike the wild type, does not start synthesizing rRNA at the gastrula stage of embryogenesis. Also in contrast to the wild type, ornithine decarboxylase activity (6) and the level of spermidine (7) do not increase after gastrulation. ^{3}H-Putrescine has been shown autoradiographically to be localized at the nucleolus in cultured *X. laevis* liver cells (8). Furthermore, in rat liver nucleoli, polyamines can raise the activity of RNA polymerase 1 (9).

There is thus considerable circumstantial evidence that polyamines are involved in rRNA metabolism. However, there is limited evidence that manipulation of polyamine levels can directly affect RNA metabolism in eukaryotes. In this chapter experiments in which the intracellular concentrations of certain polyamines are artificially raised will be described. This has been done by microinjection of ^{3}H-uridine with and without polyamines into the mature amphibian egg.

The amphibian egg is peculiarly suited to this type of experiment: it synthesizes RNA at a low rate, and yet contains all the cofactors for rapid RNA synthesis, which can be switched on simply by an external, hormonal stimulus (10), and the microinjection of polyamines into this very large, single cell avoids the difficulties of permeability barriers.

In these experiments ^{3}H-uridine, with and without putrescine, were injected into small batches of mature *X. laevis* eggs, and the newly synthesized RNA prepared and separated by agarose gel electrophoresis. There was found to be a three- to fivefold stimulation of incorporation into total cell RNA with two concentrations of putrescine, and an even greater stimulation of incorporation into rRNA.

II. MATERIALS AND METHODS

Adult female *X. laevis* toads were sacrificed and the ovaries dissected out and placed in Gurdon-modified Barth's saline (GMB) (11). Single, mature oocytes were hand dissected from the ovary and placed in sterile Petri dishes on a surface of 2% Bacto-agar in GMB. Batches of 20 to 30 eggs each were injected with 125 nl of radioactive solution. The injected eggs were left in GMB overnight, then washed, harvested, and frozen at −70°C.

The injectates were prepared by drying 100 μC aliquots of ^{3}H-uridine (specific activity 29 C/mmole, supplied as aqueous solution, 1 mC/ml by the Radiochemical Centre Amersham) in small plastic tubes. The isotope was then taken up in one-tenth of its original volume in injecting saline (12) alone, or in injecting saline containing 0.8 M or 4.0 M putrescine. This solution was drawn into a fine glass needle for microinjection.

Total cell RNA was prepared by the method of Brown and Littna (13). Each batch of eggs was thawed and placed into 0.4 ml of 0.1 M sodium acetate plus 1% sodium dodecyl sulfate plus 10 μg/ml polyvinyl sulfate, pH 5.0, and hand homogenized. The homogenate was kept at 37°C for 3 min, and the rest of the procedure carried out in an ice bath. A small aliquot of the total homogenate was counted in an Intertechnique SL40 liquid scintillation counter. Phenol extractions were carried out until there was a clear interphase between the phenol and aqueous layers, followed by three extractions with 1% isoamyl alcohol in chloroform. The RNA in the aqueous layer was precipitated with two volumes of absolute alcohol plus 0.1 volume of MNaCl.

The RNA was then taken up in 0.005 M Tris HCl plus 10^{-4} M EDTA plus 10 μg/ml polyvinyl sulfate buffer (TEP_1), for electrophoresis in 1.8% agarose gels (14, 15).

III. RESULTS

Figures 1a and 2a show two agarose gel electrophoresis profiles of RNA from eggs injected with ^{3}H-uridine in saline. The radioactivity in each slice of these gels is expressed as a percentage of the total radioactivity (c.p.m.) in the electrophoresis gel, to allow for easy comparison of the gel profiles. The positions of the major peaks are found by U.V. photoprinting of each gel before sectioning and counting. It can be seen from these two profiles that the mature eggs are making small amounts of rRNA, approximately 11 and 12% of the total RNA, respectively. It is difficult to say from these profiles whether the eggs are making 5s rRNA or 4s tRNA, because this area of the electrophoresis gel is obscured by several peaks of low molecular weight RNA. In these saline-injected controls, the radioactivity incorporated into RNA represents 3.2 and 6.1%, respectively, of the radioactivity in the total homogenates.

Figures 1b, 1c, 2b, and 2c show radioactivity profiles of RNA from eggs injected under identical conditions and at the same time as the control for each experiment, the only difference being that the injection saline contained either 0.8 M (Figs. 1b and 2b) or 4.0 M (Figs. 1c and 2c) putrescine. Assuming the mean diameter of the mature *X. laevis* egg to be about 1.25 mm, this will raise the intracellular levels of putrescine by about 98 mM and 490 mM, respectively.

This, however, is by no means the true intracellular concentration, since there is leakage of the injectate from the wound immediately after injection. Comparing the known injected radioactivity (approximately 555,000 c.p.m. allowing for counting efficiency) with the radioactivity in the total homogenate at the end of the incubation period (see Table 1), the true intracellular concentration is probably from 100 to 500 times lower than that calculated in each case. That is, in reality the intracellular concentration is enhanced in the range of .2 to 5 mM.

Several differences between the RNA profiles of the polyamine-injected eggs and the saline-injected eggs can be seen. (a) The 28s and 18s peaks are much larger in the putrescine-injected eggs. In the first experiment they represent 27 and 17%, respectively, of the newly synthesized RNA, and in the second experiment 36% and 30%. (b) There is a peak of radioactivity between the origin and the 28s rRNA peak, which is either very small or absent in the controls. In order to find the rough s-value of this peak, the molecular weights of 5s HeLa cell rRNA (16), *X. laevis* 18s and 28s rRNA (17) were taken from the literature and a graph plotted of $\log_{10}$ molecular

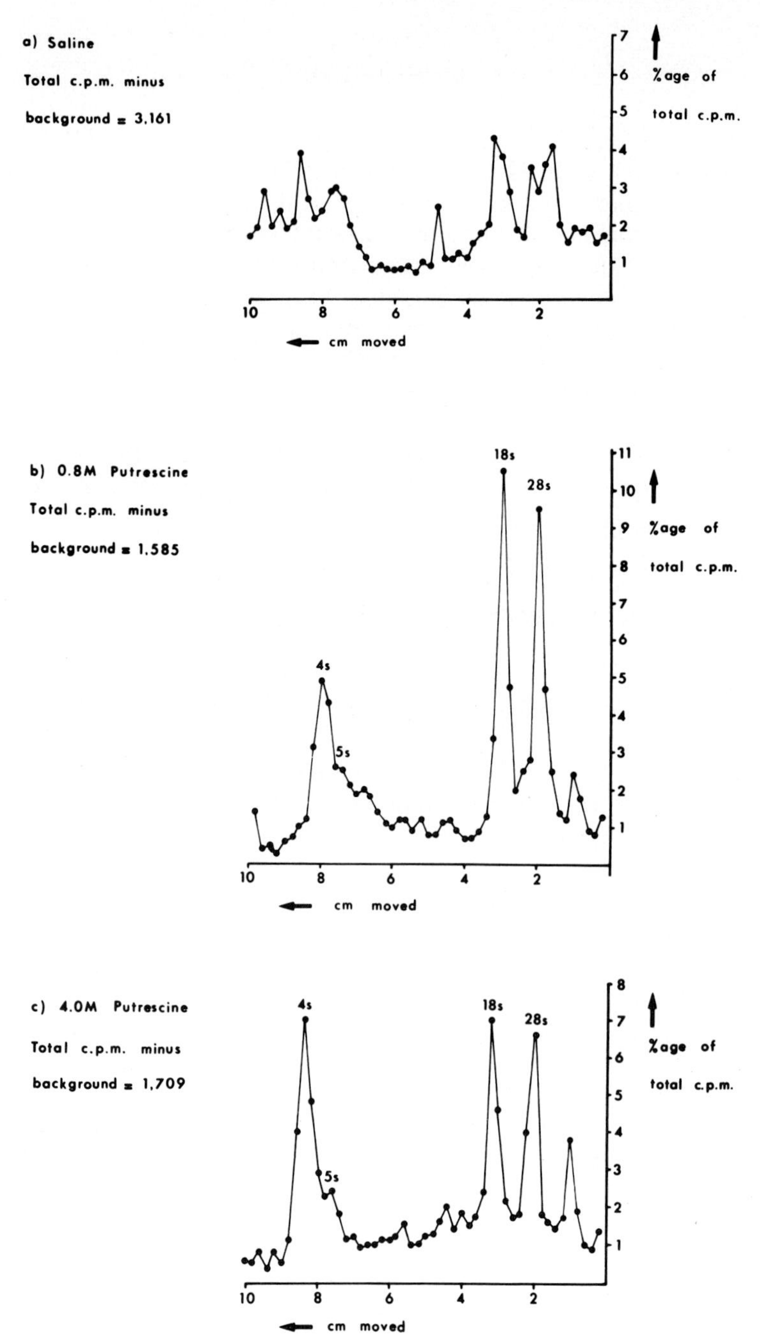

FIG. 1. Agarose gel electrophoresis profiles, showing radioactivity incorporated into the various species of total cell RNA. Each point represents the c.p.m. in 2×1 mm slices of gel, expressed as a percentage of the total c.p.m. in the gel.

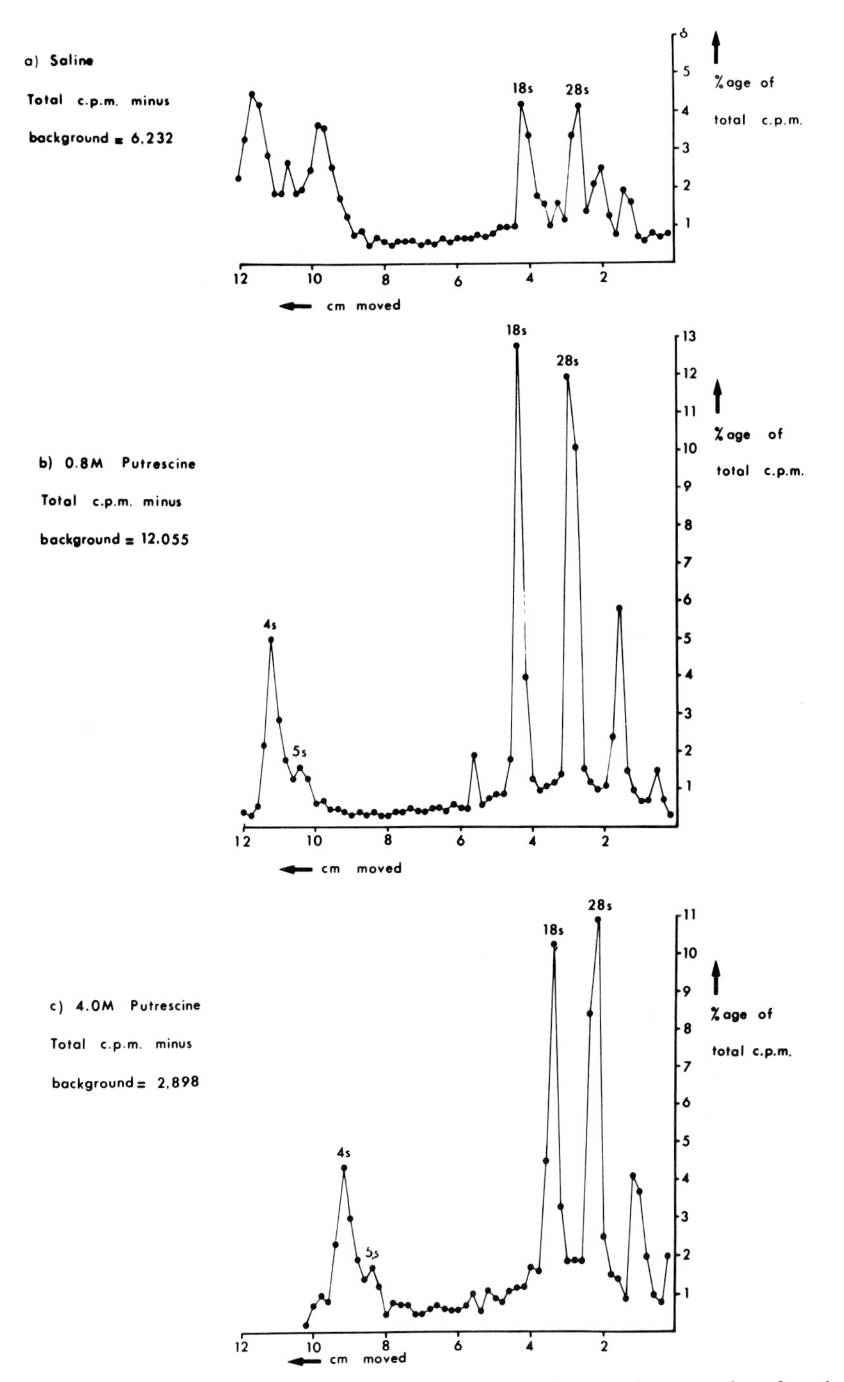

FIG. 2. A repeat of the experiment shown in Fig. 1. The conditions and presentation of results are exactly the same in both experiments.

weight versus distance moved in 1.8% agarose gels (Fig. 3). This was found to be linear. When the position of the heavy RNA peak was plotted, the observed molecular weight (2.5×10^6) was found to be exactly that of the known 40s precursor to rRNA in *X. laevis* kidney cells (18). The obvious possibility is that this peak is the heavy molecular weight precursor to 28s and 18s rRNA. One other possibility is that it could be labeled DNA. However, all efforts to label this peak with 14C thymidine have failed (3). Newly synthesized 5s and 4s RNA are far more obvious than in the controls, either due to a suppression of synthesis of the other low molecular weight components visible in the controls, or a stimulation of synthesis to levels which obscure these other peaks.

A fuller quantitative comparison of the putrescine-injected egg RNA and

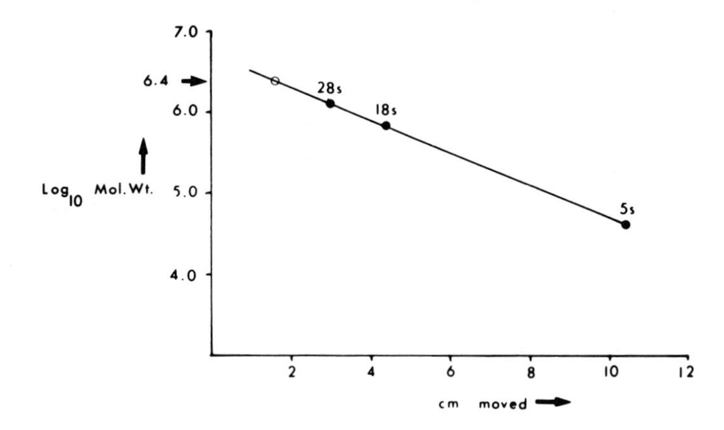

RNA Species		Mol. Wt.	Log_{10} Mol. Wt.	Reference
X. laevis	40s	2.6×10^6	6.42	Loening et al. (1969)
"	28s	1.5×10^6	6.18	Loening (1968)
"	18s	0.7×10^6	5.85	.. ..
Mammalian	5s	3.7×10^4	4.57	Knight & Darnell (1967)

FIG. 3. Graph to show the relationship between the distance moved in 1.8% agarose gels and $\log_{10}$ of the molecular weights of *X. laevis* rRNA. The distances moved are taken from Fig. 2b, and the molecular weights from the literature cited. The open circle represents the putative ribosomal precursor peak (for details see text).

the saline-injected egg RNA is shown in Table 1. From the table, the following points should be mentioned. (a) In each experiment, the percentage of the total homogenate radioactivity found in the purified RNA is stimulated more by the lower concentration of putrescine than the higher. Putrescine was found to stimulate RNA synthesis by 2 to 4.7 times that in the controls. (b) When the radioactivity in the 28s and 18s peaks alone was expressed as a percentage of the total homogenate radioactivity, in each experiment, it was found that incorporation into these components was stimulated by 4.6 to 11.7 times that found in the controls. In other words, the stimulation of RNA synthesis under the influence of putrescine is selective. As is the case with total RNA, the stimulation of incorporation into rRNA is higher with the lower concentration of putrescine in both experiments. (c) It is not clear from the experiments whether the overall stimulation of incorporation into RNA is entirely due to the stimulation of rRNA and tRNA synthesis: if the radioactivity in the stable components above the level of heterogeneous RNA is totaled and deducted from the total RNA radioactivity, then the percentage incorporation into the heterogeneous RNA alone is still slightly higher than the percentage incorporation into the total RNA of the controls (see column 7 of Table 1). It is possible, therefore, that the synthesis of at least some species of heterogeneous RNA is stimulated by putrescine, although more experiments will have to be done to establish this point fully. (d) Note that in both experiments the total homogenate radioactivity is

TABLE 1. *Quantitative data from the two experiments described in the text*

Treatment	Number of eggs	c.p.m./egg		%age of total homogenate c.p.m. in:			%age of RNA in 28s + 18s	
		total homog.	RNA	total RNA	28s + 18s RNA	heterogeneous RNA		
Saline	29	7.882	255	3.23	0.35		10.7	Expt. 1)
0.8M Put.	29	872	133	15.31	4.09	8.97	26.7	
4.0M Put.	17	2.052	235	11.47	1.96	6.70	17.1	
Saline	22	18.694	1.133	6.06	0.74		12.2	Expt. 2)
0.8M Put.	26	4.420	790	17.87	6.49	7.75	36.3	
4.0M Put.	30	1.678	193	11.52	3.41	5.69	29.6	

much lower in the putrescine-injected eggs than in the saline-injected eggs. This reflects the fact that there is a certain amount of leakage through the injection wound in the presence of putrescine, and virtually none in the presence of saline alone. This leakage occurs immediately after injection and so, presumably, does not materially influence the incorporation figures.

IV. DISCUSSION

There is considerable circumstantial evidence that RNA metabolism is influenced by the intracellular levels of certain polyamines. However, direct evidence involving the manipulation of polyamine levels in eukaryote cells is difficult to obtain. For example, in rapidly growing systems, the polyamine levels are already high [approximately 5 mM in rat uterine decidual tissue (5)], and to add polyamines in this situation would be to stimulate an already maximally stimulated system. Also, it is not known for certain whether the uptake of polyamines into cells is passive or an active process subject to intracellular control.

Growth of the frog oocyte takes place while the nucleus is arrested in the first meiotic prophase. During this period of intense metabolic activity, stores of ribosomal, transfer, and informational RNA are laid down, which enable the fertilized egg to develop through cleavage. Enough rRNA to last the embryo until the swimming tadpole stage (19, 20) is synthesized on multiple nucleoli during this period (21). In the full-grown oocyte, however, there is little RNA synthesis, the oocyte remaining in a state of near dormancy for weeks, or even months, before a hormonal stimulation from the anterior pituitary starts off a series of events known as the "maturation" of the oocyte (10). Considerable RNA synthesis has been reported to take place in *X. laevis* during the period of maturation (13, 22). The full-grown oocyte therefore represents a eukaryote cell which is potentially capable of synthesizing a great deal of RNA, both heterogeneous and ribosomal, but is, in fact, only doing so slowly. It is therefore an attractive target for the application of polyamines in a search for their role in RNA metabolism. It has the further advantage of being a very large cell into which test solutions can be injected, thus obviating any permeability barriers.

The experiments reported here, although of a preliminary nature, demonstrate that by artificially raising the levels of putrescine, a stimulation of RNA synthesis for a period of 18 hr is obtained. Furthermore, by preparing and separating the RNA species, the stimulation is shown to be selective, rRNA synthesis being stimulated more than the overall stimulation of RNA synthesis.

Obviously, further work is necessary and is in progress to find the mechanism by which this stimulation takes place. There are only two basic ways in which the transcription rate of rRNA can be raised: either each pre-existing gene is transcribed more often or the number of ribosomal genes is increased. It must be borne in mind that there are some 2,000 times the normal haploid number of ribosomal genes present in the full-grown frog oocyte (23), and it may be these genes which are reactivated. There is some evidence that a new gene amplification process may be involved (Wylie, *unpublished observations*).

An alternative mechanism by which the observed results may be obtained is by conservation of RNA by preventing its breakdown. This may be via an inhibition of nucleases or by some protection mechanism. There are several arguments for or against such a possibility; e.g., a selective action of putrescine on different species of RNA has been demonstrated, a difficult fact to reconcile with a theory of RNAase inhibition. On the other hand, the peak of ribosomal precursor RNA is rather higher than expected, which could be due to a partial inhibition of its further processing.

Whatever the mechanisms involved, these results suggest that putrescine occupies an important role in RNA metabolism, particularly rRNA, of the frog oocyte. Since this is probably true of mammalian cancer cells, among other rapidly growing systems, this may open a new avenue of research into the control of cell growth.

V. SUMMARY

There is considerable circumstantial evidence that polyamines are involved in cell growth and ribosomal RNA synthesis. However, there is little evidence of direct experimental manipulation of polyamine levels exhibiting an effect on RNA metabolism in eukaryotic cells. In this chapter we report the effect of putrescine in full-grown oocytes of *Xenopus laevis*. Putrescine, when microinjected in two concentrations together with ^{3}H-uridine, stimulates RNA synthesis at two to five times the levels seen in saline-injected controls, during an 18-hr period. Moreover, this stimulation is selective, the synthesis of 18s and 28s rRNA being stimulated five to 12 times more than that seen in controls.

ACKNOWLEDGMENTS

Thanks are due to Professor J. Z. Young and Drs. Ruth Bellairs and Martin Evans for helpful discussion in the course of this work. Excellent technical assistance was provided by Mrs. M. Reynolds and Mr. R. F. Moss.

REFERENCES

1. Ham, R. G., *Biochem. Biophys. Res. Comm.*, **14**, 34 (1964).
2. Jänne, J., and Raina, A., *Acta Chem. Scand.*, **20**, 1174 (1966).
3. Dykstra, W. G., and Herbst, E. J., *Science*, **149**, 428 (1965).
4. Caldarera, C. M., Barbiroli, B., and Moruzzi, G., *Biochem. J.*, **97**, 84 (1965).
5. Russell, D. H., and McVicker, T. A., *Biochim. Biophys. Acta*, **259**, 247 (1972).
6. Russell, D. H., *Ann. N.Y. Acad. Sci.*, **171**, 772 (1970).
7. Russell, D. H., *Proc. Nat. Acad. Sci. U.S.A.*, **68**, 523 (1971).
8. Gfeller, E., and Russell, D. H., *Z. Zellforsch.*, **120**, 321 (1971).
9. Russell, D. H., Levy, C. C., Taylor, R. L., Gfeller, E. E., and Sterns, D. N., *Fed. Proc.*, **30**, 1093 (1971).
10. Smith, L. D., and Ecker, R. E., *Curr. Top. Devel. Biol.*, **5**, 1 (1970).
11. Gurdon, J. B., *J. Embryol. Exp. Morphol.*, **20**, 401 (1968).
12. Gurdon, J. B., Lane, C. D., Woodland, H. R., and Marbaix, G., *Nature*, **233**, 177 (1971).
13. Brown, D. D., and Littna, E., *J. Mol. Biol.*, **8**, 669 (1964).
14. Evans, M. J., Ph.D. dissertation, University of London, (1969).
15. Wylie, C. C., *J. Embryol. Exp. Morphol.*, **28**, 367 (1972).
16. Knight, E., and Darnell, J., *J. Mol. Biol.*, **28**, 491 (1967).
17. Loening, U. E., *J. Mol. Biol.*, **38**, 355 (1968).
18. Loening, U. E., Jones, K., and Birnstiel, M. L., *J. Mol. Biol.*, **45**, 353 (1969).
19. Brown, D. D., and Gurdon, J. B., *Proc. Nat. Acad. Sci. U.S.A.*, **51**, 139 (1964).
20. Brown, D. D., and Gurdon, J. B., *J. Mol. Biol.*, **19**, 399 (1966).
21. Davidson, E. H., *Gene activity in early development*, Academic Press, New York, 1968.
22. Brown, D. D., and Littna, E., *J. Mol. Biol.*, **20**, 81 (1966).
23. Brown, D. D., and Dawid, I. B., *Science*, **160**, 272 (1968).

Polyamines in Normal and Neoplastic Growth. Edited by
D. H. Russell. Raven Press, New York

Interrelations of S-Adenosylmethionine and Polyamines in *Escherichia coli* K12

Ching-Hsiang Su and Seymour S. Cohen*

Department of Therapeutic Research, University of Pennsylvania School of Medicine, Philadelphia, Pennsylvania 19104

I. INTRODUCTION

The two major polyamines in *Escherichia coli* are putrescine and spermidine. The current state of knowledge about the physiological functions of these compounds has been reviewed recently (1). There are two pathways for the biosynthesis of putrescine (12): one by decarboxylation of ornithine and the other by decarboxylation of arginine followed by hydrolysis of the resultant agmatine to yield putrescine and urea. The ornithine decarboxylase which catalyzes the first pathway was shown to be regulated both by feedback and repression mechanisms (19).

In *E. coli,* as well as in mammalian cells, putrescine and methionine function as precursors for spermidine synthesis (4, 9, 10, 13, 14, 17, 18). Methionine is first activated by S-adenosylmethionine (SAM) synthetase to form SAM, which is decarboxylated to form decarboxylated SAM. The propylamine group is then transferred to putrescine to form spermidine. The last two enzymes, i.e., SAM decarboxylase and propylamine transferase, have been extensively purified and studied. Wickner et al. (20) showed that SAM decarboxylase from *E. coli* W has a molecular weight of 113,000. The enzyme requires magnesium for activity and contains a covalently linked pyruvate moiety. The purified transferase of *E. coli* W does not require any cofactor for activity (16). No information on the regulation of these enzymes has been reported. The availability of *E. coli* mutants possessing different levels of SAM synthetase activity enabled us to study the effects of the SAM concentration on spermidine biosynthesis and also on the regulation of SAM

* Present address: Department of Microbiology, University of Colorado School of Medicine, Denver, Colorado 80220

decarboxylase by the end product spermidine. As has been shown with ornithine decarboxylase (19), SAM decarboxylase is also regulated by feedback and repression mechanisms.

II. MATERIALS AND METHODS

All strains used are derivatives of *E. coli* K12 (5, 6, 15). The standard minimum medium was a modified Davis-Mingioli medium (5) containing 7 g K_2HPO_4, 3 g KH_2PO_4, 1 g $(NH_4)_2SO_4$, 0.1 g $MgSO_4 \cdot 7H_2O$, and 1 g dextrose per liter of distilled water. Cells were routinely grown at 37°C with aeration.

For enzyme assays, 1 g of twice-washed cells harvested in the early log phase was ground with 2.5 g of alumina in the cold and extracted with 10 ml of buffer containing 1 mM EDTA, 10 mM potassium phosphate, pH 7.4, and 0.1 M KCl. The extract was centrifuged at 15,000 rpm for 10 min in a Sorvall RC-2B refrigerated centrifuge. The clear supernatant was used directly for enzyme assays.

The SAM decarboxylase assay was based on measurements of the release of $^{14}CO_2$ from SAM-^{14}COOH. The standard reaction medium contained 100 μmoles of $MgCl_2$, 40 μmoles of potassium phosphate of pH 7.4, enzyme, SAM-^{14}COOH (7.3 μC/μmole), and water in a total volume of 1.0 ml. Reactions were carried out in 25-ml Erlenmeyer flasks with a center well into which 0.2 ml of 0.2 M KOH was placed to trap released CO_2. The flasks were stoppered with rubber caps. After incubation at 37°C for 30 min, 0.1 ml of 2 N $HClO_4$ was injected to stop the reaction. After an additional 20 min of incubation, the KOH solution was pipetted into 15 ml of Aquasol in a counting vial. The well was rinsed with 0.8 ml of water and the wash was added to the counting solution. The radioactivity was measured in a Packard Tri-Carb liquid scintillation spectrometer.

For spermidine and putrescine analyses, 5-ml aliquots of the log phase culture having an absorbancy of about 1.2 at 550 nm measured in a Gilford model 240 spectrophotometer (1 absorbance unit is equivalent to 1.5 mg of wet cells) were passed through a Millipore filter. The membrane was immediately placed in a vial containing 1 ml of 0.2 N perchloric acid and extracted overnight at 4°C. Two-tenths ml of the extract was used for polyamine analysis as described by Dion and Herbst (3), except that the thin-layer chromatography plates were developed twice in ethyl acetate–cyclohexane (30:60).

The protein concentrations of the crude extracts were estimated by a micro-Biuret method (21) with crystalline bovine serum as the standard. SAM-^{14}COOH was obtained from New England Nuclear Corp.; spermidine trihydrochloride, putrescine dihydrochloride, 5-dimethylamino-1-naphthalene sulfonyl chloride (dansyl chloride), and L-methionine from Calbiochem. Other reagents were the purest available grade from standard commercial sources.

III. RESULTS

A. Polyamine levels in SAM synthetase mutants

The two types of mutants of *E. coli* K12 used in this experiment have been described previously (5, 15). Both classes are derepressed for methionine biosynthetic enzymes, but the met J mutants (D8, E12, and E31) are also derepressed for SAM synthetase (15), although the mutants of other classes (E4, D7, and E40) have low levels of SAM synthetase (5). The intracellular SAM pools of met J mutants are about twice the amounts of the wild type, whereas E4 has about 50% of the wild type, and D7 and E40 have about 10% (R. C. Greene, *unpublished data*). Since SAM is the precursor for spermidine biosynthesis, the polyamine content of these mutants was analyzed to determine if there is a correlation between the level of the SAM pool and the intracellular spermidine concentration. Table 1 shows that D7 and E40, which have the smallest SAM pools, have the least amount of intracellular spermidine. E4 occupies an intermediate position with pool sizes of SAM and spermidine higher than those of D7 and E40 but still consistently lower than those of the wild type. In contrast, D8, E12, and E31, the high SAM strains, have much higher spermidine contents. In this experiment E31 had a higher spermidine concentration than that of D8 and E12. However, in other experiments, the met J mutants contained approximately the same amounts of spermidine. This result indicates that the SAM and the

TABLE 1. *Spermidine and putrescine content*

Strains	Spermidine (μmoles/g wet cells)	Putrescine (μmoles/g wet cells)
Wild type		
K12	2.7	10.3
Low SAM strains		
E4	1.3	11.4
D7	1.1	12.7
E40	0.8	14.3
High SAM strains		
D8	4.6	8.8
E12	4.4	7.5
E31	6.0	7.2

spermidine pools increase in parallel fashion. Table 1 also shows that the low SAM strains have slightly elevated levels of putrescine. A decrease in putrescine was observed in the high SAM strains, suggesting that spermidine exerts a fine control over putrescine biosynthesis by feedback and repression mechanisms, as described by Tabor and Tabor (19).

B. Spermidine and the growth of the mutants

Inouye and Pardee (8) have shown that polyamines affect cell division. Also, putrescine-deficient mutants are markedly slowed in cell growth in the absence of exogenous polyamine (2, 7). Moreover Tabor et al. (19) showed that under severe ornithine-arginine restriction, a substantial part of the ornithine or arginine entering the cell (5 to 18%) is used for polyamine synthesis. It thus appears that polyamines are important in normal bacterial growth. A positive or negative change therefore, in the polyamine contents, may significantly affect growth. Since D7 and E40 have only about 40 and 30%, respectively, of the spermidine contents of the wild type, the slow growth rate of these mutants may be related to their low spermidine content.

Experiments were carried out to test this possibility by growing cells on minimal medium supplemented with spermidine and/or methionine. Methionine was included in this experiment because it was thought that methionine may increase the SAM pool and thus the spermidine level. These exogenous nutrients immediately changed the growth rate of the cells as shown in Table 2. Either methionine or spermidine alone stimulated the growth rate

TABLE 2. *Growth rate of* E. coli *K12 and SAM synthetase mutants on various media*

	Generation time (min)			
Strain	min[a]	met[b]	spd[c]	met + spd[d]
K12	81	65	81	68
E4	86	80	85	72
D7	130	91	113	73
E40	139	112	123	78
D8	87	85	92	87
E12	85	88	91	88
E31	82	78	90	85

[a] Minimal medium.

[b] Minimal medium supplemented with 10^{-3} M L-methionine.

[c] Minimal medium supplemented with 2×10^{-3} M spermidine.

[d] Minimal medium supplemented with 10^{-3} M L$^-$ methionine and 2×10^{-3} M spermidine.

of D7 significantly, for example, but the growth rate was still much slower than in the wild type. A combination of these two supplements, however, had a much stronger effect and raised the growth rate of D7 to the wild-type level. This result shows that the slow growth of D7 is partially but not completely due to a deficiency of spermidine.

As can be seen in Table 2, spermidine stimulated the growth of low SAM strains but slightly inhibited the growth of high SAM strains. At higher spermidine concentrations, this inhibition was even more pronounced and an inhibition of the wild type was observed. In contrast to the strong stimulation of the growth of the low SAM strains by methionine and spermidine together, the growth rate of the high SAM strains was not affected by the presence of these two supplements in the medium.

C. Repression and inhibition of SAM decarboxylase

As shown in Table 3, low SAM strains, which are also low in spermidine content, have an elevated level of SAM decarboxylase. On the other hand, in high SAM strains, the enzyme was repressed, suggesting that its synthesis is regulated by the intracellular spermidine levels. This is further supported by the finding that addition of spermidine to the growth medium of the wild type causes a repression in the synthesis of SAM decarboxylase (Table 3).

TABLE 3. *SAM decarboxylase activities of* E. coli *K12 and mutants*

Strain	Growth medium	Enzyme activity (nmoles CO_2/30 min/mg protein)[a]
Wild type	min[b]	47.2
	met[c]	65.5
	spd[d]	25.6
	NB[e]	17.8
Low SAM strains		
E4	min	61.8
D7	min	60.3
E40	min	59.2
High SAM strains		
D8	min	23.5
E12	min	25.3
E31	min	22.7

[a] The incubation procedure described in the text was followed. It contained 135 nmoles of SAM-^{14}COOH and 25 μl of enzyme preparation.
[b] Minimal medium.
[c] Minimal medium supplemented with 2×10^{-3} M L-methionine.
[d] Minimal medium supplemented with 10^{-3} M spermidine.
[e] Nutrient broth (8 g/liter).

Nutrient broth, which contains substantial amounts of spermidine and spermine, also represses enzyme synthesis. In contrast to the spermidine effect, methionine stimulates the biosynthesis of this enzyme; however, it is not known whether methionine or SAM functions as a true inducer.

Mixtures of enzyme preparations from the wild type and the mutants gave within 90% of the calculated activities, suggesting that the high activities of the low SAM strains and the low activities of the high SAM strains are not due to accumulations of an activator or of an inhibitor. Experiments with mixtures of extracts of the wild type grown on minimal medium and on methionine- or spermidine-supplemented medium also led to the same conclusion. In addition to the regulation of enzyme synthesis by spermidine, the activity of SAM decarboxylase is also regulated by a feedback control mechanism. With 2×10^{-3} M spermidine, the enzyme was 50% inhibited. Dixon plots of 1/v against spermidine concentration yielded a K_i of about 2×10^{-3} M.

IV. DISCUSSION

The present results show that there is a correlation between the levels of SAM pool and spermidine content. The high SAM strains have high levels of intracellular spermidine, whereas the low SAM strains have low levels. The elevated levels of spermidine in high SAM strains exert feedback and repression controls on ornithine decarboxylase, leading to a decrease in the putrescine synthesis. The reverse was observed in the low SAM strains.

Exponentially growing cells of E40 have only 30% of the spermidine content of the wild type; this is presumably attributable to the low SAM pool. Addition of spermidine to the medium resulted in a 12% increase (from 139 to 123 min) in growth rate. This rate is still much slower than that of the wild type (81 min). Addition of methionine alone causes a similar effect. A combination of both metabolites almost restores a normal rate of growth, suggesting that SAM is needed for reactions over and above that of spermidine synthesis, perhaps for transmethylation. In any case the joint effects of methionine and spermidine on the growth of E40 and D7 suggest that these two supplements together can increase the SAM pool and thereby correct the defects in these strains. The possibility that the presence of methionine leads to higher SAM content and improved transmethylation has not yet been tested.

The formation of SAM decarboxylase is repressed by high concentrations of spermidine and stimulated by methionine. However, the mechanism of these controls is still not understood. Pegg and Williams-Ashman (13) showed that SAM decarboxylase from rat prostate requires putrescine for maximum activity and that spermidine produced a significant increase in enzyme activity. In contrast to these results, the present study shows that the *E. coli* K12 enzyme is inhibited by spermidine and that putrescine at a concentration of 3×10^{-3} M does not affect the enzyme activity (*unpublished data*). Attempts to isolate mutants completely deficient in SAM synthetase, and hence spermidine deficient, have not been successful due to the impermeability of SAM to *E. coli* (6). However, conditional putrescine-deficient mutants defective in agmatine ureohydrolase have been isolated by Hirshfield et al. (7), and were shown to grow very poorly in the presence of arginine, which represses and inhibits ornithine decarboxylase and hence inhibits putrescine synthesis. As noted earlier (2, 7), addition of putrescine or spermidine to the medium stimulates the growth of these mutants in the presence of arginine. A mutant defective in putrescine synthesis (11) which grew almost normally was shown to have increased its spermidine content and therefore appears leaky in putrescine synthesis. This organism, therefore, cannot be considered to be as polyamine deficient as is the mutant isolated by Maas (7). Thus, the low SAM mutants are possibly the second instance of an *E. coli* strain whose growth response to spermidine is clear.

V. SUMMARY

Mutants of *E. coli* K12 with low concentrations of SAM have low spermidine contents, whereas the high SAM mutants have high levels of spermidine. Putrescine biosynthesis is regulated by intracellular spermidine concentration, and an inverse relationship between levels of SAM and putrescine is observed in these organisms. The growth rates of the low SAM strains D7 and E40 are very slow. Addition of methionine or spermidine to the medium stimulates their growth significantly. The presence of both spermidine and methionine in the medium raises the growth rate of the low SAM strains to that of the wild-type level. In contrast, the growth rate of high SAM strains is not affected by exogenous methionine and spermidine, either alone or together. The synthesis of SAM decarboxylase is repressed in wild-type cells growing in medium supplemented with spermidine. The concentration of the enzyme is also low in the high SAM strains grown in minimal medium, whereas the enzyme synthesis is derepressed in the low SAM strains. Spermidine inhibits the activity of the enzyme, with a K_i of about 2×10^{-3} M.

ACKNOWLEDGMENT

This investigation was supported by U.S. Public Health Service grant AI-7005 from the National Institute of Allergy and Infectious Diseases.

REFERENCES

1. Cohen, S. S., *Introduction to the polyamines,* Prentice-Hall, Englewood Cliffs, New Jersey, 1971.
2. Dion, A. S., and Cohen, S. S., *J. Virol.,* **9,** 423 (1972).
3. Dion, A. S., and Herbst, E. J., *Ann. N.Y. Acad. Sci.,* **171,** 723 (1970).
4. Greene, R. C., *J. Amer. Chem. Soc.,* **79,** 3929 (1957).
5. Greene, R. C., Su, C. H., and Holloway, C. T., *Biochem. Biophys. Res. Comm.,* **38,** 1120 (1970).
6. Holloway, C. T., Greene, R. C., and Su, C. H., *J. Bacteriol.,* **104,** 734 (1970).
7. Hirshfield, I. N., Rosenfeld, H. J., Leifer, Z., and Maas, W. K., *J. Bacteriol.,* **101,** 725 (1970).
8. Inouye, M., and Pardee, A. B., *J. Bacteriol.,* **101,** 770 (1970).
9. Jänne, J., Schenone, A., and Williams-Ashman, H. G., *Biochem. Biophys. Res. Comm.,* **42,** 758 (1971).
10. Jänne, J., and Williams-Ashman, H. G. *Biochem. Biophys. Res. Comm.,* **42,** 222 (1971).
11. Morris, D. R., and Jorstad, C. M., *J. Bacteriol.,* **101,** 731 (1970).
12. Morris, D. R., and Pardee, A. B., *J. Biol. Chem.,* **241,** 3129 (1966).
13. Pegg, A. E., and Williams-Ashman, H. G., *J. Biol. Chem.,* **244,** 682 (1969).
14. Raina, A., *Acta Chem. Scand.,* **16,** 2463 (1962).
15. Su. C. H., and Greene, R. C., *Proc. Nat. Acad. Sci. U.S.A.,* **68,** 367 (1971).
16. Tabor, C. W., in S. P. Colowick and N. O. Kaplan, (Editors), *Propylamine transferase: Methods in enzymology, Vol. 5,* Academic Press, New York, 1962, p. 761.
17. Tabor, H., Rosenthal, S. M., and Tabor, C. W., *J. Amer. Chem. Soc.,* **79,** 2978 (1957).
18. Tabor, H., Rosenthal, S. M., and Tabor, C. W., *J. Biol. Chem.,* **233,** 907 (1958).
19. Tabor, H., and Tabor, C. W., *J. Biol. Chem.,* **244,** 2286 (1969).
20. Wickner, R. B., Tabor, C. W., and Tabor, H., *J. Biol. Chem.,* **245,** 2132 (1970).
21. Zamenhof, S., in S. P. Colowick and N. O. Kaplan (Editors), *Preparation and assay of deoxyribonucleic acid from animal tissue: Methods in enzymology, Vol. 3,* Academic Press, New York, 1957, p. 702.

Polyamines in Normal and Neoplastic Growth. Edited by D. H. Russell. Raven Press, New York © 1973.

Putrescine: A Sensitive Assay and Blockade of Its Synthesis by αHydrazino Ornithine

Sami I. Harik, Gavril W. Pasternak, and Solomon H. Snyder

Departments of Pharmacology and Experimental Therapeutics and Psychiatry and the Behavioral Sciences, The Johns Hopkins University School of Medicine, Baltimore, Maryland 21205

Putrescine and the polyamines spermidine and spermine are widely distributed in nature and their concentrations are highest in tissues undergoing rapid metabolic activity (1). The decarboxylation of ornithine to yield putrescine seems to be the rate-limiting step in the formation of the polyamines (2–5). Ornithine decarboxylase (ODC), the enzyme catalyzing this reaction, has a very short half-life (6) and is not detectable except in those few tissues undergoing rapid metabolism, such as the prostate (7) and neoplastic (5) and embryonic tissues (5, 8). Furthermore, ODC is very sensitive to conditions that provoke increased metabolic activity, such as partial hepatectomy (2, 5) and the administration of growth hormone (9, 10).

Despite these lines of evidence that link polyamines to rapid tissue growth, the precise site and mode of action of the polyamines remain unknown. In an attempt to shed more light on the biological functions of the polyamines, we sought to inhibit their synthesis. Such inhibition, if successful, would have important theoretical and practical implications. The logical site for inhibiting polyamine synthesis would be the rate-limiting step involving ODC. However, since ODC activity can be measured in only a few mammalian tissues and since putrescine is the immediate product of the reaction catalyzed by ODC, it was reasoned that putrescine tissue concentrations could be considered to reflect closely the activity of ODC in these tissues.

This communication discusses work done in our laboratory on (a) a simple, sensitive, and specific enzymatic-isotopic microassay for tissue putrescine concentrations that reflects ODC activity, (b) αhydrazino ornithine (αHO), a potent and rather specific inhibitor of *Escherichia coli* and rat prostate ODC *in vitro*, and (c) the effect of αHO on ODC activity and putrescine levels in various tissues of mice and rats.

I. ENZYMATIC-ISOTOPIC ASSAY FOR PUTRESCINE

In mammalian tissues, putrescine exists in relatively small concentrations which in some tissues reach as low as one-fiftieth the concentration of spermidine (1). The currently used methods for the determination of tissue putrescine depend mainly on its separation from other amines by chromatography (11, 12) or electrophoresis (13, 14) and its subsequent colorimetric determination. These methods are tedious and nonspecific, and some have yielded conflicting results.

Making use of the findings reported by Jänne et al. (15) that baker's yeast (*Saccharomyces cerevisiae*) S-adenosyl-L-methionine decarboxylase (SAMeDC) requires putrescine to manifest its activity, we developed an enzymatic-isotopic microassay where yeast SAMeDC activity, measured as $^{14}CO_2$ evolved from S-adenosyl-L-(^{14}C-carboxyl)-methionine (^{14}C-SAMe), would accurately measure the quantity of putrescine in the incubation mixture.

The assay procedure is briefly described as follows: tissues were homogenized in 9 volumes of cold distilled water in motor-driven glass homogenizers. The homogenates were heated in a water bath at 95°C for 20 min, centrifuged for 10 min at 20,000 × *g*, and aliquots of the supernatant fluid were used for the putrescine assay. Heating deproteinates the homogenates and destroys their content of S-adenosyl-L-methionine (SAMe) (16). Measurement of SAMeDC activity as a function of putrescine concentration in the incubation mixture was done according to the method of Jänne and Williams-Ashman (17) with some modification. The incubation mixture contained the following ingredients in a final volume of 120 μl: 8 μmoles of Tris buffer, pH 7.6; 0.6 μmole of dithiothreitol; 0.82 nmole of ^{14}C-SAMe (61 mC per mmole); 20 units of yeast SAMeDC prepared according to the procedure of Jänne et al. (15) with certain modifications and carried through the ammonium sulfate fractionation step only; and 5- to 40-μl aliquots of tissue extract (representing 0.5 to 4 mg of tissue). Control samples in which tissue extract was replaced by distilled water as well as varying amounts of a standard putrescine solution were included with each set of determinations. Incubations were carried out at 37°C for 1 hr, at which time the reaction was terminated by adding 50 μl of 10 N H_2SO_4 to the incubation medium. The evolved CO_2 was trapped and its radioactivity measured.

Putrescine concentration and evolved $^{14}CO_2$ were related linearly over a range of 20 to 400 pmoles of putrescine (Fig. 1) and with different amounts

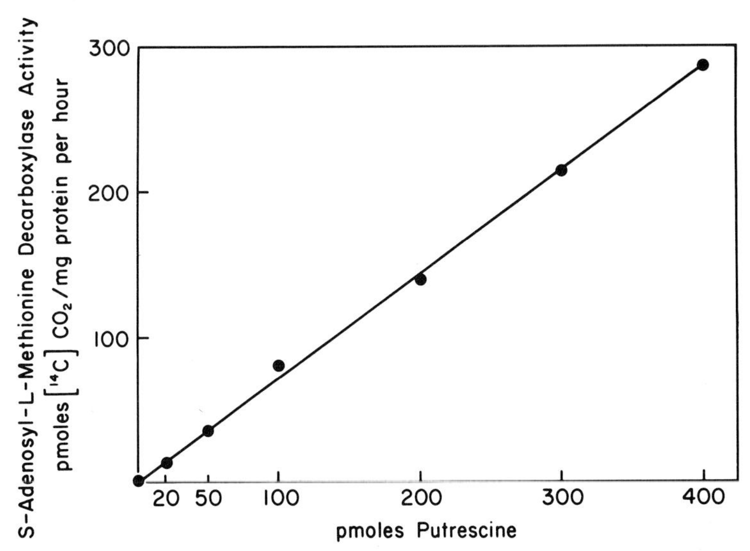

FIG. 1. Standard curve for the assay of putrescine. Putrescine (20 to 400 pmoles) was added to the assay system containing the following ingredients in a final volume of 120 μl: 8 μmoles of Tris buffer, pH 7.6; 0.6 μmole of dithiothreitol; 0.82 nmole of ^{14}C-SAMe (0.05 μC); and 20 units of yeast SAMeDC. The graph passes through the origin because the blank value of 600 cpm (amount of $^{14}CO_2$ evolved in the absence of putrescine) was subtracted from the original readings.

of tissue extract. Small amounts of $^{14}CO_2$ were evolved from the blank samples, presumably indicating that some decarboxylation occurs in the absence of putrescine. These blank values did not exceed half the amount of $^{14}CO_2$ evolved in the presence of 20 pmoles of putrescine. Tissue concentrations of putrescine, expressed in nmoles per gram, were derived from the standard curve or by calculation from an internal standard consisting of an aliquot of tissue homogenate with a known amount of exogenously added putrescine that was carried through the whole procedure to determine recovery. The mean recovery values in a variety of tissues ranged between 90 and 110% and did not vary when the amounts of exogenous putrescine added ranged from 50 to 200% of the endogenous putrescine levels.

To study the specificity of the assay, several compounds that are structurally related to putrescine were tested for their ability to stimulate yeast SAMeDC (Table 1). Agmatine was as effective as putrescine in enhancing $^{14}CO_2$ evolution. It is not clear to us at this point whether agmatine was itself active or whether it was transformed to putrescine by arginase which may be present in the yeast SAMeDC preparation. Both 1,3-diaminopropane and cadaverine showed limited activity (10 and 5% of putrescine's

TABLE 1. *Enhancement by various compounds of yeast SAMeDC activity*

Compound	Enhancement of SAMeDC activity expressed as % of that produced by an equimolar concentration of putrescine
Agmatine	100
1,3-Diaminopropane	10
Cadaverine	5.35
Spermidine	0.25
Spermine	0
γ-Aminobutyric acid	0
Histamine	0
L-Histidine	0
L-Arginine	0
L-Lysine	0
L-Ornithine	0

Compounds (20 pmoles to 100 nmoles) were added to the assay system containing the following ingredients in a final volume of 120 μl: 8 μmoles of Tris buffer, pH 7.6; 0.6 μmole of dithiothreitol; 0.82 nmole of ^{14}C-SAMe (0.05 μC); and 20 units of yeast SAMeDC. The amount of $^{14}CO_2$ evolved was compared to that produced by putrescine and expressed as percent of putrescine enhancement. Zero percent activity means that 100 nmoles of the compound was incapable of evolving an amount of $^{14}CO_2$ higher than blank samples containing no added amine.

activity, respectively), whereas spermidine was only 0.25% as active as putrescine. The remainder of the compounds tested were inactive.

To ascertain if agmatine, 1,3-diaminopropane, and cadaverine are present in animal tissues, samples of mouse brain and liver were processed by the amino acid analyzer technique of Kremzner (18), which quantitatively resolves these compounds from each other (Fig. 2A). Agmatine and cadaverine were not detected in the brain and liver of the mouse. However, liver extracts revealed a slight deflection in the region of 1,3-diaminopropane which amounted to a small fraction of the putrescine peak. No such peak could be detected when brain extracts were analyzed. Closely similar data regarding the presence of agmatine, cadaverine, and 1,3-diaminopropane in rat brain and liver when analyzed by this amino acid analyzer technique have been obtained by Dr. L. Kremzner (*personal communication*).

Spermidine concentrations in most mammalian tissues are up to 50 times those of putrescine and accordingly can interfere with the assay for putrescine even if its activity was only a few percent of that of putrescine. Fortunately, however, we observed that during heating and centrifugation of

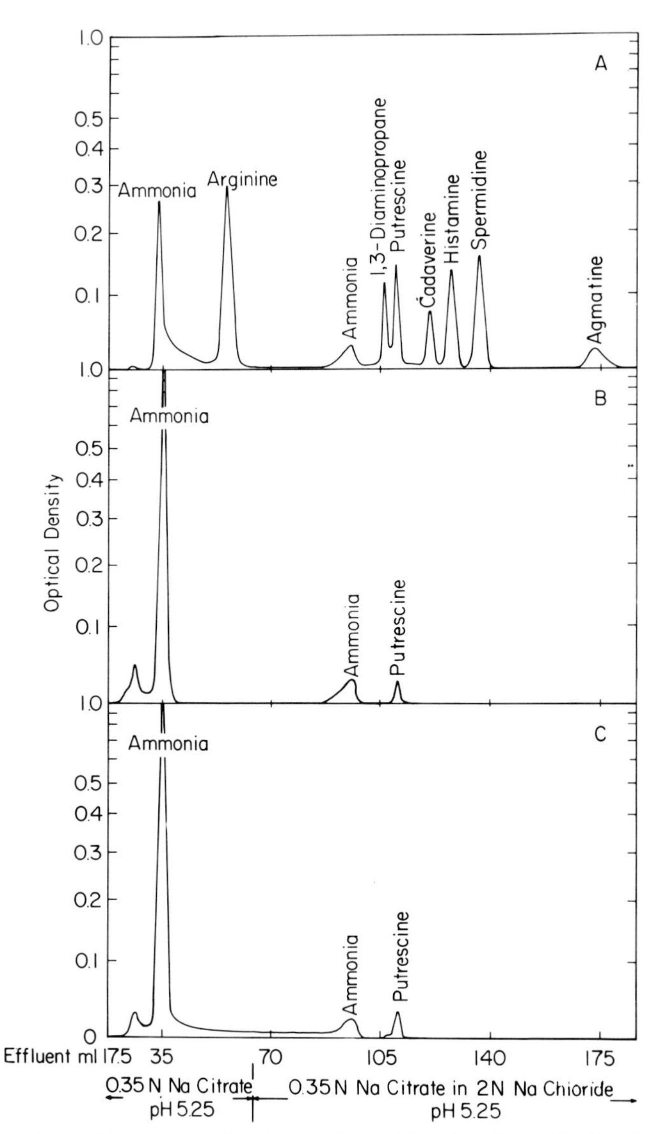

FIG. 2. Chromatographic analysis of amines performed by a Beckman 120 C amino acid analyzer. A sulfonic acid exchanger (Beckman PA-35 resin) was utilized in a column 6.3 cm long and 0.9 cm in diameter. The temperature was maintained at 39°C and the effluent pumping rate at 70 ml/hr. A: Mixture containing 20 nmoles L-arginine, 10 nmoles 1,3-diaminopropane, 10 nmoles putrescine, 10 nmoles cadaverine, 20 nmoles histamine, 10 nmoles spermidine, and 10 nmoles agmatine. B: The 0.06 N HCl eluate obtained from a phosphorylated cellulose column loaded with tissue extract from 200 mg of mouse brain. The putrescine deflection represents 2.3 nmoles (11.5 nmoles/g) as calculated by comparing the area encompassed by the curve to that produced by 10 nmoles of authentic putrescine in Fig. 2A. C: The 0.06 N HCl fraction obtained from a phosphorylated cellulose column loaded with tissue extract from 200 mg of mouse liver. The putrescine deflection represents 2.43 nmoles (12.15 nmoles/g) calculated in an identical manner to that of brain eluate.

tissue homogenates in preparation for the putrescine assay, 70% of ^{14}C-spermidine added to the tissues was precipitated and hence unavailable for the enzymatic-isotopic assay. This, coupled with the very low activity of spermidine in enhancing yeast SAMeDC activity, being only 0.25% of that of putrescine, makes any serious interference by spermidine with the putrescine assay very unlikely. To assess further the possibility of interference with spermidine, we separated putrescine from spermidine by phosphorylated cellulose column chromatography (12). Putrescine tissue concentrations were the same with or without the separation procedure.

Putrescine levels in rat and mouse brain and liver, assayed by the enzymatic-isotopic method are listed in Table 2 and compared to those obtained by other techniques. The results obtained by the enzymatic-isotopic method differ from those obtained by phosphorylated cellulose column chromatography (12) followed by the 2,4-dinitro-1-fluorobenzene (DNFB) colorimetric determination (19) and by ninhydrin colorimetric determination after electrophoresis (13, 14). However, putrescine levels determined by enzymatic-isotopic assay were quite similar to those obtained by fluorimetric assay of the 1-dimethylaminonaphthalene-5-sulfonyl derivative of putrescine (20).

To check the validity of our results, we assayed the same samples of mouse brain and liver for putrescine by the enzymatic-isotopic technique as well as by the DNFB method after phosphorylated cellulose chromatography and also by the amino acid analyzer procedure (Table 3 and Fig. 2). The

TABLE 2. *Tissue putrescine content measured by various methods*

Tissue	n	Enzymatic-isotopic method	Method A	Method B	Method C
Mouse brain	13	19.0 ± 1.0	30–50		10.9 (20)
Mouse liver	14	35.9 ± 3.3	40–70	171 (24)	25.2 (20)
Rat brain	17	19.0 ± 1.7		91 (27)	
Rat liver	13	25.4 ± 3.4	70–85	100–250 (14)	

Putrescine concentrations, expressed in nmoles/g, in the brain and liver of mice and rats were determined by the indicated methods. Data obtained by the enzymatic-isotopic assay are mean values ± S.E.M. Data obtained by other methods are quoted from the literature, and the reference is indicated in parentheses. The data reported under Method A were obtained in our laboratory.

Method A: Separation of putrescine by phosphorylated cellulose column chromatography (12) followed by DNFB colorimetric determination (19).

Method B: Separation of putrescine by electrophoresis followed by ninhydrin colorimetric determination (13, 14).

Method C: Separation by thin-layer chromatography of 1-dimethylaminonaphthalene-5-sulfonyl derivative of putrescine followed by its fluorimetric determination (20).

TABLE 3. *Mouse brain and liver putrescine content measured by different methods*

	Putrescine tissue contents (nmoles/g tissue)		
Tissue	Enzymatic-isotopic method	Phosphorylated cellulose column chromatography followed by DNFB determination	Amino acid analyzer
Brain	13.8	34.0	11.5
Liver	15.6	50.6	12.1

Brains and livers from two mice were pooled, homogenized, heated, and centrifuged. Aliquots of the supernatant fluid of each were assayed for their putrescine content by the three indicated procedures. A comparison of the results obtained by the different methods for single samples of each tissue is presented.

results obtained by the enzymatic-isotopic assay were corroborated with those obtained by the amino acid analyzer, a totally unrelated technique, while the DNFB method gave much higher results.

We then studied the relationship between ODC activity and putrescine tissue levels in the regenerating liver of the rat where a marked increase in ODC activity is manifested (Table 4). There seems to be good correlation between the enhanced ODC activity and putrescine tissue levels measured by the enzymatic-isotopic assay. The same samples were also assayed for putrescine by the DNFB method which gave values which were 40 to 50 nmoles/g higher than those obtained by the enzymatic-isotopic assay. This constant difference suggested to us that the DNFB method overestimates the levels of tissue putrescine by reacting with a substance present in tissue extracts that fails to increase during hepatic regeneration.

II. αHO: A POTENT INHIBITOR OF ODC

Numerous ornithine analogues were screened in an attempt to find an inhibitor of ODC. Only one, αHO, was found to be a potent inhibitor of both *E. coli* and rat prostate ODC. The *E. coli* ODC was purified threefold by a modification of the method of Karpetsky and Talalay (21), and the rat prostate ODC was purified three- to fourfold according to a modification of the method of Jänne and Williams-Ashman (22). ODC activity was measured by estimating the amount of $^{14}CO_2$ evolved upon the incubation of (1-^{14}C)-ornithine with the enzyme preparation in the presence of 0.01 mM pyridoxal phosphate and 5 mM dithiothreitol. ODC activity was also measured in the presence of varying concentrations of αHO ranging from 10^{-4} to 10^{-7} M. The ID_{50} of the *E. coli* enzyme was approximately 5×10^{-7} M;

TABLE 4. *ODC activity and putrescine levels in normal and regenerating rat liver*

Groups	ODC activity		Putrescine levels: Enzymatic-isotopic method		Putrescine levels: Phosphorylated cellulose column chromatography followed by DNFB determination	
	pmoles $^{14}CO_2$/mg protein/hr	% of control	nmoles/g tissue	% of control	nmoles/g tissue	% of control
Nonoperated rats	82 ± 38	100	22.5 ± 5.5	100	74.0 ± 3.6	100
Sham-operated rats	897 ± 35*	1,094	57.2 ± 3.6*	254	93.5 ± 5.5**	126
Partially hepatectomized rats	2,317 ± 200*	2,826	143.0 ± 7.5*	636	196.5 ± 13.4*	266

Groups of four rats underwent partial hepatectomy, were sham operated, or were not operated. The rats were killed 4 hr later, their livers quickly removed and assayed for ODC activity by a modification of the method of Jänne and Williams-Ashman (22) and for putrescine. Data presented are the mean values ± S.E.M.

Experimental values differ from control $*p < .005$, $**p < .05$.

the ID_{50} of the rat prostate enzyme was 5×10^{-6} M. The enzymes' kinetics in the presence of two concentrations of αHO and in its absence are shown in Figs. 3 and 4. This inhibition by αHO seems to be competitive in nature. The discrepancies between the K_m and the K_i, especially in the case of the *E. coli* ODC (Fig. 3), speak of the potency of inhibition. Assessing *E. coli* and rat prostate ODC activity in the presence of a constant concentration

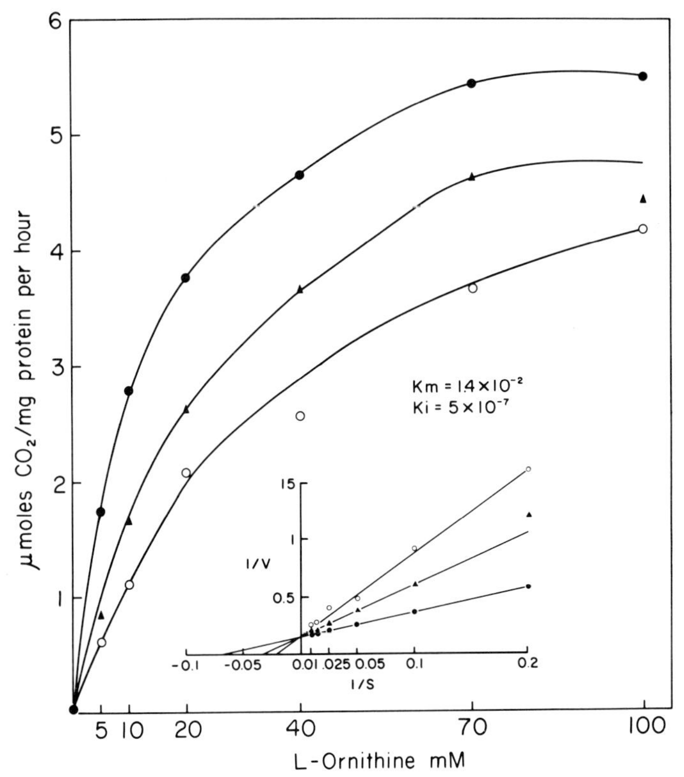

FIG. 3. Effects of substrate concentration on *E. coli* ornithine decarboxylase activity and on its inhibition by αhydrazino ornithine. L-Ornithine, in various concentrations, was added to the assay system containing in a final volume of 300 μl the following ingredients: 30 μmoles of Tris buffer, pH 7.2; 1.5 μmole of dithiothreitol; 3 nmoles of pyridoxal phosphate; threefold purified *E. coli* enzyme preparation (86 μg of protein per assay). The amount of DL(1-^{14}C)-ornithine was kept constant at 0.2 mM (0.28 μC), but the specific radioactivity was varied by adding suitable quantities of L-ornithine. αHO was added at all concentrations of L-ornithine in two final concentrations of 5×10^{-7} M and 1×10^{-6} M. The mixture was incubated for 1 hr at 37°C. The reaction was terminated by adding 100 μl of 10 N H_2SO_4 to the incubation mixture. The radioactivity of the evolved $^{14}CO_2$ was measured and converted to μmoles CO_2/mg protein/hr. No inhibitor (●—●). αHO concentration 5×10^{-7} M (▲—▲). αHO concentration 1×10^{-6} M (○—○).

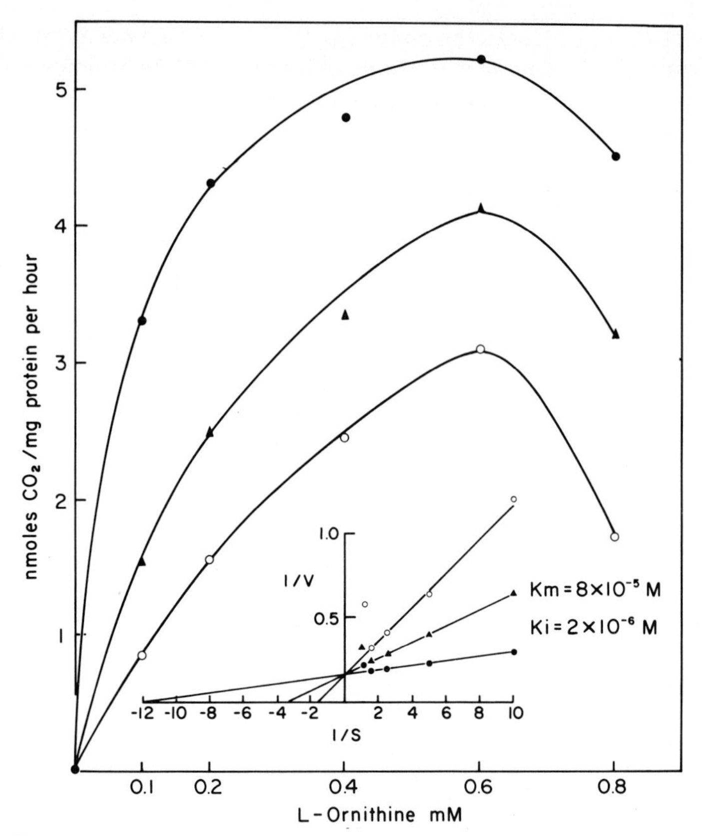

FIG. 4. Effects of substrate concentration on rat prostate ODC activity and on its inhibition by αHO. L-Ornithine, in various concentrations, was added to the assay system containing in a final volume of 300 μl the following ingredients: 30 μmoles of Tris buffer, pH 7.2; 1.5 μmole of dithiothreitol; 3 nmoles of pyridoxal phosphate; three- to fourfold purified rat prostate enzyme preparation (150 μg of protein per assay). The amount of DL(1-^{14}C)-ornithine was kept constant at 0.2 mM (0.28 μC), but the specific radioactivity was varied by adding suitable quantities of L-ornithine. αHo was added at all concentrations of L-ornithine in two final concentrations of 5×10^{-6} M and 1×10^{-5} M. The mixture was incubated for 1 hr at 37°C. The reaction was terminated by adding 100 μl of 10 N H_2SO_4 to the incubation mixture. The radioactivity of the evolved $^{14}CO_2$ was measured and converted to nmoles CO_2/mg protein/hr. No inhibitor (●—●). αHO concentration 5×10^{-6} M (▲—▲). αHO concentration 1×10^{-5} M (○—○).

of αHO while varying the quantity of enzyme added (Ackerman-Potter plots) showed the inhibition to be completely reversible. The inhibition by αHO of either enzyme, however, was abolished in the presence of large concentrations of pyridoxal phosphate (0.1 mM or more).

To determine the specificity of αHO as an inhibitor of ODC, we studied its efficacy in inhibiting two of the pyridoxal-dependent amino acid decar-

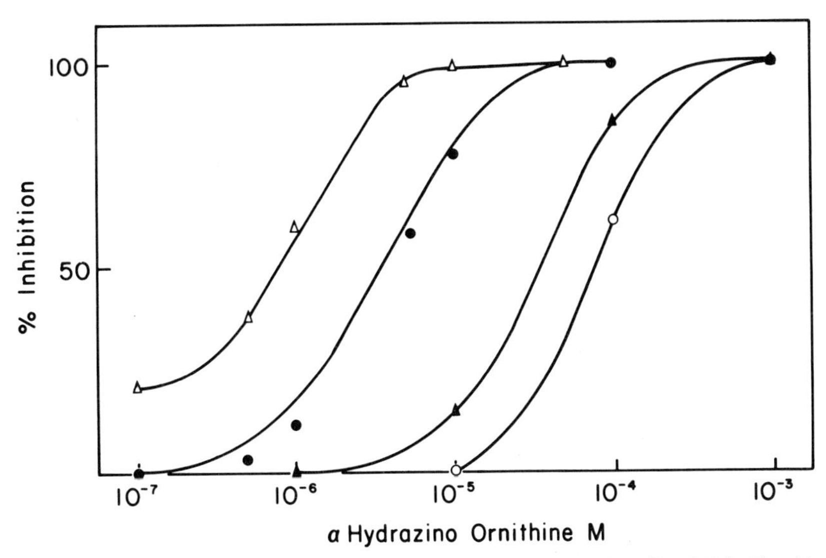

FIG. 5. Inhibitory effects of various concentrations of αHO on *E. coli* L-ODC (△—△), rat prostate L-ODC (●—●), rat brain L-glutamic acid decarboxylase (▲—▲), and rat brain L-DOPA decarboxylase (○—○). αHO was added in various concentrations to the assay system containing in a final volume of 300 μl the following ingredients: 30 μmoles of Tris buffer, pH 7.2; 1.5 μmoles of dithiothreitol; 3 nmoles of pyridoxal phosphate; the enzyme preparation indicated; and the appropriate (1-^{14}C)-amino acid (in the case of *E. coli* L-ODC, 3 μmoles of L-ornithine was used, but for the rat prostate L-ODC, rat brain L-glutamic acid decarboxylase, and rat brain L-DOPA decarboxylase, 30 nmoles of the appropriate L-amino acid was used). The mixture was incubated for 1 hr at 37°C and the radioactivity of the evolved $^{14}CO_2$ measured. Percent inhibition was calculated by comparing the amount of $^{14}CO_2$ evolved at that concentration of αHO to that evolved in the absence of the inhibitor.

boxylases: L-DOPA decarboxylase and L-glutamic acid decarboxylase. These two enzymes were less sensitive to inhibition by αHO by about one and two orders of magnitude than rat prostate and *E. coli* ODC, respectively (Fig. 5). On the other hand, 4-bromo-3-hydroxybenzyloxyamine (NSD-1055), a potent inhibitor of L-histidine decarboxylase (23), another pyridoxal-dependent amino acid decarboxylase, was one and two orders of magnitude less effective than αHO in inhibiting rat prostate and *E. coli* ODC, respectively, under comparable conditions.

III. EFFECTS OF THE ADMINISTRATION OF αHO ON MICE AND RATS

The potent *in vitro* inhibitory effects of αHO on *E. coli* and rat prostate ODC prompted us to determine its efficacy when administered to animals on tissue putrescine levels and ODC activity.

TABLE 5. *Effects of hydrocortisone, αHO, and NSD-1055 on prostatic and hepatic ODC activity and on hepatic putrescine levels of rats*

Treatment	Prostatic ODC activity		Hepatic ODC activity		Hepatic putrescine levels	
	pmoles $^{14}CO_2$/mg protein/hr	% of control	pmoles $^{14}CO_2$/mg protein/hr	% of control	nmoles/g tissue	% of control
Controls	3,343 ± 271	100	15.6 ± 0.3	100	25.3 ± 3.2	100
Hydrocortisone (62 mg/kg)	3,991 ± 1,066	119	439.3 ± 98.1**	2,816	142.7 ± 20.7**	564
Hydrocortisone (62 mg/kg) αHO (400 mg/kg)	9,984 ± 2,169*	299	2,265.3 ± 536.5**	14,521	153.0 ± 12.5**	605
NSD-1055 (200 mg/kg)	4,474 ± 756	134	263.3 ± 107.6	1,688	62.7 ± 10.8*	248

Groups of three rats received the indicated agents intraperitoneally 4 hr prior to sacrifice (only αHO was administered in two doses, 200 mg/kg each, at 4 and 2 hr prior to sacrifice). Control rats received 0.5 ml of normal saline. After decapitation the prostate and liver were quickly removed and assayed for ODC activity by a modification of the method of Jänne and Williams-Ashman (22) and for putrescine by the enzymatic-isotopic methods. Data presented are the mean values ± S.E.M.

Experimental values differ from control * $p < 0.05$, ** $p < 0.025$.

In preliminary experiments, αHO was injected into mice in doses of 200 to 2,000 mg/kg body weight. Our choice of mice was dictated by the meager amount of αHO in our possession. The putrescine content of various organs was monitored for periods of 30 to 120 min after a single injection of αHO, but no significant changes in putrescine tissue levels occurred. Unfortunately most of our supply of αHO was exhausted in these preliminary experiments on mice before we realized that mice are probably not the best animal model to study in this case because of their rather slow polyamine turnover rate (24). As an example of this, the increase in ODC activity in the liver remnants after partial hepatectomy in mice is much less intense than that occurring in rats (24).

Whatever was left of our αHO was then used in one experiment on three rats: these rats were given hydrocortisone 4 hr prior to sacrifice, which is a known powerful stimulant of hepatic ODC activity (25, 26), and then αHO in two equal doses, each dose 200 mg/kg, was administered at 4 and 2 hr prior to sacrifice. The rationale for administering αHO to animals also receiving hydrocortisone was that αHO might block the hydrocortisone-induced increase in hepatic ODC activity as well as the resultant liver putrescine elevation even if it were ineffective in significantly changing the resting ODC activity and putrescine levels. ODC activity was measured in the liver and prostate, and putrescine tissue levels were determined in the brain, liver, and kidney of these rats and compared to other groups of rats from the same litter who received hydrocortisone, NSD-1055, or saline. There were no significant differences between the various groups in their brain and kidney putrescine content. However, definite effects were evident when the prostatic and hepatic ODC activities as well as the hepatic putrescine levels were compared (Table 5). Hydrocortisone, as expected, caused a 28-fold increase in hepatic ODC activity and a five- to sixfold increase in putrescine levels; αHO, however, instead of blocking these effects of hydrocortisone by inhibiting ODC as was projected, caused a remarkable 145-fold increase in hepatic and a threefold increase in prostatic ODC activity. NSD-1055 also unexpectedly increased hepatic ODC activity and putrescine levels.

The nature of this unexpected action of αHO and NSD-1055 on ODC activity *in vivo* is unclear, and will require extensive study.

IV. CONCLUSIONS

A highly sensitive and specific enzymatic-isotopic microassay procedure for the measurement of putrescine is described. The method depends on the enhancement by putrescine of the decarboxylation of ^{14}C-SAMe by baker's yeast SAMeDC. The quantity of $^{14}CO_2$ evolved is a linear function of the amount of putrescine present in the incubation mixture. The method can reliably measure 20 pmoles of putrescine, only small amounts of tissue (0.5 to 4mg) are required for the assay, and the technique is simple to perform so that 100 samples can be assayed in one working day.

This method was used to measure the putrescine content of various tissues. Data collected by this method were much lower than those obtained by other workers for the rat and mouse brain and liver, but our values were corroborated by an amino acid analyzer technique. Putrescine tissue levels correlated well with ODC activity and may serve as an indicator of the activity of this enzyme in tissues where ODC activity cannot be detected readily.

αHO was found to be a potent and specific *in vitro* inhibitor of *E. coli* and rat prostate ODCs. However, when given to rats it greatly enhanced the activity of hepatic and prostatic ODC.

ACKNOWLEDGMENTS

S. I. Harik is supported by the Commonwealth Fund Exchange Fellowship Program at The Johns Hopkins University. G. W. Pasternak is supported by the Mutual of Omaha Insurance Company and United Benefit Life Insurance Company. S. H. Snyder is recipient of N.I.M.H. Research Scientist Development Award MH-33128. We would like to thank Drs. J. G. Johansson and W. B. Skinner for their generous gift of αhydrazino ornithine.

REFERENCES

1. Tabor, H., and Tabor, C. W., *Pharmacol. Rev.*, **16,** 245 (1964).
2. Jänne, J., and Raina, A., *Acta Chem. Scand.*, **22,** 1349 (1968).
3. Williams-Ashman, H. G., and Pegg, A. E., *Adv. Enzyme Regul.*, **7,** 291 (1969).
4. Pegg, A. E., and Williams-Ashman, H. G., *J. Biol. Chem.*, **244,** 682 (1969).
5. Russell, D. H., and Snyder, S. H., *Proc. Nat. Acad. Sci. U.S.A.*, **60,** 1420 (1968).
6. Russell, D. H., and Snyder, S. H., *Mol. Pharmacol.*, **5,** 253 (1969).
7. Pegg, A. E., and Williams-Ashman, H. G., *Biochem. J.*, **108,** 533 (1968).
8. Anderson, T. R., and Schanberg, S. M., *J. Neurochem.*, **19,** 1471 (1972).
9. Russell, D. H., and Snyder, S. H., *Endocrinology,* **84,** 223 (1969).
10. Russell, D. H., Snyder, S. H., and Medina, V. J., *Endocrinology,* **86,** 1414 (1970).
11. Hammond, J. E., and Herbst, E. J., *Anal. Biochem.*, **22,** 474 (1968).
12. Kremzner, L. T., Barrett, R. E., and Terrano, M. J., *Ann. N.Y. Acad. Sci.*, **171,** 735 (1970).
13. Raina, A., *Acta Physiol. Scand.*, Suppl. **218,** 1 (1963).
14. Jänne, J., *Acta Physiol. Scand.*, Suppl. **300,** 1 (1967).
15. Jänne, J., Williams-Ashman, H. G., and Schenone, A., *Biochem. Biophys. Res. Comm.*, **43,** 1362 (1971).
16. Snyder, S. H., Baldessarini, R. J., and Axelrod, J., *J. Pharmacol. Exp. Ther.*, **153,** 544 (1966).

17. Jänne, J., and Williams-Ashman, H. G., *Biochem. Biophys. Res. Comm.*, **42,** 222 (1971).
18. Kremzner, L. T., this volume.
19. Dubin, D. T., *J. Biol. Chem.*, **235,** 783 (1960).
20. Seiler, N., and Askar, A., *J. Chromatogr.*, **62,** 121 (1971).
21. Karpetsky, T., and Talalay, P., in preparation.
22. Jänne, J., and Williams-Ashman, H. G., *J. Biol. Chem.*, **246,** 1725 (1971).
23. Leinweber, F. J., *Mol. Pharmacol.*, **4,** 337 (1968).
24. Russell, D. H., and McVicker, T. A., *Biochim. Biophys. Acta,* **244,** 85 (1971).
25. Medina, V. J., Russell, D. H., and Snyder, S. H., *unpublished observations.*
26. Panko, W. B., and Kenney, F. T., *Biochem. Biophys. Res. Comm.*, **43,** 346 (1971).
27. Russell, D. H., Medina, V. J., and Snyder, S. H., *J. Biol. Chem.*, **245,** 6732 (1970).

Polyamines in Normal and Neoplastic Growth. Edited by
D. H. Russell. Raven Press, New York © 1973.

Cation Requirement for RNA- and DNA-Templated DNA Polymerase Activities of B-Type Oncogenic RNA Viruses (MuMTV)

Arnold S. Dion and Dan H. Moore

Institute for Medical Research, Copewood Street, Camden, New Jersey 08103

I. INTRODUCTION

Temin and Mizutani (1) and, independently, Baltimore (2) reported the presence of an endogenous enzyme (reverse transcriptase) catalyzing the formation of DNA polymers from RNA templates by disrupted oncogenic RNA viruses. Later reports (3–5) also indicated the presence of an endogenous, viral DNA-dependent DNA polymerase in these preparations. These observations laid the theoretical basis for genetic information transfer from oncogenic RNA viruses to DNA and the possible integration of the latter into the host genome. Polyamine enhancement of DNA-directed RNA polymerase has been amply demonstrated in both *in vivo* and *in vitro* studies (6), and it was of obvious importance to test these ubiquitous organic cations for possible effects on viral RNA and DNA-dependent DNA polymerase.

Murine mammary tumor virus (MuMTV) is a B-type, oncogenic RNA virus (7), with a demonstrated ability to induce mammary tumors in mice (8). Although these viruses differ morphologically (9) from other oncogenic RNA viruses of the C-type, both groups share the common attributes of a 60-70S RNA (10), an endogenous RNA-directed DNA polymerase (11, 12), and extensive polyadenylate sequences (13, 14).

Rapidly proliferating normal cells have been demonstrated to contain higher spermidine than spermine concentrations, a situation not unlike that found in tumor cells (15, 16). In contrast, normal cells contain higher concentrations of spermine than spermidine. Of particular importance for the present investigation are the early studies of Rosenthal and Tabor (17) which demonstrated higher ratios of spermidine to spermine in mouse mammary carcinoma, sarcoma, and hepatoma than in normal tissues.

II. METHODS AND MATERIALS

A. Preparation of purified virus

MuMTV was prepared from RIII mouse milk according to the unpublished procedure of Dr. N. H. Sarkar of this institute. Four ml of milk was skimmed by adding 16 ml of 0.15 M EDTA (pH 7.4) and 10 ml of phosphate-buffered saline (PBS, pH 7.4) and centrifuging for 5 min at 5,000 rpm (SW 25.1). After discarding the cream pellicle and pellet, the skim-milk fraction was centrifuged for 60 min at 21,000 rpm (SW 25.1). PBS (0.5 ml) was added to the viral pellet, allowed to stand overnight at 4°C, and resuspended. The resuspended pellet was then applied to a 0 to 60% sucrose (PBS) gradient and centrifuged at 32,000 rpm (SW 50.1) for 60 min. The virus band at a density of 1.18 was collected, diluted with PBS, and centrifuged for 60 min at 32,000 rpm (SW 50.1). This final pellet was stored at 4°C after the addition of 0.2 ml PBS.

B. DNA polymerase assay

Virus pellets were resuspended in PBS, diluted with T.15NE (0.01 M Tris, 0.15 M NaCl, 0.002 M disodium EDTA, pH 8.3) and centrifuged 30 min at 25,000 rpm (SW 27). After the supernate was discarded, the virus pellet was resuspended in 0.01 M Tris (pH 8.3). Each assay contained 50 μl of the resuspended virus (approximately 1 mg protein) which was disrupted (0°C for 10 min) by the addition of 6 μl of 1 M dithiothreitol (Sigma) and 3 μl of 5% NP-40 (Shell Chemical Co.). Each reaction mixture contained, in μmoles per 125 μl: 6.25 Tris HCl (pH 8.3), 1.25 NaCl, and 0.2 μmoles of each of the unlabeled deoxyribonucleoside triphosphates (Sigma). Where designated, $MgCl_2$, manganese acetate, or spermidine trihydrochloride (Calbiochem) was added at the indicated concentrations. Thymidine-^{3}H-methyl 5′-triphosphate (^{3}H-TTP) (New England Nuclear, Boston, Mass.) was employed at a specific activity of 2,520 cpm/picomole. The reactions were terminated after 30 min at 37°C by the addition of 10 μl of 4 M NaCl, 10 μl of 10% sodium dodecyl sulfate, and dilution with 0.5 ml T.15 NE (pH 8.3). After phenol/cresol extraction (11) at room temperature for 5 min, trichloroacetic acid-precipitable incorporation (3) was determined from 100 to 200 μl of the aqueous phase.

III. RESULTS

A. Endogenous RNA-templated DNA synthesis

1. *Inorganic Cation Requirement.* Figure 1 demonstrates the effect of Mg^{++} concentration on the incorporation of ^{3}H-TTP by detergent-disrupted MuMTV. Maximal incorporation was observed at a Mg^{++} concentration of 5.25 mM, resulting in a 26-fold enhancement. Higher Mg^{++} concentrations were inhibitory. The optimal Mn^{++} concentration was determined to be approximately 0.7 mM; however, Mn^{++} was only 30 to 50% as effective as Mg^{++}. Higher Mn^{++} concentrations were also observed to be inhibitory. Ca^{++} concentrations between 3.5 and 21.0 mM were ineffective in substituting for Mg^{++}. In summary, the order is $Mg^{++} > Mn^{++} >> Ca^{++}$ in fulfilling the requirement for a divalent cation for the endogenously templated reac-

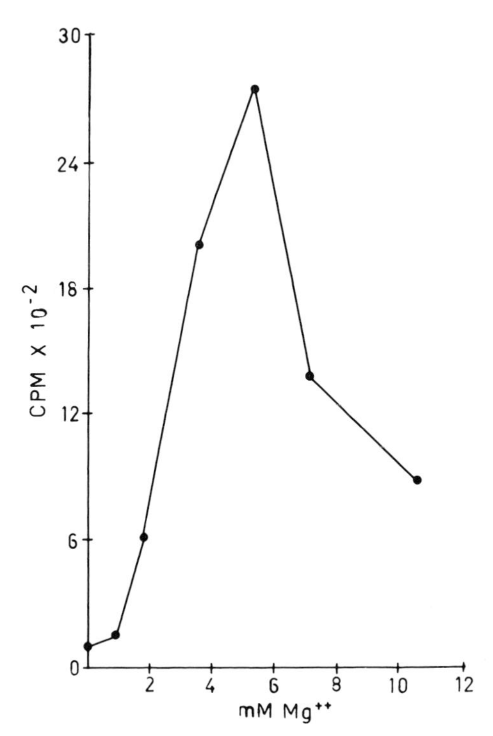

FIG. 1. Dependence of endogenous RNA-templated DNA synthesis by disrupted RIII MuMTV on magnesium concentration. Virions were purified and assayed for DNA polymerase activity as described in Methods and Materials.

tion. Interestingly, stimulation of the endogenous reaction was inversely proportional to the ionic radii, i.e., 0.65, 0.80, and 0.99 angstrom units for Mg^{++}, Mn^{++} and Ca^{++}, respectively.

2. *Effect of Spermidine.* The effect of spermidine added at various concentrations was assayed at two concentrations of Mg^{++}, at the optimum concentration (5.25 mM) and at 2.63 mM. As shown in Fig. 2, the addition of spermidine at final concentrations of 0.66 to 2.64 mM, in the presence of 5.25 mM Mg^{++}, resulted in a 60 to 70% inhibition of ^{3}H-TTP incorporation.

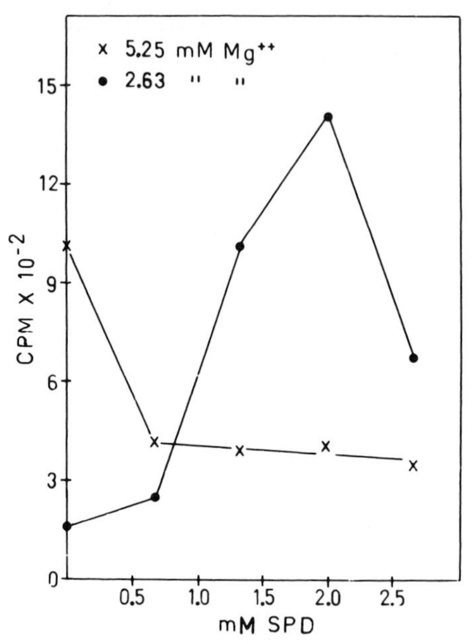

FIG. 2. Effect of spermidine on endogenous RNA-templated DNA synthesis in the presence of optimal and suboptimal magnesium concentrations. Details of virus purification and DNA polymerase assay are given in Methods and Materials.

However, the addition of the same concentrations of spermidine effected a marked stimulation (more than ninefold) of ^{3}H-TTP incorporation in the presence of 2.63 mM Mg^{++}. Maximum incorporation was observed in the presence of 2.63 mM Mg^{++} and 1.98 mM spermidine, exceeding the incorporation in the presence of 5.25 mM Mg^{++} by 34%. That the same concentrations of spermidine cannot totally replace Mg^{++} is shown in Table 1. Little or no incorporation of ^{3}H-TTP was observed in the absence of added Mg^{++}. Similar results were reported by Lazarus et al. (18) for the RNA-dependent RNA polymerase activity of foot-and-mouth disease virus.

TABLE 1. *Lack of spermidine enhancement of polymerase activity in the absence of Mg^{++}*

Mg^{++} (mM)	Spermidine (mM)	CPM(^{3}H-TTP)
–	–	177
3.52	–	1,971
8.80	–	1,248
14.08	–	819
–	0.33	60
–	0.66	54
–	1.65	69
–	2.64	66

B. DNA-templated DNA synthesis

1. *Inorganic Cation Requirement.* As previously noted, disrupted oncogenic RNA viruses also utilize various DNAs as templates for DNA synthesis (3–5). In the present assays, each reaction mixture was supplemented with 1 μg of native calf thymus DNA (Sigma Chemical Co., St. Louis, Mo.), resulting in a four- to fivefold stimulation of ^{3}H-TTP incorporation as compared to the endogenously templated reaction. Optimal Mg^{++} and Mn^{++} concentrations for these DNA-dependent reactions are demonstrated in Fig. 3. Maximal incorporation was observed at a Mg^{++} concentration of 7.0 mM, effecting a 25-fold stimulation. A similar enhancement of incorporation was found for Mn^{++} at an optimal concentration of 3.5 mM.

2. *Effect of Spermidine.* The effect of spermidine concentration on the DNA-templated reaction was investigated with three concentrations of Mg^{++}, as shown in Fig. 4. In the presence of 10.5 mM Mg^{++}, all concentrations of spermidine (0.66 to 2.64 mM) inhibited ^{3}H-TTP incorporation. At the highest concentration of spermidine (2.64 mM), a 62% inhibition of incorporation was observed. A similar inhibition by spermidine was noted in the presence of 4.4 mM Mg^{++}. In the latter series, 7.92 mM spermidine inhibited the polymerase reaction by 84%. In the presence of the lowest concentration of Mg^{++} (1.7 mM), the addition of spermidine had little or no effect.

C. Poly rA·poly dT—templated DNA synthesis

1. *Inorganic Cation Requirement.* Spiegelman and collaborators (19) reported that a number of synthetic duplexes greatly stimulated incorporation when added to the endogenous reaction. In the experiments to be de-

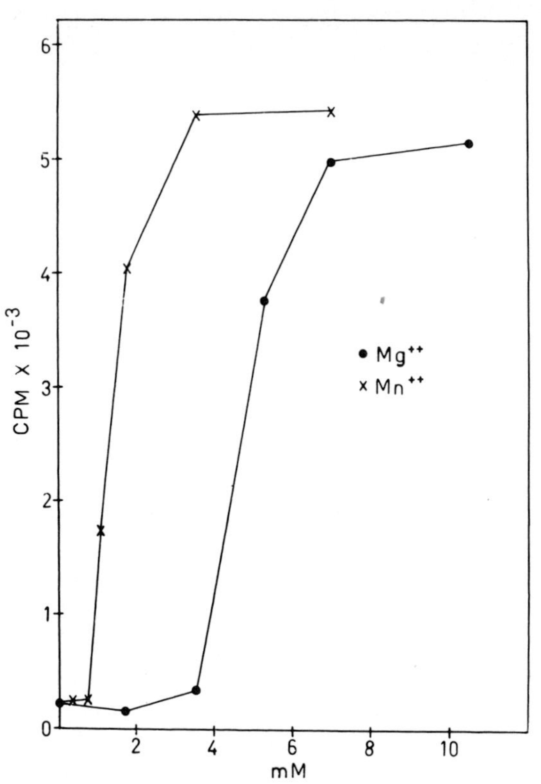

FIG. 3. Dependence of DNA-dependent DNA synthesis on magnesium or manganese concentrations. Each reaction mixture was supplemented with 1 μg native calf thymus DNA (0.15 M NaCl) after detergent disruption in the presence of dithiothreitol. Assay conditions are given in Methods and Materials.

scribed, poly rA·poly dT (Miles Laboratories, Elkhart, Ind.) addition effected a 30 to 40-fold increase in ^{3}H-TTP incorporation as compared to the endogenously templated reaction. Optimal concentrations of Mg^{++} and Mn^{++} for the poly rA·poly dT-templated reaction are shown in Fig. 5. Maximum incorporation was obtained in the presence of 7.0 mM Mg^{++}, resulting in a 264-fold increase in incorporation as compared to the reaction performed in the absence of Mg^{++}. Inhibition by higher Mg^{++} concentrations was observed.

The optimal concentration of Mn^{++} was determined to be 3.5 mM, increasing ^{3}H-TTP incorporation by 380-fold as compared to the incorporation noted in the absence of divalent cations. Higher concentrations of Mn^{++} were also inhibitory. Mn^{++}, especially at suboptimal concentrations, was

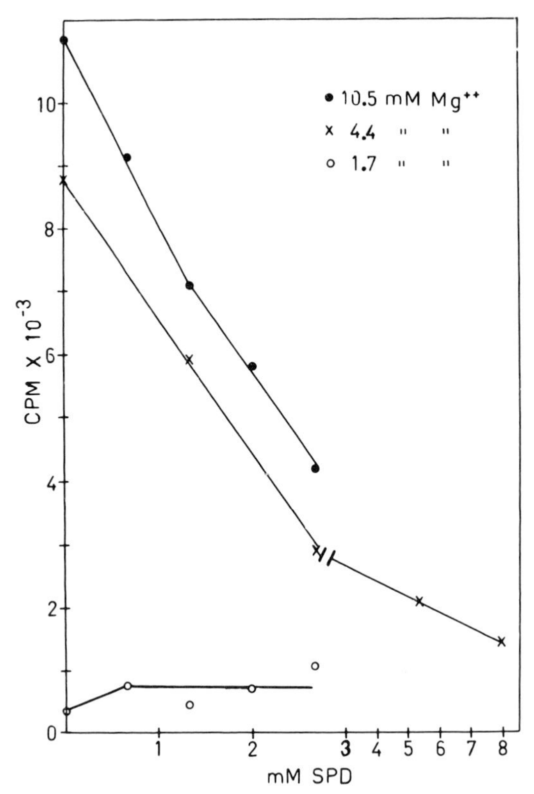

FIG. 4. The effect of spermidine concentration on DNA-dependent DNA synthesis in the presence of various concentrations of magnesium. Conditions for the assay of DNA polymerase activity are given in the legend to Fig. 3 and Methods and Materials.

found to be much more effective than Mg^{++} in satisfying the requirement for a divalent cation. Similar results have been reported by Scolnick et al. (20) for a number of oncogenic RNA viruses; however, in their investigation poly rA·poly rU was employed as exogenous template and Mg^{++} was found to be nearly ineffective.

2. *Effect of Spermidine.* As in the previous studies, the effect of spermidine addition was examined at two concentrations of Mg^{++}, 7.0 and 3.5 mM (Fig. 6). In the presence of optimal Mg^{++} concentration (7.0 mM), the addition of spermidine (1.32 mM) resulted in enhanced ^{3}H-TTP incorporation amounting to a 55% increase above the level found with Mg^{++} alone. Higher spermidine concentrations in the presence cf 7.0 mM Mg^{++} were inhibitory. At suboptimal Mg^{++} concentrations (3.5 mM), spermidine addition effected a marked stimulation of ^{3}H-TTP incorporation, i.e., 200

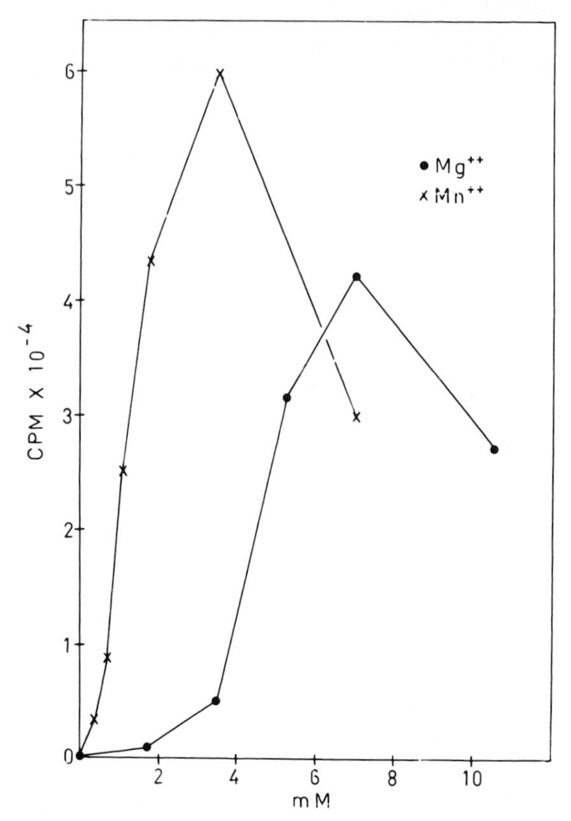

FIG. 5. Optimal magnesium and manganese concentrations for DNA synthesis by RIII MuMTV, employing poly rA· poly dT as exogenous template. Each reaction mixture was supplemented with 1.25×10^{-2} O.D. (260 nm) units [$E(p)_{260} = 5.97 \times 10^3$]. See Methods and Materials for assay conditions.

to 250% over the control containing only Mg^{++}. The level of incorporation in the presence of 3.5 mM Mg^{++} and 3.3 mM spermidine was nearly commensurate to that observed in the presence of 7.0 mM Mg^{++} and 1.32 mM spermidine.

IV. SUMMARY AND DISCUSSION

A summary of the data obtained in the present investigation is presented in Table 2. The endogenous reaction, directed by viral RNA, is stimulated by spermidine addition only at suboptimal Mg^{++} concentrations. In the absence of Mg^{++}, spermidine has no effect. The addition of spermidine in the presence of optimal Mg^{++} concentration results in inhibition. Similar results

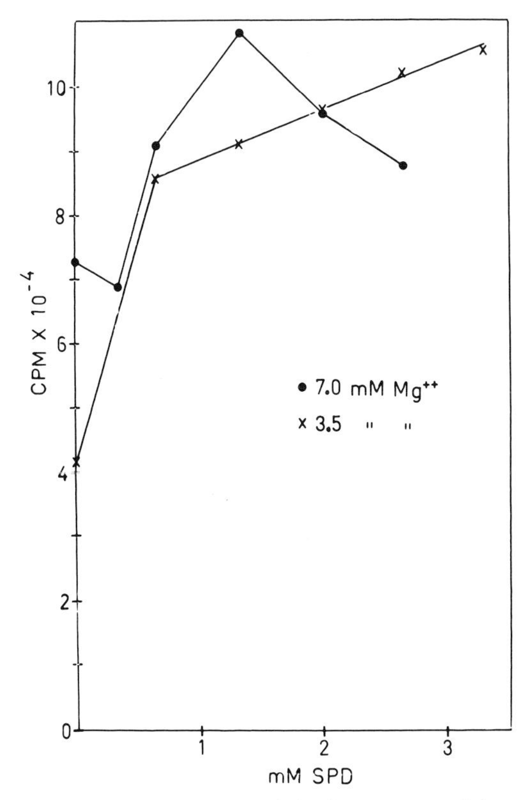

FIG. 6. Effect of spermidine on DNA synthesis in the presence of the exogenous template, poly rA· poly dT. Assay conditions are given in the legend to Fig. 5 and Methods and Materials.

have been reported for the RNA-dependent RNA polymerase of foot-and-mouth disease virus (18). From these data, it is suggested that suboptimal Mg^{++} and spermidine act in a cooperative manner to optimize the conformation of the RNA template-primer complex and excess concentrations of these cations inhibit DNA polymerase activity, possibly by effecting an unfavorably restricted template-primer conformation.

Table 3 presents a compilation of published reports relating to polyamine studies on the *in vitro* synthesis of nucleic acids. DNA-dependent synthesis of RNA has by far been most studied in a number of systems (21–29). These polymerases share a number of common properties, e.g., cofactor requirements, reaction kinetics, and the utilization of native and denatured DNA as well as RNA templates. Since these polymerases apparently bind to both DNA and RNA, it is possible that DNA polymerase binding to product

TABLE 2. *Summary of data: Mg^{++} and Mn^{++} optima and effect of spermidine*

Reaction	Exogenous template	Optimal conc. (mM)		Effect of spermidine	
		Mg^{++}	Mn^{++}	Optimal Mg^{++}	Subopt. Mg^{++}
Endogenous	–	5.25	0.7	Inhibited	Enhanced
Endogenous	Native calf thymus DNA	7.0	3.5	Inhibited	Inhibited
Endogenous	Poly rA·Poly dT	7.0	3.5	Enhanced	Enhanced

RNA results in product inhibition. Enzyme-RNA complex formation and its reversal by spermidine has been demonstrated by Gumport (27), and serves as an explanation for polyamine enhancement of DNA-dependent RNA synthesis. That a similar mechanism of polyamine action may be operative regarding the RNA-template DNA product hybrid formed as a result of RNA-dependent DNA synthesis, is currently being investigated.

Only two brief reports (30, 31) have appeared concerning the effect of polyamine addition on *in vitro* DNA-dependent DNA synthesis. In these studies, polyamine addition either had little effect at low concentrations or severely inhibited polymerase activity at higher concentrations. We have observed a similar phenomenon of spermidine inhibition for DNA-dependent DNA synthesis by disrupted MuMTV. In only one instance has polyamine stimulation of DNA-dependent DNA synthesis been reported (31), i.e., when nucleohistone was employed as template.

When the synthetic hybrid poly rA· poly dT was employed as exogenous template, spermidine addition effected enhanced incorporation of ^{3}H-TTP

TABLE 3. *Polyamine studies on the* in vitro *synthesis of nucleic acids*

Nucleic Acid	Source	Reference
RNA polymerases		
DNA-dependent	*Escherichia coli* B	21
	" " MRE600	22, 23
	Azotobacter vinelandii	24
	Micrococcus lysodeikticus	25–27
	Pseudomonas indigofera	28
	Rat testis	29
RNA-dependent	Foot-and-mouth disease virus	18
DNA polymerases		
DNA-dependent	*Escherichia coli* K12	30, 31
	MuMTV	Present study
RNA-dependent	MuMTV	Present study

with both optimal and suboptimal Mg^{++} concentrations. It appears that in this instance spermidine is functioning more efficiently than Mg^{++} in solely a "cation role" or that spermidine is performing a function not realized by Mg^{++} alone. It is likely that the synthetic duplex poly rA· poly dT is not a complete hybrid, but consists as well of single-stranded regions of poly rA and poly dT. In the present study, employing ^{3}H-TTP as label, the assayed template would consist of single-stranded poly rA regions and the complexed poly dT as primer. It has been demonstrated (32) that single-stranded poly rA exhibits considerable secondary helical structure under the proper ionic conditions. From the data reported here, it is suggested that spermidine is better able to maintain an optimal template structure than is Mg^{++}, this effect being reflected in greater incorporation in the presence of spermidine. This phenomenon could have important implications for the *in vitro* synthesis of DNA complementary to natural m-RNAs, since DNA formation by the RNA-dependent DNA polymerase proceeds only in the presence of added oligo dT. Oligo dT probably acts as primer complexed to poly A sequences located at the 3′-OH termini of natural m-RNAs (33). Of course, polyamine enhancement of this type synthesis would occur only if polyamines did not restrict the conformation of the remainder of the m-RNA.

The relatively narrow range of favorable spermidine concentrations for RNA- and DNA-dependent DNA synthesis by disrupted MuMTV, and possibly other oncogenic RNA viruses, may reflect important control mechanisms for the intracellular expression of the oncogenic viral genome. It would be of interest to determine the concentrations of polyamines and divalent cations during the process of infection leading to virus-induced transformation, and the effect of exogenous polyamines. As has been noted, spermidine increases rather dramatically in malignant tissues. These polyamine increases are especially impressive when one considers the reports of Russell et al. (34, 35), who observed elevated urinary excretion of conjugated polyamines in a variety of cancer patients, and its diminution as a result of surgical removal of the tumor mass or chemotherapy. The ubiquity of the polyamines and their intimate association with the processes of growth and development establish these compounds as prime candidates for the possible management of malignant cells.

ACKNOWLEDGMENTS

The authors wish to express their appreciation for expert technical assistance rendered by the following: polymerase assays, Garland S. Fout and Patricia Totten; purification of RIII MuMTV, Evelyn Craig.

This investigation was supported by contract PH 43-68-1000 and grant CA-08740 from the National Cancer Institute, General Research Support grant FR-5582 from National Institutes of Health, and Grant-in-Aid M-43 from the State of New Jersey.

REFERENCES

1. Temin, H. M., and Mizutani, S., *Nature,* **226,** 1211 (1970).
2. Baltimore, D., *Nature,* **226,** 1209 (1970).
3. Spiegelman, S., Burny, A., Das, M. R., Keydar, J., Schlom, J., Travnicek, M., and Watson, K., *Nature,* **227,** 1029 (1970).
4. Mizutani, S., Boettiger, D., and Temin, H. M., *Nature,* **228,** 424 (1970).
5. Riman, J., and Beaudreau, G. S., *Nature,* **228,** 427 (1970).
6. Cohen, S. S., *Introduction to the polyamines,* Prentice-Hall, Englewood Cliffs, N.J., 1971.
7. Nowinski, R. C., Old, L. J., Sarkar, N. H., and Moore, D. H., *Virology,* **42,** 1152 (1970).
8. Moore, D. H., Pillsbury, N., and Pullinger, B. D., *J. Nat. Cancer Inst.,* **43,** 1263 (1969).
9. Sarkar, N. H., Charney, J., and Moore, D. H., *J. Nat. Cancer Inst.,* **43,** 1275 (1969).
10. Duesberg, P. H., and Cardiff, R. D., *Virology,* **36,** 696 (1968).
11. Schlom, J., and Spiegelman, S., *Science,* **174,** 840 (1971).
12. Dion, A. S., and Moore, D. H., *Nature, New Biol.,* **240,** 17 (1972).
13. Gillespie, D., Marshall, S., and Gallo, R. C., *Nature New Biol.,* **236,** 227 (1972).
14. Dion, A. S., Sarkar, N. H., and Moore, D. H., Sixth Miles International Symposium on Molecular Biology, Cellular Modification and Genetic Transformation by Exogenous Nucleic Acids, in press.
15. Bachrach, U., Bekierkunst, A., and Abzug, S., *Israel J. Med. Sci.,* **3,** 474 (1967).
16. Neish, W. J. P., and Key, L., *Int. J. Cancer,* **2,** 69 (1967).
17. Rosenthal, S. M., and Tabor, C. W., *J. Pharm. Exp. Ther.,* **116,** 131 (1956).
18. Lazarus, L. H., Popescu, M., Barzilai, R., and Goldblum, N., *Arch. ges. Virusforsch.,* **36,** 311 (1972).
19. Spiegelman, S., Burny, A., Das, M. R., Keydar, J., Schlom, J., Travnicek, M., and Watson, K., *Nature,* **228,** 430 (1970).
20. Scolnick, E., Rands, E., Aaronson, S. A., and Todaro, G. J., *Proc. Nat. Acad. Sci. U.S.A.,* **67,** 1789 (1970).
21. Petersen, E. E., Kroger, H., and Hagen, U., *Biochim. Biophys. Acta,* **161,** 325 (1968).
22. So, A. G., Davie, E. W., Epstein, R., and Tissieres, A., *Proc. Nat. Acad. Sci. U.S.A.,* **58,** 1739 (1967).
23. Abraham, K. A., *Eur. J. Biochem.,* **5,** 143 (1968).
24. Krakow, J. S., *Biochim. Biophys. Acta,* **72,** 566 (1963).
25. Fox, C. F., and Weiss, S. B., *J. Biol. Chem.,* **239,** 175 (1964).
26. Fox, C. F., Robinson, W. S., Haselkorn, R., and Weiss, S. B., *J. Biol. Chem.,* **239,** 186 (1964).
27. Gumport, R. I., *Ann. N.Y. Acad. Sci.,* **171,** 915 (1970).
28. Tani, T., McFadden, B. A., Homann, H. R., and Shishiyama, J., *Biochim. Biophys. Acta,* **161,** 309 (1968).
29. Ballard, P., and Williams-Ashman, H. G., *Nature,* **203,** 150 (1964).
30. O'Brien, R. L., Olenick, J. G., and Hahn, F. E., *Proc. Nat. Acad. Sci. U.S.A.,* **55,** 1511 (1966).
31. Schwimmer, S., *Biochim. Biophys. Acta,* **166,** 251 (1968).
32. Eisenberg, H., and Felsenfeld, G. J., *J. Mol. Biol.,* **30,** 17 (1967).
33. Verma, I. M., Temple, G. F., Fan, H., and Baltimore, D., *Nature New Biol.,* **235,** 163 (1972).
34. Russell, D. H., *Nature New Biol.,* **233,** 144 (1971).
35. Russell, D. H., Levy, C. C., Schimpff, S. C., and Hawk, I. A., *Cancer Res.,* **31,** 1555 (1971).

Polyamines in Normal and Neoplastic Growth. Edited by
D. H. Russell. Raven Press, New York © 1973.

Structural and Biochemical Changes in the Nucleolus in Response to Polyamines

Eduard Gfeller*, Carl C. Levy, and Diane H. Russell

* Department of Anatomy, Johns Hopkins University School of Medicine,
725 N. Wolfe Street, Baltimore, Maryland 21205, and Baltimore Cancer Research Center,
National Cancer Institute, Baltimore, Maryland 21211

Elucidating the subcellular distribution of polyamines has been particularly difficult. *In vitro,* polyamines bind to nucleic acids and membrane systems, and they enhance ribosomal aggregation (1). As polycations, they bind to most cellular components in tissue homogenates, and it has been difficult to establish definitively their subcellular localization. Raina and Telaranta (2) found about 60% of radiolabeled spermidine associated with the microsomal fraction. However, after more exogenous labeled spermidine was added, the same percentage appeared in the microsomal fraction, and redistribution seemed to occur during homogenization.

We studied the intracellular distribution of polyamines by means of autoradiography. Twenty-four hr after intracisternal injection of ^{3}H-putrescine in rats, we found a homogeneous distribution of the label in cerebellar cortex (Table 1). No tissue component showed any preferential accumulation of the label. The almost complete turnover of putrescine in rat brain which occurs within 24 hr seemed ideal for studies of this kind, but neither distribution nor movement of label during the first 24 hr could be studied. In the hope of achieving better resolution in a rapidly growing system, which at the same time had cells of considerably larger size than mammalian neurons, and which consisted moreover of a homogeneous cell population, we turned to *Xenopus* liver cell cultures during logarithmic growth phase (3). This study showed that ^{3}H-putrescine was initially (Fig. 1), i.e., at 1 hr after incubation, associated primarily with the nucleus; at 4 hr it was associated mostly with the nucleolus. Cytoplasmic label only increased slowly and showed its highest level at 24 hr. This movement of label is similar to that of ^{3}H-uridine except for the continued rise of ^{3}H-uridine in nucleoli up to 24 hr after incubation. Pulse-labeling was used in this study, and ^{3}H-putres-

TABLE 1. *Grain distribution in autoradiographs from rat cerebellum 24 hr after intracisternal injection of* ^{3}H-putrescine

	Surface area (%)	Grains (%)
Molecular layer		
Neuropil	93	94
Stellate cells	4	1.5
Blood vessels and glia	3	4.5
Granular layer and Purkinje cell layer		
Neuropil	59	56
Purkinje cells	8	11.2
Basket cells	6	3.5
Granule cells	17	20.2
Blood vessels	6	6.2

cine was "chased" with 10 times the concentration of unlabeled putrescine. At 24 hr after incubation, 30% of the putrescine had been converted to spermidine, whereas less than 1% was converted to spermine.

The absence of labeled mitoses was particularly noteworthy. Since there is little nuclear RNA present during interphase, this finding lends additional support to the notion of an association of polyamines with RNA. Also, basic aniline dyes fail to stain nuclear RNA during interphase, and Harris (4) has proposed that the binding sites to which aniline dyes attach may not be available during interphase, and it is possible that polyamines may be associated with the same binding sites.

Although the localization of polyamines in this study was of limited resolution, and although there may still be redistribution of label during fixation, this study was the first demonstration of the subcellular distribution of putrescine as a function of time after pulse-labeling. The results indicated the strong possibility of an association of putrescine with nuclear RNA.

This latter possibility was further studied in rat liver nuclei (5, 6). Biochemical studies indicated that isolation of rat liver nuclei using polyamines was possible and that this isolation procedure led to a dramatic increase in nucleolar DNA content, as well as to an activation of RNA polymerase I. Table 2 shows that there are considerable differences between fractions isolated with 5 mM spermidine and those isolated with 5 mM magnesium. The spermidine-isolated nucleoli contained twice as much DNA as did those isolated with magnesium; the nucleoli contained equal amounts of protein and RNA under both conditions. The amount of DNA found in association with magnesium-isolated nucleoli is in good agreement with data from other laboratories (13). The amount of DNA found in spermidine-isolated nucleoli

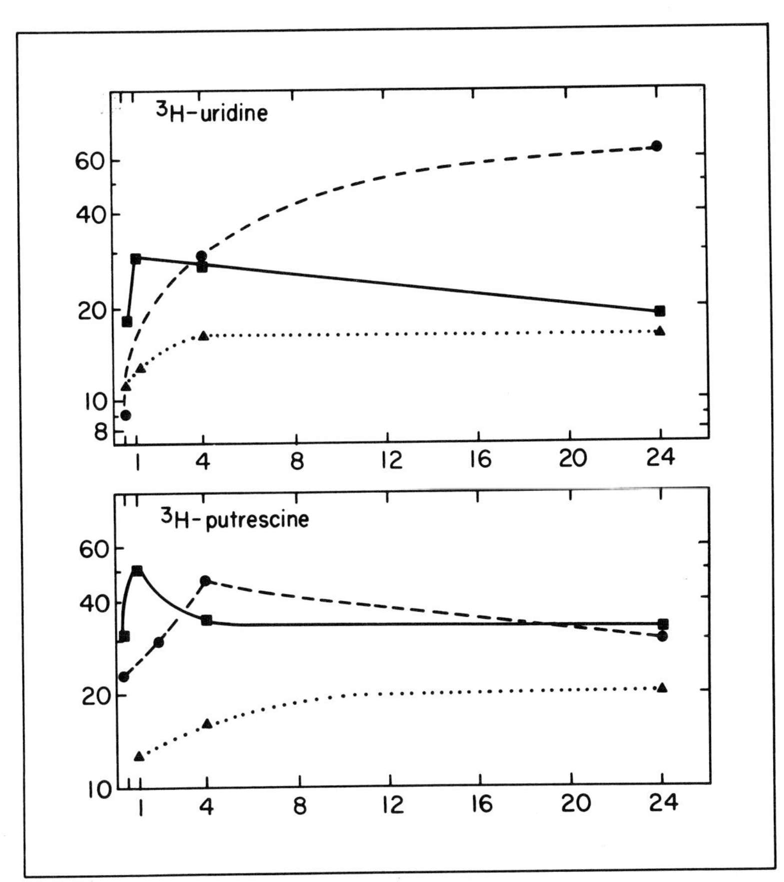

FIG. 1. Grain densities over different cell compartments after ^{3}H-uridine. Abscissae: time in hr; ordinates: grains per unit surface area. Squares connected by solid line: nuclear label. Solid circles connected by broken line: nucleolar label. Triangles connected by dotted line: cytoplasmic label. [From Gfeller and Russell (3).]

is, on the other hand, similar to amounts found in Walker tumor cell nucleoli (13).

RNA polymerase I was 10-fold more active in nucleoli isolated with spermidine than in those isolated with magnesium (Table 1). This activity was insensitive to inhibition by 1 μM α-amanatin. Neither spermidine nor magnesium was used in the actual assay for RNA polymerase activity. Moreover, after partial purification of the enzyme, addition of magnesium or spermidine did not affect enzyme activity. These data were interpreted to mean that spermidine was not affecting the polymerase activity by inter-

TABLE 2. *Effects of isolation of rat liver nucleoli with spermidine upon their RNA Polymerase I activity, DNA, RNA, and protein content*

Fraction	RNA Polymerase I Activity (cpm/10^6 nucleoli)	pg protein/ nucleolus	pg RNA/ nucleolus	pg DNA/ nucleolus
Nucleoli isolated with 5 mM spermidine	486	8.2	1.2	1.0
Nucleoli isolated with 5 mM magnesium	54	7.5	0.99	0.41

RNA polymerase assayed according to the method of Reoder and Rutter (7). Nucleoli were isolated by the method of Ro and Busch (8). DNA content was assayed according to the method of Burton (9), RNA by the method of Schneider (10), and protein by the method of Lowry et al. (11), except in the spermidine fractions in which the microbiuret method was used (12). pg = picograms.

action with the enzyme itself, and that we had purified away the factors upon which spermidine was acting in order to cause enhanced nucleolar RNA polymerase activation in the intact nucleoli.

In an attempt to find morphological correlates of these events, we studied nucleolar ultrastructure in rat liver nucleoli as a function of polyamine concentration during isolation. Figure 2 shows that magnesium at concentrations of 2.5, 5.0 and 10.0 mM had little or no effect on nuclear morphology, whereas putrescine at the same concentrations showed slightly increased nucleolar size. Particularly in 5.0 mM putrescine, occasional dense bodies of finely granular, homogeneous appearance were observed. These microspherules, however, were not numerous.

Using spermidine for isolating nuclei produced numerous lacunae in the nucleoli. Lacunae measured from 0.02 to 0.05 μ at 2.5 mM concentration, and from 0.1 to 0.2 μ at 10.0 mM spermidine. The granular component of the nucleoli was much more dispersed in spermidine-treated nucleoli than in those treated with putrescine or magnesium.

The most prominent dispersion of the granular component of the nucleolus occurred with spermine. Lacunae measuring up to 0.3 μ were frequent, particularly at 10.0 mM. Extranucleolar chromatin was usually condensed around the enlarged nucleolus, forming chromatin caps. These caps give the nucleoli a ring-shaped appearance in the light microscope. At 10.0 mM spermine, single fibrillar strands of the fibrillar component of the nucleolus could be observed in some lacunae. These strands measured 50×300 to 400 Å.

The presence of chromatin caps correlates well with the finding of elevated levels of DNA associated with nucleoli after isolation with polyamines. This DNA could represent either DNA synthesized *de novo* or DNA condensed around the nucleolus from extranucleolar sources. We studied the incorporation of dTTP-^{32}P into DNA, and found that the addition of spermidine produced a two- to threefold increase in the rate of incorporation. It is therefore possible that the chromatin caps represent rapidly synthesized DNA. ^{3}H-Thymidine incorporation or incorporation of deoxynucleotides is also enhanced by spermidine. Moreover, spermidine appears to overcome the inhibitory effects of the supernatant fraction upon incorporation of thymidine or deoxynucleotides into nuclear preparations. We have some preliminary evidence that this inhibition may be due to a DNAase present in the supernatant, and it is tempting to assume that spermidine might inhibit this DNAase. However, these experiments are preliminary and further work is necessary to establish their validity.

The structural changes observed in the nucleoli are similar to those seen after long-term treatment with thioacetamide. We therefore studied the possibility of a similar direct effect of thioacetamide on nucleoli *in vitro*. Thioacetamide-treated nucleoli isolated with magnesium or polyamines showed neither changes of their own nor changes which would be a modification beyond the effect of polyamines alone. Arnold et al. (14) recently described similar effects of polyanions on nuclear and nucleolar structure. As Table 3 shows, there were marked differences in the effectiveness of various

TABLE 3. *Effect of polyanions on DNA synthesis and nuclear morphology*

Anion	DNA synthesis[a] (pmoles)	Morphology
Control	30	
Polycytidylic acid	35	No changes
Polyadenylic acid	40	Slight nucleolar changes
Polyuridylic acid	60	Chromatin dispersion, condensation of fibrillar nucleolonema, small defects in nuclear membrane
Polyinosinic acid	250	Dispersion of chromatin and appearance of microspherules
Polyxanthine and heparin	350	Marked chromatin dispersion, appearance of single strands of chromatin, large number of microspherules, lysis of nuclear membranes, complete nucleolar dispersion

From (14).
[a] dTMP-^{3}H incorporated into acid-insoluble DNA in 30 min at 37°.

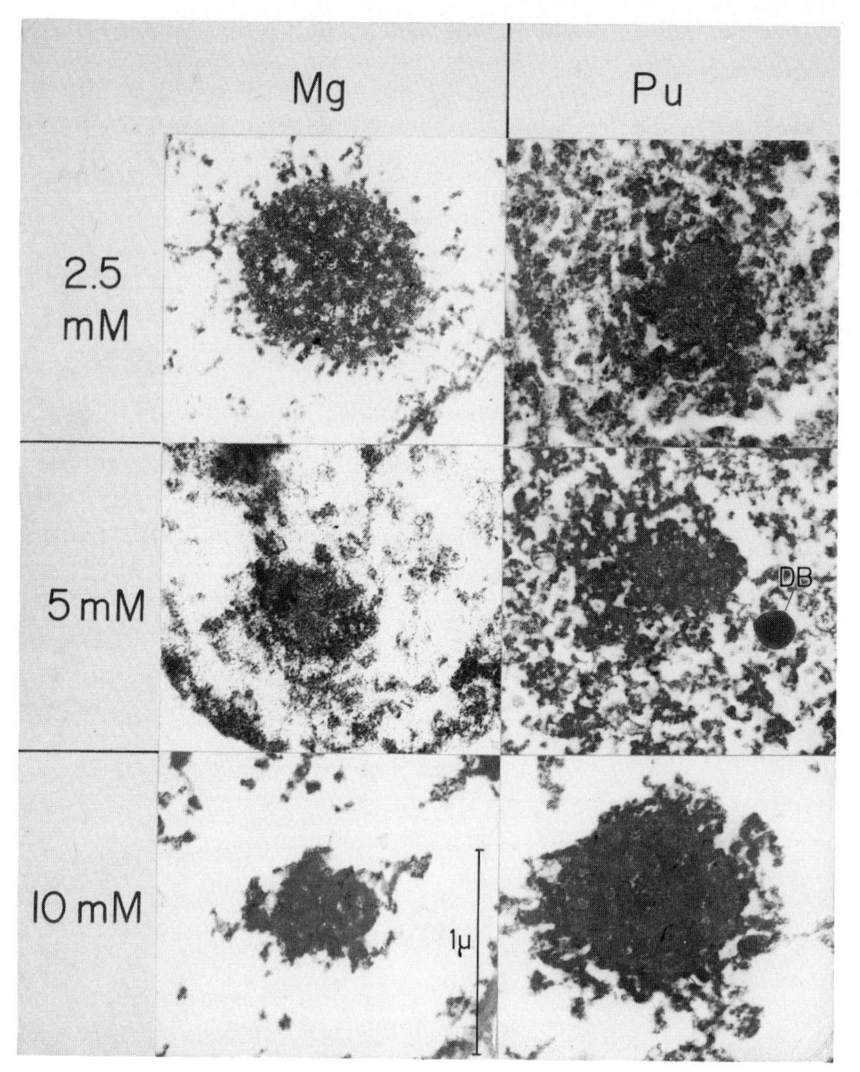

FIG. 2. Nucleolar changes observed after *in vitro* isolation of rat liver nucleoli in 2.4, 5, and 10 mM magnesium (Mg), putrescine (Pu), spermidine (Spd) or spermine (Spm). The magnification is approximately × 20,000. Note the large lacunae, particularly at 5 and 10 mM spermidine

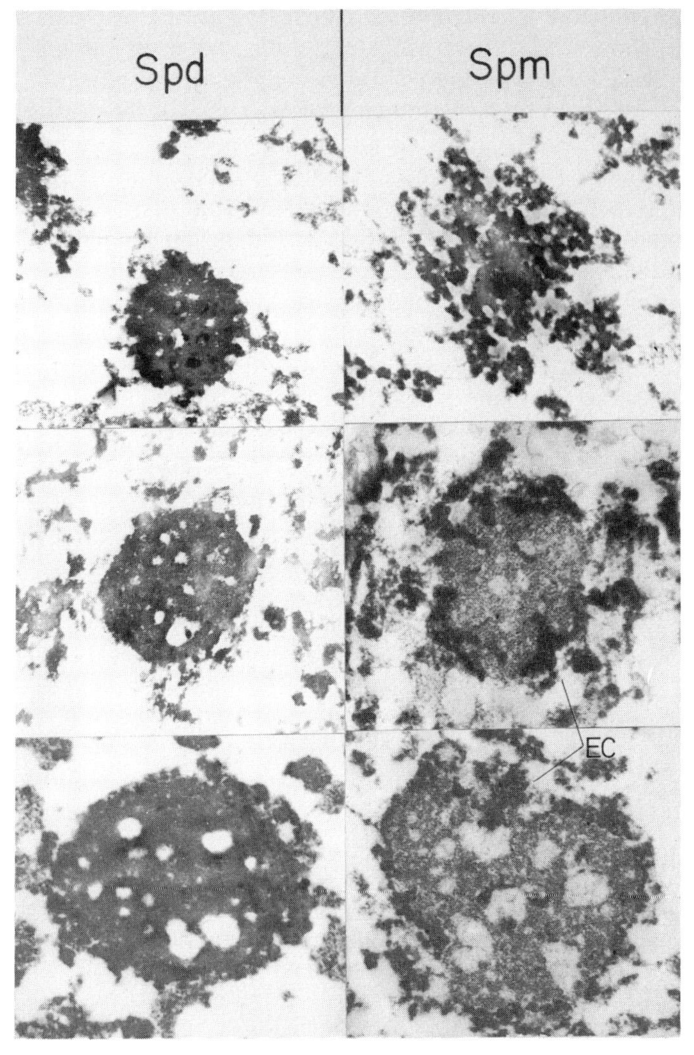

and spermine. Extrachromatic caps (EC) are particularly obvious at 5 and 10 mM spermine. DB = dense body; these were found only in putrescine-isolated fractions. [From Gfeller et al. (5).]

polyanions, polyxanthine and heparin being the most effective. Their study, moreover, showed that there was a correlation between the ability of polyanions to increase DNA synthesis, and their structural effects.

There is no doubt that isolation of rat liver nuclei in the presence of polyamines leads to what is known as nucleolar hypertrophy. Such hypertrophy has been observed in cancer cells, embryonic cells, and regenerating liver cells, but it has also been seen after stimulation of salivary glands with isoproterenol and in livers of nutritionally deficient animals (15). Hypertrophy of the nucleolus, therefore, can be induced not only by polyamines, but in certain experimental conditions it is associated with increased endogenous polyamine levels. Biochemical evidence clearly implicates polyamines as stimulators of RNA synthesis, and it is tempting to assume that the effects on DNA might be specific for the nucleolar organizer. Absence of labeled mitoses in our autoradiographic study and the appearance of chromatin caps are certainly compatible with such an assumption.

ACKNOWLEDGMENTS

This investigation was supported by grant NS07934 from the National Institute of Neurological Diseases and Stroke.

REFERENCES

1. Tabor, H., and Tabor, C. W., *Pharmacol. Rev.*, **16,** 245 (1964).
2. Raina, A., and Telaranta T., *Biochim. Biophys. Acta,* **138,** 200 (1967).
3. Gfeller, E., and Russell, D. H., *Z. Zellforsch.,* **120,** 321 (1971).
4. Harris, H., *Progr. Nucleic Acid Res.,* **2,** 19 (1963).
5. Gfeller, E., Stern, D. N., Russell, D. H., Levy, C. C., and Taylor, R. L., *Z. Zellforsch.,* **129,** 447 (1972).
6. Russell, D. H., Levy, C. C., and Taylor, R. L., *Biochem. Biophys. Res. Comm.,* **47,** 212 (1972).
7. Roeder, R. G., and Rutter, W. J., *Nature,* **224,** 234 (1969).
8. Ro, T. S., and Busch, H., *Cancer Res.,* **24,** 1630 (1964).
9. Burton, K., *Biochem. J.,* **62,** 315 (1956).
10. Schneider, W. C., *J. Biol. Chem.,* **161,** 293 (1945).
11. Lowry, O. H., Rosebrough, N. J., Farr, A. L., and Randall, R. J., *J. Biol. Chem.,* **193,** 265 (1951).
12. Zamenhof, S., in S. P. Colowick and N. O. Kaplan (Editors), *Methods in enzymology,* Vol. *3,* 1957, p. 696.
13. Busch, H., and Smetana, K., *The nucleolus,* Academic Press, New York, 1970, p. 174.
14. Arnold, E. A., Yawn, D. H., Brown, D. G., Wyllie, R. C., and Coffey, D. S., *J. Cell Biol.,* **53,** 737 (1972).
15. Simard, R., *Int. Rev. Cytol.,* **28,** 169 (1970).

Polyamines in Normal and Neoplastic Growth. Edited by D. H. Russell. Raven Press, New York © 1973.

The Determination of Polyamines in Urine by Gas-Liquid Chromatography

Charles W. Gehrke, Kenneth C. Kuo, Robert W. Zumwalt, and T. Phillip Waalkes*

*Experiment Station Chemical Laboratories, University of Missouri-Columbia, Columbia, Missouri 65201, and *National Cancer Institute, Bethesda, Maryland 20014*

I. INTRODUCTION

The polyamines putrescine, spermidine, and spermine have become the subject of intense research interest. They have been found to occur in many bacteria, plants, and bacteriophages as well as in animal tissue and body fluids. An extensive review by Tabor and Tabor (1) included the occurrence, biologic and pharmacologic effects, and biosynthesis of the polyamines. More recently, the increased synthesis and accumulation of polyamines after tissue stimulation have been found to be characteristic of certain rapid growth systems (2–6), both normal and neoplastic.

The possible role of polyamines as clinical biochemical markers for malignancy was first suggested by the report of Russell et al. (7) indicating that polyamines are present in increased amounts in the urine of cancer patients. Steps in the assay included butanol extraction, high-voltage paper electrophoresis, and ninhydrin staining reaction for final quantitative determination. The procedure, however, lacked the sensitivity necessary to determine the range of values for urine levels in normal control subjects.

In order to obtain the desired sensitivity and to gain increased specificity, gas-liquid chromatography (GLC) was considered, and the following procedure developed. Substances present in urine which interfere with the accurate determination of the polyamines had to be removed prior to assay by GLC. To do so, polyamines were first isolated by means of a cation-exchange clean-up procedure, then converted to the volatile N-trifluoroacetyl derivatives and quantitatively analyzed by GLC. This chapter is concerned primarily with the experimental details involved in the assay method. Initial data are presented on amounts of polyamines found in the urine of normal subjects with a limited number of preliminary analyses carried out on urine samples from patients with malignant disease. The ultimate goal of this work is to evaluate whether polyamine levels in urine can be utilized as a sensitive indicator for changes in total tumor cell number in cancer patients undergoing therapy.

II. EXPERIMENTAL

A. Apparatus

A Bendix 2500 series gas chromatograph with a four-column oven bath, four hydrogen flame detectors, two differential electrometers, and a linear temperature programmer, and equipped with a Varian Model A-20 dual pen strip chart recorder was used. The GLC columns were composed of 2% OV-17 and 1% SP-2401 coated on 100/120 mesh Gas-Chrom Q, 1 m × 2 mm I.D. glass. Peak areas were determined with Infotronics Corp. CRS-104 digital integrators.

Solvents were removed from the samples with a Calab rotary evaporator, and "cold finger" condenser filled with dry ice in methyl cellosolve, and a Welch Duo-Seal vacuum pump. During evaporation the samples were heated in a 60°C constant temperature water bath.

Pyrex 16 × 75 mm glass, screw-top culture tubes with Teflon-lined caps (Corning No. 9826) were used as the reaction vessels for the acylation reactions.

Filters containing shell-type charcoal and indicating molecular sieve (Type 13X) were placed in the lines for removal of trace hydrocarbons and water in the nitrogen, hydrogen, and air to the chromatograph. For removal of particulate matter from the gas streams, "F" series inline 7-μ sintered stainless steel filters were obtained from the Nupro Company (Cleveland, Ohio) and fitted into the lines.

The ion-exchange columns for cleanup of the samples 6 mm I.D. glass columns with Teflon stopcocks were obtained from Fischer-Porter Co. (Warminster, Pa.). The cation-exchange resin (Dowex 50 × 8, 50/100 mesh, H^+ form) was obtained from BioRad Laboratories (Richmond, Calif.).

The various solutions were passed through the ion-exchange columns with the aid of a proportioning pump obtained from Technicon, Inc. (Tarrytown, N.Y.).

B. Reagents and materials

The polyamines putrescine, spermidine, and spermine were obtained from Calbiochem (Los Angeles, Calif.).

The liquid phase SP-2401 was obtained from Supelco, Inc. (Bellefonte, Pa.), and OV-17 and Gas Chrom Q were obtained from Applied Science Laboratories (State College, Pa.).

Trifluoroacetic anhydride (TFAA) was purchased from Eastman Kodak Co. (Rochester, N.Y.).

The following reagents were obtained from commercial sources, and the solutions were prepared as listed: tri-*n*-butyl citrate (antifoamant); hydrochloric acid, 6 N and 0.50 N; ammonium hydroxide, 3.0 N and 4.0 N; ammonium hydroxide, pH 10.5; EDTA, ammonium form, 5%; ethanol, 80%; and acetonitrile.

The general techniques used on derivatization and gas-liquid chromatography followed the methods described by Gehrke et al. (9–11). For purpose of comparison with GLC, classical ion-exchange analyses (CIE), after hydrolysis and cation-exchange cleanup, were carried out on a BC 200 amino acid analyzer with some modifications as outlined by Bremer and Kohne (12). Buffer A was pH 5.28 citrate buffer, 1.35 M in sodium chloride; Buffer B was pH 5.28 and 2.35 M in sodium chloride; and Buffer C was pH 5.28 citrate buffer, 2.85 M in sodium chloride.

C. Samples, collection, and storage

The 24-hr urine samples were collected at ice temperature. Aliquot samples were frozen and stored at −20°C. Some of the normal control urines were from laboratory personnel, ages 18 to 55. Another group of eight normal control subjects, ages 5 to 52, had a second collection either the following 24 hr after the first collection or at a later date. In general, the cancer patients selected had advanced malignant disease, and at the time of the urine collection the majority of the patients were not receiving antineoplastic drugs or other antitumor therapy. The 24-hr urine samples from the cancer patients were obtained through the courtesy of the following hospital services: (a) the Johns Hopkins University Medical School, Oncology Division, (b) the National Cancer Institute Solid Tumor Service, and (c) the Cancer Research Center, Columbia, Missouri.

D. Procedure

Because polyamines are excreted in a conjugated form, hydrolysis is necessary in order to assay them as the free amines (7).

1. *Hydrolysis and Cation-Exchange Cleanup.* A 5- to 25-ml aliquot of urine was evaporated to near dryness using a rotary evaporator and 60°C water bath; 5 ml of 6 N HCl was added, and then hydrolyzed in a 110°C

oven for 16 hr. The hydrolysate was evaporated to dryness to remove HCl, basicity adjusted, and interferences removed (8) by cleanup on a small Dowex 50 W × 8, 50/100 mesh, H^+ form cation-exchange column.

2. *Chromatographic and Instrumental.* The sample eluate from the cation-exchange clean-up column was evaporated to approximately 1 to 2 ml by rotary evaporation, transferred to a 16 × 75 mm Pyrex glass screw-top culture tube, then evaporated to dryness under a stream of highly pure N_2 gas at 95°C. Equal portions (0.5 ml) of CH_3CN and of TFAA were added, the tube capped, and the mixture sonicated for 1 min. Derivatization of the polyamines was accomplished by placing the sample tube in an oil bath at 100°C for 5 min. The CH_3CN and TFAA were removed by evaporation with N_2 gas at room temperature. The sample was then redissolved in 0.50 ml of diethyl ether containing an exact amount of phenanthrene as internal standard. The amount of internal standard added should be at a level of the expected amounts of polyamines in the sample. Three to 5 μl was injected on a mixed-phase GLC column of 2 w/w% OV-17 and 1 w/w% SP-2401 on 100/120 mesh Gas Chrom Q, 1 m × 2 mm I.D. glass. The initial column temperature was 80°C with a program rate of 8°C/min and a final temperature of 255°C. Phenanthrene was used as a chromatographic internal standard to determine the response ratios and for calculation of the concentrations of polyamines in the sample.

III. RESULTS AND DISCUSSION

The chromatogram presented in Fig. 1 shows the GLC analysis of a standard solution of polyamines. Each peak represents about 120 ng of polyamines. (The instrumental and chromatographic parameters are presented on this and all subsequent figures.) Most of the interfering materials were removed by the ion-exchange clean-up method, resulting in a good chromatogram and adequate separation of the polyamines.

Straight-line response curves were obtained for putrescine, spermidine, and spermine over the range investigated of 10 to 600 ng injected. The w/w response values for each of the polyamines were essentially the same (Fig. 2).

Figure 3 shows a chromatogram for putrescine (N_2), spermidine (N_3), and spermine (N_4) in normal human urine. Only a trace of spermine was found for this individual.

The polyamine content of 24-hr urine samples from normal control subjects is presented in Tables 1 and 2. For the latter, a second 24-hr collection was analyzed for each subject. The range of values for the putrescine, spermidine, and spermine levels in the control urines expressed as mg/24 hr were as follows, for N_2 (0.29 to 2.42), N_3 (0.37 to 2.7), and N_4 (a trace to

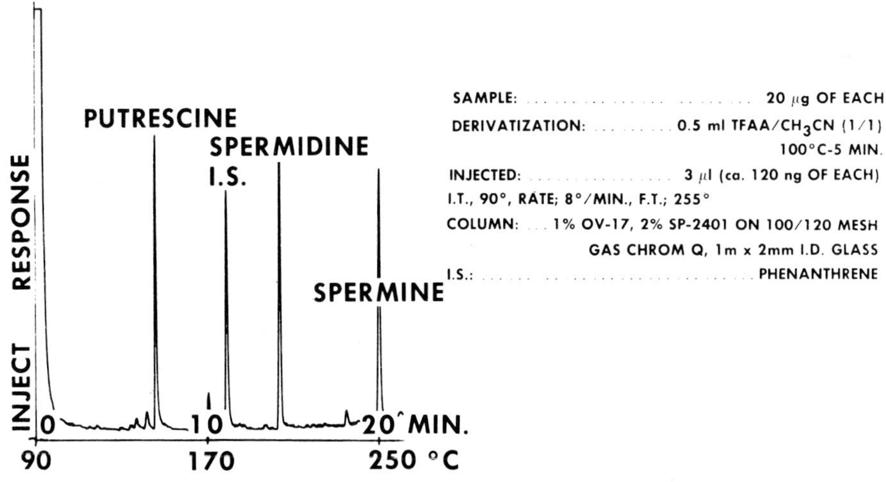

FIG. 1. GLC analysis of polyamines.

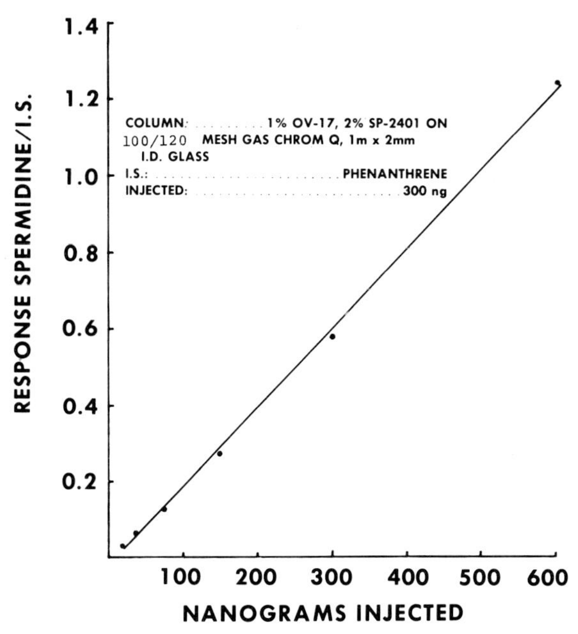

FIG. 2. Chromatographic response of spermidine N-TFA derivative.

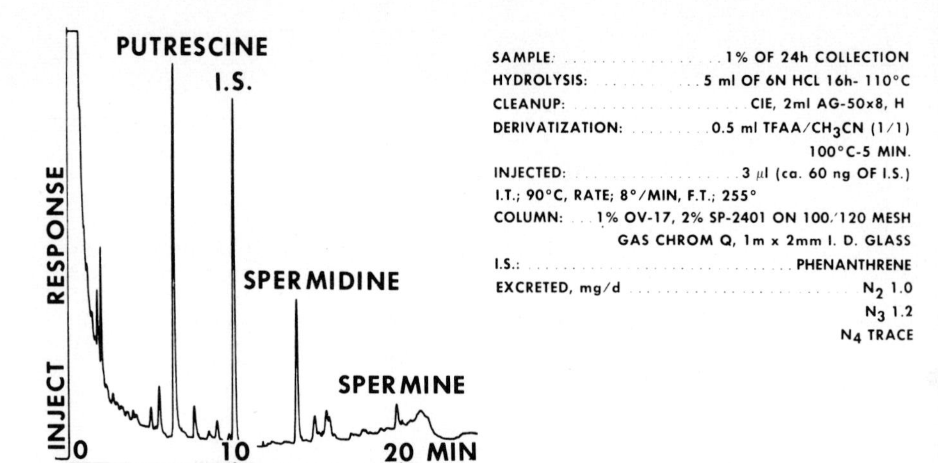

FIG. 3. GLC analysis of polyamines in human urine; normal.

TABLE 1. *Gas-liquid chromatographic and classical ion-exchange analysis of polyamines in normal urine*

Age of normals (yr)	mg/24 hr[a] Putrescine GLC	Putrescine CIE	Spermidine GLC	Spermidine CIE	Spermine GLC	Spermine CIE	Added µg	% recovery GLC N_2	N_3	N_4
27	2.4	2.4	1.4	1.5	0.3	0.1	20	77	66	40
26	1.8	1.6	2.7	2.2	1.5	0.6	20	97	64	39
26	0.9	0.7	1.1	1.1	0.2	0.1	20	107	70	45
24	1.3		1.3		0.5		50	–	89	62
55	1.2	1.1	1.3	1.2	0.3	0.3	100	107	94	74
30	1.2		1.7		0.3		100	91	103	74
26	1.1		0.8		0.1		100	87	101	72
29	1.4		1.1		0.7		100	77	80	44
26	1.3	1.1	1.1	1.1	trace	0.2	100	80	93	55
31	1.7		0.8		0.2		–	–	–	–
18	1.4		1.4		0.2		–	–	–	–
20	1.4		1.1		0.3		–	87	108	45
Average	1.4		1.3		0.4		Average	90	87	55

[a] Approximately 25-ml urine, evaporated, hydrolyzed with 5 ml 6 N HCl, 110°C–16 hr. Corrections were made for recovery.

TABLE 2. *Gas-liquid chromatographic analysis of polyamines in normal urine*

Age (yr) and sex of normals	mg/24 hr[a]					
	Putrescine		Spermidine		Spermine	
	GLC	% recovery	GLC	% recovery	GLC	% recovery
10 F	0.7		0.4		trace	
	0.6		0.5		trace	
5 F	0.3		0.6		trace	
	0.6		0.4		0.04	
24 F	1.1		0.5		trace	
	0.8	83	1.0	89	trace	79
19 M	0.5		0.9		0.2	
	0.6		0.8		0.1	
27 M	1.1		1.4		0.1	
	0.8	92	0.9	120	trace	40
15 M	0.9		1.3		trace	
	0.9		0.9		0.02	
52 M	1.3	74	1.1	94	trace	46
	0.8		0.8		0.1	
48 F	2.4		1.3		trace	
	1.7	83	0.9	88	trace	36
Average	0.94		0.86			

[a] Approximately 25-ml urine, evaporated, hydrolyzed with 5 ml 6 N HCl, 110°C–16 hr. Corrections were made for recovery.

1.5 mg/24 hr). Table 3 gives the values for polyamine content in 24-hr urine collections for 26 cancer patients, of whom 13 had lung cancer. The analyses were made by GLC and confirmed for some of the samples by CIE liquid chromatography (12).

In addition, a series of recovery experiments with normal control urine was made to confirm the integrity of the chromatographic methods used (Tables 1 and 2). The polyamine standards were added to the urine prior to hydrolysis. As shown in Table 1, the recovery for N_2 averaged 90%, for N_3 84%, and for N_4 56%. Further work (unpublished data) indicates that the recovery for N_4 can be substantially improved. In addition, a number of recovery experiments were made in which 20, 50, and 100 μg of each polyamine were added to aliquots of urine from different cancer patients. As before, the polyamines in the urine were hydrolyzed, interferences removed by cation-exchange cleanup, and the polyamines eluted and then analyzed by both GLC and CIE. In 14 experiments in which polyamines were added to cancer urine, the recoveries by GLC averaged about 90% for N_2, greater than 90% for N_3, and 50 to 60% for N_4. The recovery of added polyamines to urine of cancer patients by CIE averaged 100, 100, and 86%, respectively, for N_2, N_3, and N_4.

TABLE 3. *GLC and CIE analysis of polyamines in urine of cancer patients*[a]

Patient number	Diagnosis	mg/24 hr[b]					
		Putrescine		Spermidine		Spermine	
		GLC	CIE	GLC	CIE	GLC	CIE
1	Malignant lymphoma	2.7	2.6	8.2	6.6	1.9	1.8
2	Reticulum cell sarcoma	1.6		2.2		1.0	
3	Lymphosarcoma	2.3		1.2		<0.1	
4	Hodgkin's disease	1.7		1.2		<0.1	
5	Multiple myeloma	1.4		1.8		0.5	
6	Acute myelocytic leukemia or aplastic anemia	5.7		1.8		1.3	
7	Chronic myelocytic leukemia	1.6	1.2	1.9	1.7	0.1	0.1
8	Bronchogenic carcinoma	1.1	1.1	2.4	2.1	0.6	0.5
9	" "	2.1	1.8	2.2	1.8	0.4	0.3
10	" "	3.5	3.1	5.2	4.5	0.8	0.6
11	" "	0.5	0.5	1.0	1.0	trace	0.1
12	" "	4.2	3.9	4.3	4.0	0.1	0.2
13	" "	1.9	1.6	1.1	1.1	0.1	0.2
14	" "	1.9		1.6		0.4	
15	" "	1.5	1.6	1.4	1.7	trace	0.2
16	" "	1.1	0.9	2.7	2.0	0.2	0.2
17	" "	0.8	0.7	1.0	0.9	trace	0.1
18	" "	2.5		2.4		1.0	
19	" "	2.2	2.0		3.4	trace	0.3
20	" "	1.2	1.5	2.5	3.8	0.2	0.5
21	Ovarian carcinoma	1.1		0.5		<0.1	
22	Adenocarcinoma of the colon	1.0	1.2	1.2	1.4	trace	0.4
23	Adenocarcinoma of the rectum	1.2	1.4	1.5	1.3	0.2	0.3
24	Adenocarcinoma of the pancreas	1.8	1.1	2.2	1.8	0.1	0.1
25	Adenocarcinoma of the pancreas	1.8	1.7	0.6	0.6	0.1	0.04
26	Probable carcinoma of breast	0.6	0.6	0.9	0.9	0.2	0.07

[a] 25-ml aliquots, evaporated, hydrolyzed with 6 N HCl, 110°C–16 hr.
[b] Analyses made by GLC and CIE on same hydrolysate. Corrections were made for recovery.

A chromatogram is shown for polyamines in normal human urine to which 100 μg of each polyamine was added prior to hydrolysis of the sample. The sample was then taken through the hydrolysis and ion-exchange clean-up steps, derivatized, and analyzed. The response and recoveries are given in Fig. 4. A corresponding chromatogram for polyamines in human urine of a patient with Hodgkin's disease was used in a recovery study (Fig. 5). The response for polyamines in urine of a patient with reticulum cell sarcoma is given in Fig. 6.

The derivatization reaction conditions should be carefully controlled as

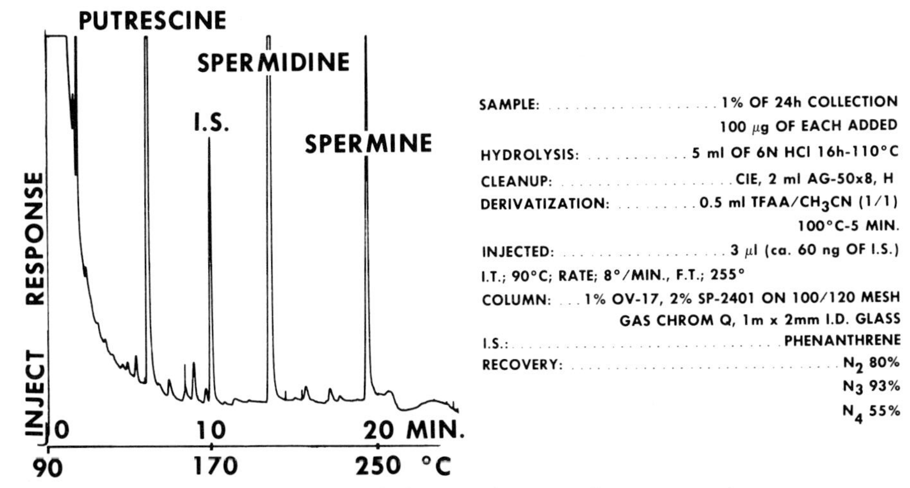

FIG. 4. GLC analysis of polyamines in human urine; normal recovery study.

lower temperatures and shorter times caused low yields for spermidine and spermine. Higher temperatures and longer times gave double peaks for putrescine and spermidine. Reproducible data were obtained on derivatization at 100 ± 5°C for 5 ± 0.5 min.

Another internal standard that may be used is 3,3′-diaminodipropylamine. This compound can be added to the urine sample prior to hydrolysis and carried through the entire analytical and instrumental method.

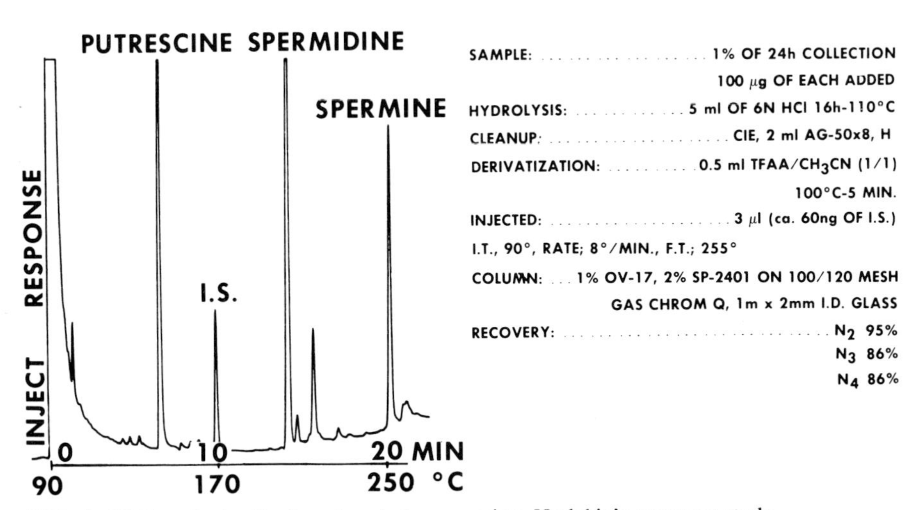

FIG. 5. GLC analysis of polyamines in human urine; Hodgkin's recovery study.

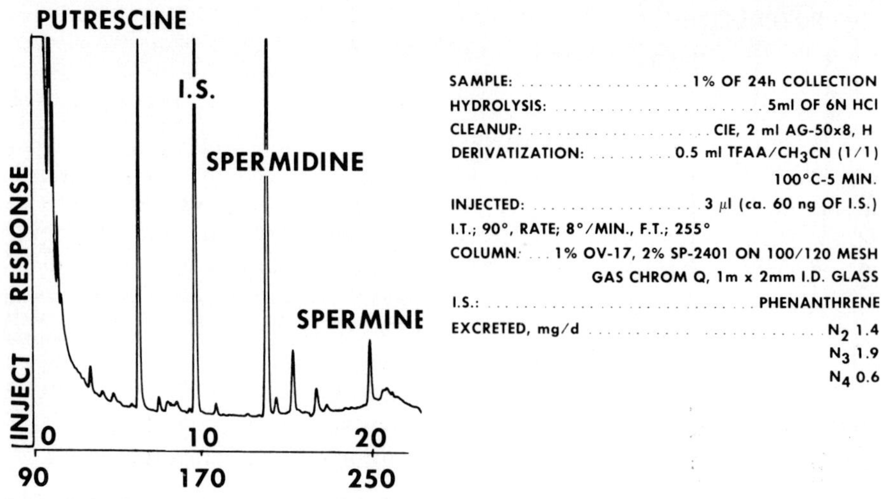

FIG. 6. GLC analysis of polyamines in human urine; reticulum cell sarcoma.

During derivatization, CH_3CN dissolves the small amount of residue; however, after prolonged use of the chromatographic column, tailing of the peaks occurs. Also, with CH_3CN as the solvent, some samples gave interfering peaks on the chromatogram after spermidine. Acetone as solvent also dissolved all of the remaining residue in the derivatization vial, but did not cause peak tailing, although some samples gave interfering peaks after spermidine with acetone.

The solvents methylene chloride and chloroform did not completely dissolve all of the brownish residue and this resulted in a low GLC response for spermidine and spermine. However, no interference peaks or peak tailing was evident. Diethyl ether is a good solvent and dissolved most of the residue. The polyamine derivatives were completely dissolved in ether, giving good peak shapes and no interfering peaks.

IV. SUMMARY AND CONCLUSIONS

In summary, a sensitive and specific method for analysis of polyamines in urine by GLC has been developed. Recovery of standard compounds added to the urine samples prior to assay is satisfactory and lends validity to the method. CIE gives essentially the same values as GLC when used following the hydrolysis and clean-up steps, further substantiating the GLC method and results. The procedure is sufficiently sensitive to determine the levels for polyamines in the urine of normal subjects, except on occasion with spermine where only trace quantities (<0.2 mg/24 hr) are found. In general the amounts of the individual polyamines found in 24-hr urines of cancer patients were below those previously reported (7). The lack of an adequate number of patients for each specific type of cancer, with the possible exception of bronchogenic carcinoma, prevents as yet a valid conclusion regarding the average levels or range obtained by GLC for any single type of malignant disease. In addition, further studies to clarify the effects of different storage conditions on polyamine stability with time must be completed prior to continuing the desired clinical evaluation and correlations.

ACKNOWLEDGMENTS

Contribution from Missouri Agricultural Experiment Station. Journal Series No. 6572. The work reported here was supported in part by a contract from the National Institutes of Health, National Cancer Institute (NIH 71 2323).

REFERENCES

1. Tabor, H., and Tabor, C. W., *Pharmacol. Rev.*, **16,** 245 (1964).
2. Dykstra, W. J., Jr., and Herbst, E. J., *Science,* **19,** 428 (1965).
3. Russell, D. H., and Snyder, S. H., *Proc. Nat. Acad. Sci. U.S.A.*, **8,** 1420 (1968).
4. Jänne, J., *Acta Physiol. Scand.*, Suppl. **300,** 1 (1967).
5. Russell, D. H., *Fed. Proc.*, **29,** 669 (1970).
6. Russell, D. H., *Ann. N.Y. Acad. Sci.*, **171,** 772 (1970).
7. Russell, D. H., Levy, C. C., Schimpff, S. C., and Hawk, I. A., *Cancer Res.*, **31,** 1555 (1971).
8. Gehrke, C. W., Kuo, K. C., Zumwalt, R. W., and Waalkes, T. P., *J. Chromatogr.*, submitted.
9. Gehrke, C. W., and Roach, D., *J. Chromatogr.*, **43,** 303 (1969).
10. Gehrke, C. W., Zumwalt, R. W., and Kuo, K. C., *J. Agric. Food Chem.*, **19,** 605 (1971).
11. Gehrke, C. W., and Takeda, H., *J. Chromatogr.*, **76,** 63 (1973).
12. Bremer, H. J., and Kohne, E., *Clin. Chem.*, **32,** 407 (1971).

Polyamines in Normal and Neoplastic Growth. Edited by D. H. Russell. Raven Press, New York © 1973.

Gas Chromatography–Mass Spectrometry of Di- and Polyamines in Human Urine: Identification of Monoacetylspermidine as a Major Metabolic Product of Spermidine in a Patient with Acute Myelocytic Leukemia

Thomas Walle

Department of Pharmacology, College of Medicine, Medical University of South Carolina, Charleston, South Carolina 29401

There has been in general a great lack of sensitivity in previously used methods for quantitative measurements of di- and polyamines in biological fluids, but also, and perhaps more important, there has been a great lack of specificity. These deficiencies have been due to the fairly nonspecific and moderately sensitive detection methods used, such as colorimetry and fluorimetry, in combination with poor separation systems such as paper chromatography, thin-layer chromatography, and paper high-voltage electrophoresis.

For identification of metabolites of this class of compounds, infrared, nuclear magnetic resonance, and mass spectrometry have been of very little use partially because of the limited sample sizes available but mainly because of the lack of adequate and reliable separation techniques for the isolation of these compounds from the biological material in a pure form.

Gas chromatography, a superior separation technique with more sensitive and specific detection devices, was applied to the analysis of aliphatic diamines for the first time in 1961 (1). It was not very successful at that time because of the high polarity and basicity of this class of compounds which adversely affected their peak symmetry and separation. In 1969 (2), however, putrescine, spermidine, and spermine were well separated with reasonably good peak symmetry as their trifluoroacetylated derivatives.

In our laboratories we have used this approach to develop a quantitative and specific assay for putrescine, spermidine, and spermine in human urine with flame ionization detection (3). We have also demonstrated the application of electron-capture gas chromatography to detect small picogram quantities of trifluoroacetylated spermidine and spermine (3), which may be the ultimate technique for the determination of these two compounds at very low concentrations in serum, plasma, cerebrospinal, and other fluids.

In this chapter, we are introducing for the first time the combined gas chromatograph-mass spectrometer (GC–MS) as an analytical tool for qualitative and quantitative studies of di- and polyamines and their metabolites in biological systems. Mass spectral characteristics are presented for some of these compounds as well as their identification in the urine of cancer patients. Two metabolites previously not reported in cancer patients were identified. The application of mass spectrometry as a highly sensitive and specific detector system for these compounds in very crude urine extracts is demonstrated.

I. EXPERIMENTAL

A. Materials

Trifluoroacetic anhydride was purchased from Pierce Chemical Co. Putrescine, cadaverine, spermidine, and spermine were purchased as their hydrochlorides from Calbiochem Co.

B. Gas chromatography—Mass spectrometry

Trifluoroacetyl derivatives of reference compounds and of compounds isolated from urine by alkaline butanol extraction were prepared as earlier described (3). Some of the urine extracts were purified prior to derivatization by ion-exchange chromatography as described elsewhere (M. D. Denton and H. S. Glazer, *unpublished data*).

The combination instrument was an LKB 9000 used at an accelerating voltage of 3.5 kV, electron energy of 20 eV, and trap current of 65 μA.

The gas chromatographic column was of Pyrex glass (100 cm × 2 mm I.D.) and was packed with 2% OV-17 on Chromosorb W HP 80/100 mesh. The carrier gas was He (10 ml/min).

C. Patient samples

Twenty-four-hr urine specimens were collected under toluene and under refrigeration from hospitalized patients. The samples were frozen until analysis.

II. RESULTS

Figure 1 demonstrates the total ion-current recording obtained by GC–MS of a mixture of about 200 ng each of trifluoroacetylated putrescine, spermidine, and spermine. Only a portion of the total amount of ions formed in the ion source of each individual compound is recorded by this detection system, which is mainly used as a guide for the operator indicating when compounds are entering the ion source from the gas chromatographic column for spectra scanning. The sensitivity and specificity of this recording system are comparable to a flame ionization detector. The figure demonstrates excellent peak symmetry and separation of all three trifluoroacetylated amines in approximately 10 min under temperature-programmed conditions.

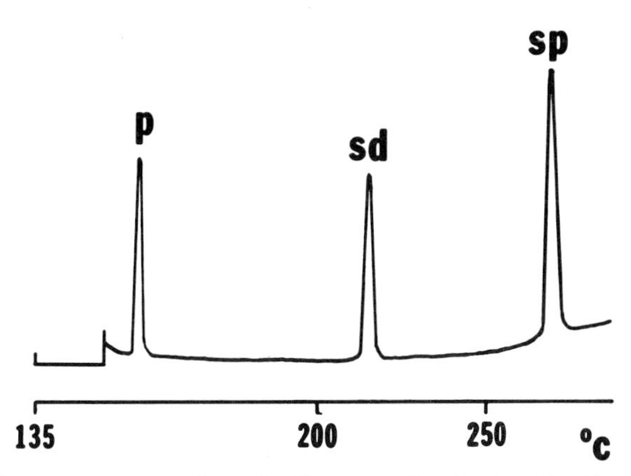

FIG. 1. Total ion-current recording of trifluoroacetylated putrescine(p), spermidine (sd), and spermine(sp). Temperature programming 15°C/min. Each peak represents approximately 200 ng.

The derivatives formed of these three amines are shown in Fig. 2. It is essential to assure complete acylation of all amino groups for quantitative work. Two trifluoroacetyl groups must be incorporated in putrescine, three in spermidine, and four in spermine. Partially acylated amines will chromatograph very badly or not at all.

The mass spectra of these three compounds are shown in Fig. 3. Weak but quite visible molecular ions (M^+) are present for all compounds: at *m/e* 280 for putrescine, *m/e* 433 for spermidine, and *m/e* 586 for spermine. Base peaks, the most intense fragment ions, are obtained at *m/e* 167 for putrescine, *m/e* 336 for spermidine, and *m/e* 517 for spermine. As might be

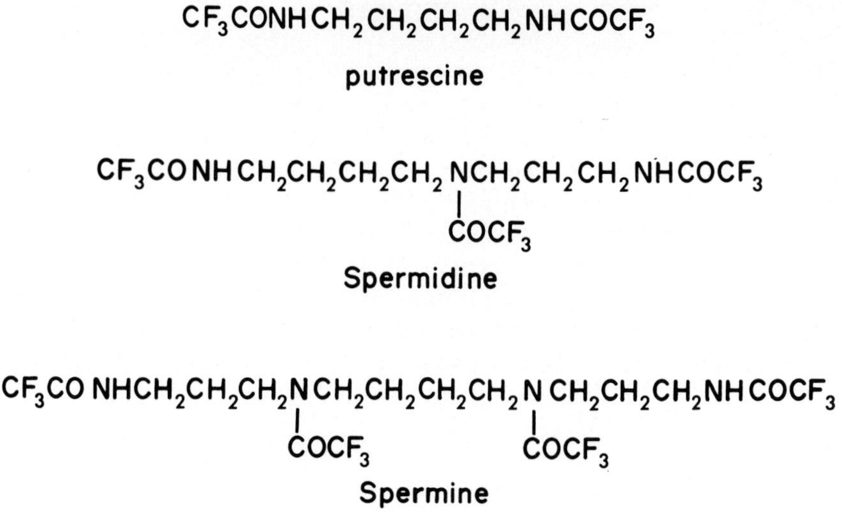

$$CF_3CONHCH_2CH_2CH_2CH_2NHCOCF_3$$

putrescine

$$CF_3CONHCH_2CH_2CH_2CH_2N(COCF_3)CH_2CH_2CH_2NHCOCF_3$$

Spermidine

$$CF_3CONHCH_2CH_2CH_2N(COCF_3)CH_2CH_2CH_2CH_2N(COCF_3)CH_2CH_2CH_2NHCOCF_3$$

Spermine

FIG. 2. Trifluoroacetylated putrescine, spermidine, and spermine.

expected from the close structural similarity of these compounds, their mass spectral fragmentation patterns are also closely related. The molecular ions all lose CF_3 (69) to form fragment ions of *m/e* 211, putrescine, *m/e* 364, spermidine, and *m/e* 517, spermine, and CF_3CO (97) to form fragment ions of *m/e* 183, putrescine, *m/e* 336, spermidine, and *m/e* 489, spermine. Other

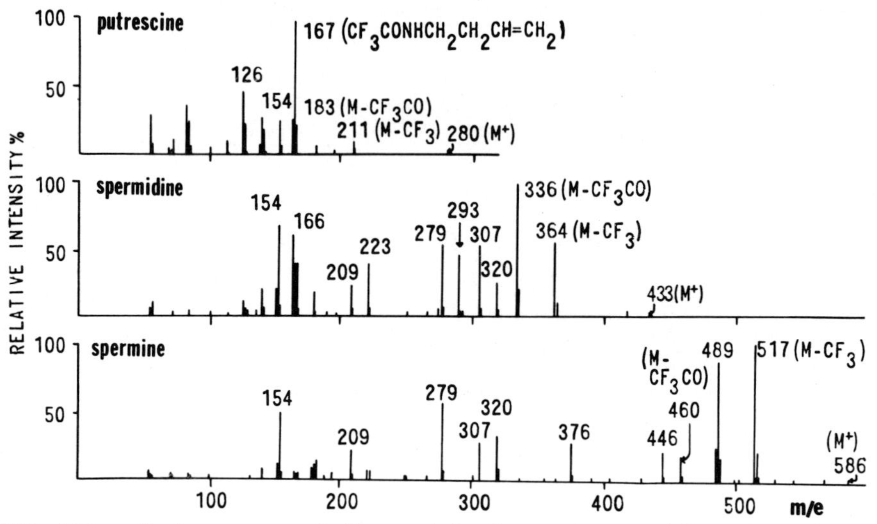

FIG. 3. Normalized mass spectra of trifluoroacetylated putrescine, spermidine, and spermine.

fragment ions arise from cleavages at different positions of the carbon chains. The presence of numerous and characteristic fragment ions and the relative intensity of these ions give structural information that can prove unequivocally the identity of these compounds as they are eluting from the gas chromatographic column even in submicrogram quantities.

Figure 4 demonstrates the total ion-current recording of a trifluoroacetylated extract of urine from a patient with acute myelocytic leukemia. Prior to extraction the urine is hydrolyzed with 6 N hydrochloric acid, the usual procedure for urine polyamine analysis (5, 6). The urine extract (pH 13, butanol) is cleaned up on an ion-exchange column (Denton and Glazer, *unpublished data*) before trifluoroacetylation. By scanning each individual peak in the total ion-current recording and comparing the obtained mass spectra with the mass spectra of reference compounds, putrescine, spermidine, and spermine could easily be identified. The small peak designated *c* in

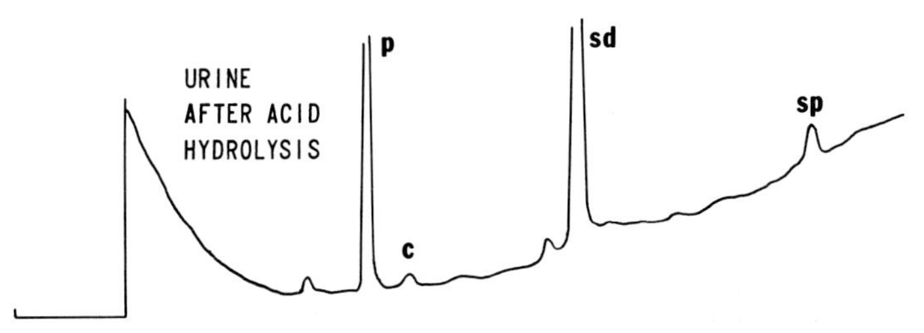

FIG. 4. Total ion-current recording of a trifluoroacetylated extract of urine from a leukemic patient. Hydrolyzed urine. Temperature programming 15°C/min from 80°C to 280°C. p, putrescine; sd, spermidine; c, cadaverine; sp, spermine.

Fig. 4 could be identified from its mass spectrum as cadaverine in spite of the fact that the peak represents only about 50 ng of material.

In Fig. 5 the mass spectrum of this compound is shown with the mass spectrum of the trifluoroacetylated reference compound cadaverine. The complete identity of the two rather complex mass spectra is obvious with reasonably intense molecular ions at *m/e* 294 and base peaks, the most intense fragment ions, at *m/e* 68. All fragment ions for the reference compound are present in the compound isolated from the human urine, and they have nearly the same intensity ratios.

Figure 6 demonstrates the total ion-current recording of a trifluoroacetylated extract of urine from the same patient but without acid hydrolysis. By scanning the mass spectra of the eluting peaks, it could easily be concluded that both putrescine and spermidine are excreted to some extent in a nonconjugated form. However, no free spermine was detected (see Fig. 4).

The largest peak in the total ion-current recording of the nonhydrolyzed

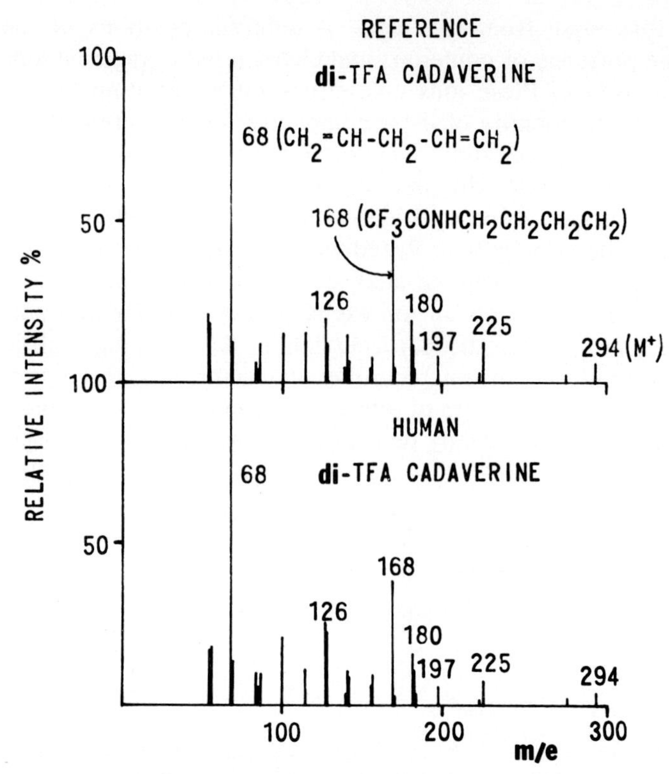

FIG. 5. Mass spectra of the trifluoroacetylated reference compound cadaverine and human cadaverine. Amount of material approximately 50 ng.

urine could be identified as a monoacetylspermidine in spite of the fact that no synthetic monoacetylspermidine was available. By comparing Fig. 6 with Fig. 4, it can easily be seen that all of this acetylated spermidine is hydrolyzed to free spermidine by hydrochloric acid. The mass spectrum of the trifluoroacetylated monoacetylspermidine from the human urine is shown in Fig. 7 together with the trifluoroacetylated spermidine.

The molecular ion for the unknown human metabolite is at *m/e* 379 which is 54 mass units downscales compared to spermidine ($M^+ = 433$), representing the difference in mass between a trifluoroacetyl group (97) and an acetyl group (43) or a di-trifluoroacetyl-monoacetylspermidine and tri-trifluoroacetylspermidine. The loss of CF_3 (69) and CF_3CO (97) during the formation of fragment ions for spermidine at *m/e* 364 and 336 gives corresponding and intense fragment ions for acetylspermidine at *m/e* 310 and 282. Monoacetylspermidine also loses CH_3 (15) and CH_3CO (43) to form

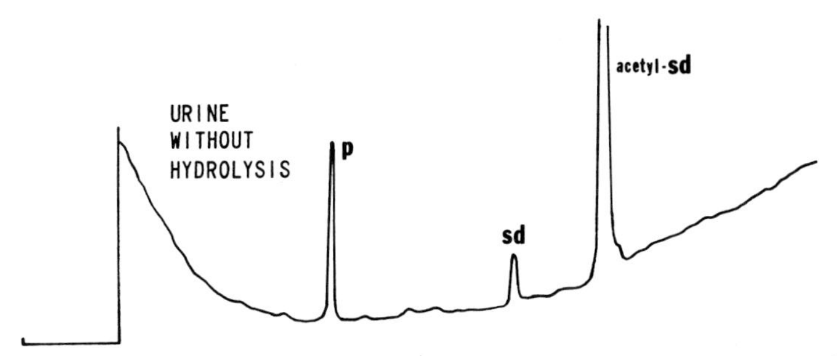

FIG. 6. Total ion-current recording of a trifluoroacetylated extract of urine from a leukemic patient. Nonhydrolyzed urine. Temperature programming 15°C/min from 80 to 280°C. p, putrescine; sd, spermidine; acetyl-sd, monoacetylspermidine.

fragment ions of *m/e* 364 and 336. In addition, *m/e* 140 and 154 are characteristic fragment ions for a trifluoroacetylated chain ($CF_3CONHCH_2CH_2$ and $CF_3CONHCH_2CH_2CH_2$, respectively, in spermidine) while *m/e* 86 and 100 are the corresponding fragment ions for an acetylated chain ($CH_3CONHCH_2CH_2$ and $CH_3CONHCH_2CH_2CH_2$, respectively, in monoacetylspermidine). More detailed studies of the fragmentation pattern

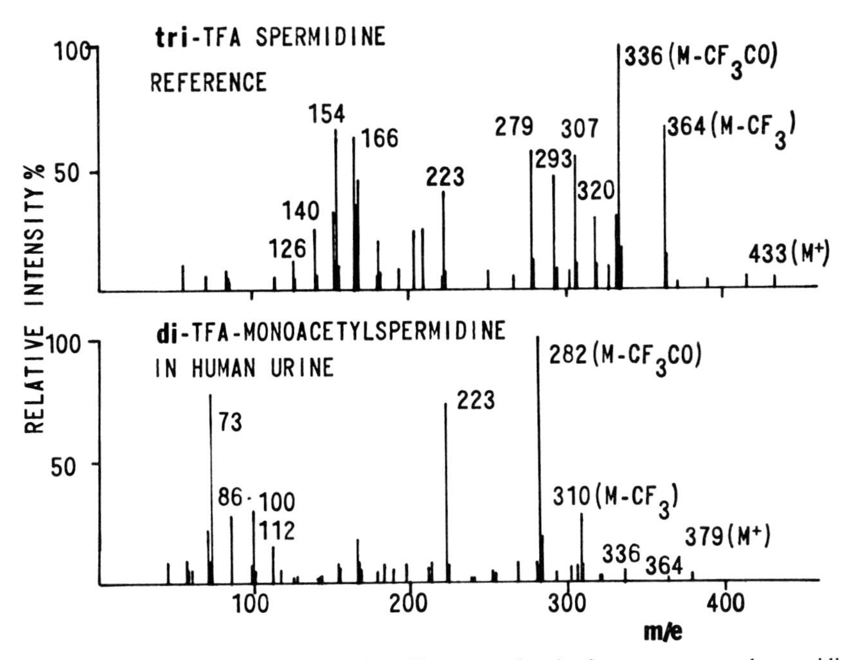

FIG. 7. Normalized mass spectra of the trifluoroacetylated reference compound spermidine and human monoacetylspermidine.

strongly suggest that monoacetylspermidine is monoacetylspermidine B. However, the absolute proof can only be obtained by comparison with pure reference compounds, which are still under investigation.

As discussed earlier, reasonably good, complete mass spectra can be obtained from as little as 50 to 100 ng of a compound eluting from a gas chromatographic column. If only partial mass spectra are recorded, the sensitivity can be enhanced up to approximately 1,000 times. The mass spectrometer can be used in that mode as an extremely sensitive and specific detector (4). This was applied to the analysis of di- and polyamines in a very crude extract of urine obtained from a cancer patient. The urine was extracted without prior hydrolysis with butanol (pH 13). The butanol phase was then extracted with a small volume of hydrochloric acid. The aqueous phase was evaporated to dryness, derivatized with trifluoroacetic anhydride, and injected without any further cleanup into the GC–MS. The top half of Fig. 8 shows the resulting total ion-current recording which demonstrates a large number of peaks, many of which are not resolved from each other. Monoacetylspermidine can easily be recognized, and possibly cadaverine and spermidine. In the bottom half of Fig. 8 is shown a single-ion (*m/e* 154) mass

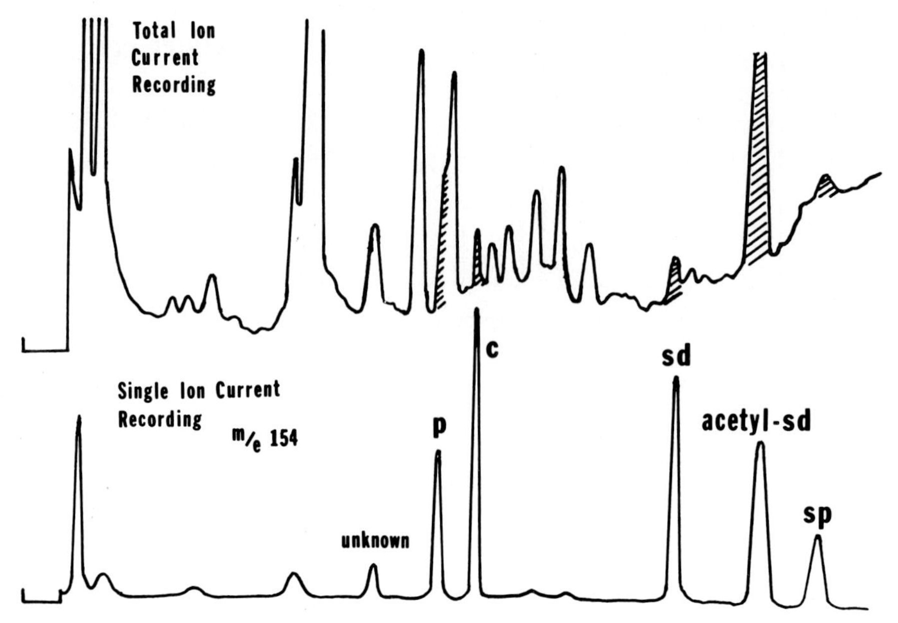

FIG. 8. GC–MS of a trifluoroacetylated crude extract of urine from a cancer patient. Temperature programming 15°C/min from 80 to 280°C. p, putrescine; sd, spermidine; sp, spermine; c, cadaverine; acetyl-sd, monoacetylspermidine.

fragmentogram (4) of the same sample. The *m/e* 154 is a characteristic fragment ion of trifluoroacetylated di- and polyamines ($CF_3CONHCH_2CH_2CH_2$). The number of peaks in this chromatogram is restricted to those compounds that can give rise to fragment ions of *m/e* 154. Putrescine, cadaverine, spermidine, monoacetylspermidine, and also spermine were all completely resolved. Further confirmation of the identity of these compounds was established by focusing the mass spectrometer not only on *m/e* 154 but also on several other fragment ions characteristic of each individual compound and on the corresponding molecular ions. There are also additional peaks containing *m/e* 154 fragment ions, but these do not necessarily have to be related to di- or polyamines. One of these peaks, however, the one labeled "unknown," has recently been identified in our laboratories as 1,3-diaminopropane, which has been part of a separate study.

III. DISCUSSION

This is the first reported application of GC–MS to studies of the biosynthesis and metabolism of di- and polyamines. Complete mass spectra, which in general means unequivocal structural proof, could be obtained from as little as 50–100 ng of putrescine, cadaverine, spermidine, and spermine eluting from the gas chromatographic column after injection of extracts of hydrolyzed urine. The identity was established by comparing the mass spectra of these compounds as isolated from urine with the mass spectra of the corresponding reference compounds. The specificity of the flame ionization detection method earlier cited (3) and further reported on in the next chapter by Denton (7) was confirmed in the same way; the same samples run with the flame ionization detector were run under identical column conditions with the mass spectrometer as the detector.

Urine from cancer patients was to our knowledge analyzed for the first time without prior acid hydrolysis. Both putrescine and spermidine were excreted in free form to a considerable extent (Fig. 6) but only small quantities of free cadaverine and spermine (Fig. 8). Monoacetylspermidine was identified in nonhydrolyzed urine of a leukemic patient. No reference compound was available, but the identity could still be established by comparing the fragmentation pattern of this urinary metabolite with the fragmentation pattern of the reference compound spermidine.

The finding of large quantities of monoacetylspermidine in the urine of a leukemic patient suggests that it is a major metabolic product of spermidine. Earlier (8) monoacetylspermidine was suggested to be a normal urinary constituent in children. It has also been reported to be formed in large quantities by *Escherichia coli* (9). Since oxidative deamination was suggested as the major metabolic pathway of spermidine, this finding in man deserves further careful investigation in a larger number of patients. Where this acetylation takes place is not known, but it certainly is of great importance at this point to investigate not only urine specimens but also what is actually present of this metabolite in the circulating blood. It is also emphasized that extensive acid hydrolysis of any samples prior to analysis, which has hitherto been the case with human urines, will give an incomplete picture of the many forms in which polyamines may be excreted.

The finding of 1,3-diaminopropane in fairly large quantities in the urine of a cancer patient suggests that it is related to the polyamine metabolism. Bachrach (10, 11) demonstrated that 1,3-diaminopropane is a metabolic product of spermidine that is formed by an enzyme isolated from *Serratia marcescens*. Not to our knowledge has it been proved to be formed by mammalian tissues.

The use of mass spectrometry as a highly sensitive and specific detection system was shown. By monitoring just one single-ion characteristic of trifluoroacetylated di- and polyamines, 1,3-diaminopropane, putrescine, cadaverine, spermidine, monoacetylspermidine, and spermine could be completely resolved from a large cluster of unidentified interfering compounds in a very crude urine extract. By simultaneous monitoring of several characteristic ions, multiple ion detection, or mass fragmentography (4), the specificity of this detection system can be greatly increased and still a high degree of sensitivity maintained; less than 50 pg can be detected. A further extension of multiple-ion detection may be toward quantitative determinations using stable, labeled internal standards as demonstrated by Gaffney and co-workers (12).

It can finally be concluded that GC–MS with single- or multiple-ion detection may be the major analytical tool in the future for a better understanding of the biosynthesis and metabolism of the di- and polyamines as they relate to growth and development.

IV. SUMMARY

For the first time GC–MS has been applied to studies of the polyamine biosynthesis and metabolism.

Complete mass spectra could be obtained from as little as 50 to 100 ng of material eluting from the gas chromatographic column. Structural proof of putrescine, spermidine, and spermine and also cadaverine in hydrolyzed human urine extracts could easily be obtained. Free putrescine and spermidine were both detected in nonhydrolyzed urine.

Monoacetylspermidine was identified as a major urinary metabolite of spermidine in nonhydrolyzed urine from a leukemic patient. A second new metabolite, 1,3-diaminopropane, may also be related to spermidine metabolism.

Using GC–MS with single-ion monitoring, 1,3-diaminopropane, putrescine, cadaverine, spermidine, monoacetylspermidine, and spermine could all simultaneously be detected in a very crude extract of urine from a cancer patient.

ACKNOWLEDGMENT

Dr. Drue Denton (Department of Internal Medicine, University of Cincinnati, College of Medicine) is gratefully acknowledged for many valuable discussions. The urine specimens were kindly provided by Dr. Denton.

REFERENCES

1. Smith, E. D., and Radford, R. D., *Anal. Chem.*, **33,** 1160 (1961).
2. Brooks, J. B., and Moore, W. E. C., *J. Microbiol.*, **15,** 1433 (1969).
3. Walle, T., and Denton, M. D., submitted for publication.
4. Hammar, C.–G., Holmstedt, B., and Ryhage, R., *Anal. Biochem.*, **25,** 532 (1968).
5. Russell, D. H., *Nature New Biology*, **233,** 144 (1971).
6. Russell, D. H., Levy, C. C., Schimpff, S. C., and Hawk, I. A., *Cancer Res.*, **31,** 1555 (1971).
7. Denton, M. D., Glazer, H. S., Walle, T., Zellner, D. C., and Smith, F. G., *this volume.*
8. Nakajima, T., Zack, J. F., Jr., and Wolfgram, F., *Biochem. Biophys. Acta*, **184,** 651 (1969).
9. Dubin, D. T., and Rosenthal, S. M., *J. Biol. Chem.*, **235,** 776 (1960).
10. Bachrach, U., *Nature*, **194,** 377 (1962).
11. Bachrach, U., *J. Biol. Chem.*, **237,** 3443 (1962).
12. Gaffney, T. E., Hammar, C.–G., Holmstedt, B., and McMahon, R. E., *Anal. Chem.*, **43,** 307 (1971).

Polyamines in Normal and Neoplastic Growth. Edited by D. H. Russell. Raven Press, New York © 1973.

Elevated Polyamine Levels in Serum and Urine of Cancer Patients: Detection by a Rapid Automated Technique Utilizing an Amino Acid Analyzer

Laurence J. Marton, James G. Vaughn*, Inez A. Hawk, Carl C. Levy, and Diane H. Russell

Laboratory of Pharmacology, Baltimore Cancer Research Center, National Cancer Institute, Baltimore, Maryland 21211

It has now been over 2 years since the first data were collected indicating that patients with diagnosed cancers excrete elevated levels of polyamines in their urine (1, 2). The initial data indicated that an elevated level of polyamines in the urine was not specific for any particular kind of cancer but was exhibited by almost all of the patients with diagnosed cancer prior to therapy. It was found that the surgical removal of a portion of the tumor mass led to a marked depression in the excretion of polyamines. Data on patients with acute nonlymphocytic leukemia indicated that these patients exhibited lower urinary polyamine levels when they obtained a medical remission.

The initial data on urinary excretion levels in cancer patients was obtained by utilizing (a) hydrolysis of the urine in 6 N HCl to ensure that no polyamines remained in a conjugated form, (b) extraction of the polyamines into alkaline-1-butanol, (c) separation by high-voltage electrophoresis, and (d) quantification utilizing ninhydrin staining and spectrophotometric determination. Although this method is perfectly adequate and reliable when used with tissue samples which do not contain diverse kinds of amines and other metabolites, there have been problems with this method in fluids. One of the problems in analyzing samples of urine has been an unknown compound that runs with spermine on the electrophoretigram, which results in

* Beckman Instruments, Inc., Palo Alto, California 94304

the overestimation of this compound by this method. Although this unknown seems to be specific for cancer, at this time it is important to be able to evaluate precisely the concentrations of polyamines in the fluids of cancer patients. An additional problem is its lack of sensitivity. It is necessary to have approximately 10 nmoles of amine present to obtain accurate determinations by the electrophoretic method. It is necessary in order to screen large numbers of samples of urine, serum, cerebral spinal fluid, and other appropriate fluids, to have a convenient, rapid, and sensitive assay method.

This chapter reports on a rapid automated technique for polyamine determination utilizing a Beckman Model 121 amino acid analyzer. At the present time, as little as 1 nmole of amine can be detected by this method, and it is both rapid and automated. Currently, it does not appear that there are any other substances present in fluids that interfere with the accurate quantification of the polyamines, putrescine, spermidine, and spermine by this method. We are presently investigating the use of a fluorimetric indicator, to replace ninhydrin, that should increase the sensitivity of this assay. Sensitivity in the range of 10^{-11} moles seems feasible.

Preliminary data are presented which indicate that elevated spermidine, and sometimes also putrescine and spermine, levels can be detected in the serum of diagnosed cancer patients by the use of this method. Its adaptation to urine has been possible, and the elevation of polyamines in the urine of cancer patients is corroborated by this method. It is postulated that screening serum for the elevation of spermidine alone might serve as an adequate diagnostic tool.

I. MATERIALS AND METHODS

A. Preparation of samples

Serum samples were prepared by hydrolyzing in 6 N HCl for 14 to 16 hr at 110°C, the adjusting the pH to between 11 and 12, and extracting into l-butanol in the presence of a mixture of Na_2SO_4 and Na_3PO_4 salts (3). Following this extraction, a known quantity of the butanol phase is removed, acidified to pH 2 with concentrated HCl, and evaporated to dryness. The dried sample is then brought up in 0.1 N HCl, washed twice with ether, and applied to the amino acid analyzer column for analysis.

Urine samples were prepared in a simpler fashion, as it is possible to eliminate the butanol extraction. Hydrolysis is still necessary in order to obtain the amines in their free form, but following hydrolysis it is only necessary to evaporate the HCl, bring the sample up in distilled water, filter through a Millipore filter, and apply to the column. This method is not applicable to serum samples due to their high protein content. If serum samples are deproteinated prior to hydrolysis, we have found that a portion of the polyamines is lost.

B. Analysis

A Beckman Model 121 automatic amino acid analyzer was used to assay polyamine levels in urine and serum. Two 12-mm cuvettes were used in the colorimeter to detect the ninhydrin-polyamine absorbance at 440 and at 570 nm. An Infotronics Model 210 digital integrator measured the polyamine peak areas. Two 0.9 × 23-cm columns were packed to a height of 5 cm with PA-35 custom sperical resin. The temperature of the columns was maintained at 65°C. For chromatographic separation of the polyamines, a modification of the buffers described by Bremer and Kohne and others (4–6) was employed. Buffer A, pH 5.06, was 0.35 M sodium citrate buffer made 0.7 M in respect to sodium; that is, 81.8 g NaCl and 400 ml of a 3.5 M Beckman sodium citrate buffer concentrate were combined in a final volume of 4 liters. Buffer B, pH 4.08, was 0.35 M sodium citrate buffer adjusted to 2.35 M in respect to sodium; that is, 467.5 g sodium chloride and 400 ml of a 3.5 M Beckman sodium citrate buffer concentrate were combined in a final volume of 4 1. A flow rate of 120 ml/hr for the buffers was maintained. The ninhydrin flow rate was 60 ml/hr.

A column elution schedule was programmed to eliminate the interference of amino acids on the resultant polyamine chromatogram. The free amino acids were eluted from the column with buffer A for 20 min followed by a 15-min elution with buffer B. During this 35-min interval the column elution is directed to drain, and the components are not detected by the ninhydrin reaction. From 35 to 95 min after sample injection, the column effluent is valved to the ninhydrin system for colorimetric detection of the polyamines. The utilization of dual columns and alternating sample injection allows for one sample determination per hour. After polyamine elution, each column is stripped for 5 min with 0.2 N NaOH and then equilibrated with buffer A for 25 min. Since the rotating sample tray on this instrument can be loaded with 72 samples, and all handling of the samples thereafter is automated, more than 20 samples can be analyzed daily without technical supervision.

Polyamine hydrochloride standards were obtained from Calbiochem Co. and recrystallized three times from ethanol before use. The ninhydrin color response of standard polyamine solutions was found to be linear over the range of 1 to 25 nmoles. Recovery of ^{14}C-labeled polyamines from the columns was essentially 100%. Figure 1 shows a representative chromatograph obtained from a standard solution of polyamines and amino acids.

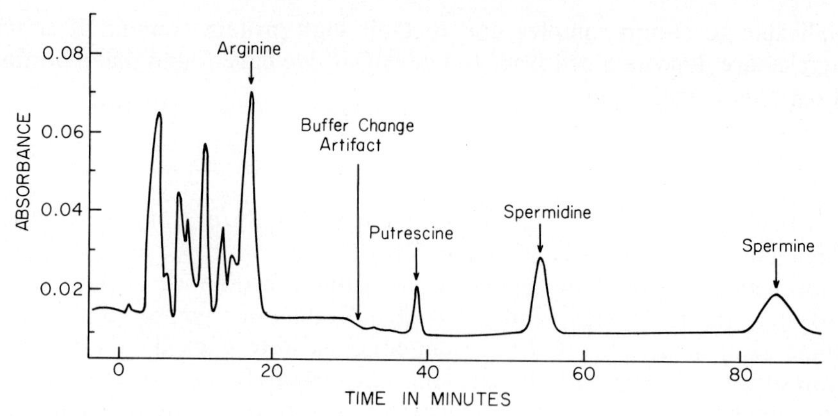

FIG. 1. Representative chromatograph obtained from a standard solution of polyamines and amino acids.

TABLE 1. *Urinary levels of polyamines in normals versus diagnosed and suspected cancer patients*

Diagnosis	Putrescine	Spermidine (mg/24 hr)	Spermine
Normal	2.5	2.1	<0.4
	2.0	2.1	<0.4
	2.1	1.4	<0.4
	1.8	1.2	<0.4
	1.6	0.9	<0.4
Mean value	2.0	1.5	<0.4
Tumors			
Adenocarcinoma	3.4	5.1	<0.4
	4.0	2.3	<0.4
	3.5	3.1	0.4
Anaplastic carcinoma (metastatic)	7.5	5.7	1.6
Anaplastic carcinoma	7.0	8.0	0.8
Suspected lung cancer	5.8	2.8	0.7
	5.2	3.1	0.7
	3.7	2.5	0.7
Hodgkin's disease	9.1	4.2	<0.4
	4.0	13.0	<0.4
	5.3	0.5	<0.4
Testicular embryonal carcinoma	4.3	4.9	<0.4

II. RESULTS AND DISCUSSION

Table 1 tabulates the results of polyamine assays on the 24-hr urinary collections of five normal laboratory volunteers and on twelve hospitalized patients with diagnosed or suspected cancer. Significant elevations of putrescine and/or spermidine could be noted in all the hospitalized cancer patients. Spermine levels were elevated in some of the hospitalized patients, but the high levels originally reported using the high-voltage electrophoretic method were not found. Nevertheless, it is evident from these results that the original finding of elevated levels of putrescine and spermidine in the urines of patients with cancer is corroborated by this method.

Table 2 illustrates the results of polyamine assays on the sera of normal volunteers and on patients with diagnosed cancer. Each patient with diagnosed cancer exhibited an elevated spermidine level, and several also had elevations in putrescine and spermine levels. Sample size was varied in

TABLE 2. *Serum levels of polyamines in diagnosed cancer patients and in normals*

Diagnosis	Sample volume (ml)	Putrescine	Spermidine	Spermine
		(nmoles/ml serum)		
Normal	1	—	—	—
	1	—	—	—
	1	—	—	—
	1	—	—	—
	1	—	—	—
	3	—	0.3	—
	3	—	0.2	—
Tumors				
Mediastinal choriocarcinoma				
pretreatment	1	—	3.3	—
during chemotherapy	1	—	2.5	—
during chemotherapy	1	—	2.4	—
preterminal	1	—	3.1	—
Pancreatic carcinoma	1	—	1.2	—
Acute nonlymphocytic leukemia	2	4.4	0.6	—
Hodgkin's disease	2	1.5	1.1	+
	2	—	1.1	—
	2	—	1.5	—
Breast carcinoma	2	—	0.7	+

—, nondetectable
+, present but not quantifiable

carrying out these analyses, and it is evident that it is an important factor. In the normal volunteers, no polyamines were detectable with a sample size of 1 ml; however, when the sample size is increased to 3 ml, spermidine is detectable. Even with the elevation of levels of the polyamines in cancer patients, the amounts are in the lower range of sensitivity of the method. It is probably best to set a lower limit of sample size at 2 to 3 ml, and to work with larger volumes if possible.

The advantages of a serum test over a test requiring the collection of a 24-hr urine sample are evident. These preliminary results indicate the feasibility of such a test. Screening sera for the elevation of spermidine alone might serve as an adequate diagnostic tool. We are presently carrying out the appropriate clinical studies to verify this postulation.

The methodology outlined in this chapter allows for the sensitivity, reproducibility, and speed necessary to conduct clinical investigations on physiological fluids, and at the same time is applicable to small tissue samples as described in the next chapter.

REFERENCES

1. Russell, D. H., *Nature,* **233,** 144 (1971).
2. Russell, D. H., Levy, C. C., Schimpff, S. C., and Hawk, I. A., *Cancer Res.,* **31,** 1555 (1971).
3. Raina, A., *Acta Physiol. Scand.,* **60,** Suppl., 218 (1963).
4. Shull, K. H., McConomy, J., Vogt, M., Castillo, A., and Farber, E., *J. Biol. Chem.,* **241,** 5060 (1966).
5. Morris, D. R., Koffron, K. L., and Okstein, C. J., *Anal. Biochem.,* **30,** 449 (1969).
6. Bremer, H. J., and Kohne, E., *Clin. Chim. Acta,* **32,** 407 (1971).

Polyamines in Normal and Neoplastic Growth. Edited by D. H. Russell. Raven Press, New York © 1973.

Clinical Application of New Methods of Polyamine Analysis

M. Drue Denton, Helen S. Glazer, Thomas Walle*, David C. Zellner, and Frank G. Smith

*University of Cincinnati College of Medicine, Cincinnati, Ohio 45229, and *Medical University of South Carolina, Charleston, South Carolina 29401*

The polyamines have been the object of extensive investigations since the early decades of this century. Their widespread occurrence in nucleated cells has been the subject of several excellent reviews (1, 2). Many data have been accumulated to relate polyamine levels and composition to growth and development (3, 4). The elegant studies which have correlated polyamine levels and composition with nucleic acid structure and function and with many other biological processes have been adequately covered by others in this volume. Our own studies have dealt primarily with the application of newer methods of polyamine analysis to the study of polyamine excretion in humans, particularly in patients with cancer. These studies were initiated because of interest generated by the reports by Diane Russell of increased excretion of polyamines in the urine of cancer patients compared to normal controls and non-cancer patients (5). It was apparent to us that one of the major limitations of extending these studies in patients was the lack of analytical methods for biological fluids of sufficient reproducibility, sensitivity and specificity which could be automated so that reliable studies of large numbers of patients could be accomplished.

In order to pursue these investigations we developed a gas-chromatographic method for the measurement of polyamines in urine. In this chapter I will discuss some of the features of this method and present some of the data we have collected. As will be apparent, some aspects of our studies are in an incomplete form but are presented to demonstrate the clinical and research capabilities possible with this method. In view of the goals of this volume, presentation of these early data is appropriate because of the relevance of our studies to understanding the role of polyamines in cancer patients. The preceding chapter (6) discussed the theoretical and experimental basis for the studies I will discuss. The data in this presentation will be concerned primarily with quantitative studies of polyamines and derivatives of polyamines in the urine of cancer patients.

I. EXPERIMENTAL PROCEDURE

A. Materials

Trifluoroacetic anhydride was purchased from Eastman Kodak Co. The column packing material was purchased from Varian Aerograph Co. 1,7-Diaminoheptane was purchased from K and K Laboratories and was recrystallized as the hydrochloride salt from ethanol three times before use. 3,3′-Diaminodipropylamine was purchased from Eastman Kodak Co. and was recrystallized as the hydrochloride salt from ethanol prior to use. Putrescine dihydrochloride, spermidine trihydrochloride, and spermine tetrahydrochloride were purchased from Calbiochem Co.

B. Gas chromatography

Trifluoroacetyl derivatives of isolated urinary compounds and reference compounds were prepared as previously described (T. Walle and M. D. Denton, *submitted*) and were injected directly into dual columns of a Varian Aerograph Model 2700 Gas Chromatograph equipped with dual flame ionization detectors and electrometers. The dual glass columns (3 ft × 0.25 in, O.D.) were packed with 5% OV-17 on 100/120 mesh Chromosorb WHP, and the ends of the columns were plugged with silane-treated glass wool (Applied Science Laboratories). Temperature programming from 75 to 280°C at 15 °C/min with 75 ml/min nitrogen carrier gas flow were used. Prior to derivatization, the polyamines from urine samples were isolated using alkaline butanol solvent extraction and column purification.[1]

C. Patient samples

Urine samples were collected from the patients while they were hospitalized on the General Clinical Research Center at the University of Cincinnati Medical Center. Urine samples were promptly refrigerated in glass bottles under toluene during collection. After the 24-hr collection was completed, the volume was measured and an aliquot was promptly frozen at −20°C.

[1] Unpublished procedure developed by M. Drue Denton and Helen S. Glazer. A detailed description is beyond the scope of this brief chapter but will be the subject of a separate publication.

II. RESULTS

A. Quantitation and recovery

Representative standard curves used for calibrating the derivatization and chromatographic procedure are shown in Fig. 1. The linearity and precision of the procedure are quite good. Because of the excellent chromatographic properties under these conditions, as can be seen in the representative chromatogram on the right side of Fig. 1, peak height is quite satisfactory for quantitation. Each peak in the representative chromatogram represents approximately 1 μg of the respective diamine or polyamine base. Shown are both naturally occurring polyamine bases and compounds being evaluated for use as internal standards. The gas-chromatographic detector used for these chromatograms was a flame ionization detector. Under these conditions there is increasing peak height associated with increasing length of the carbon chain, as is best appreciated by observing the relative positions of the diamine curves on the left side of Fig. 1. In the case of the polyamines shown in the middle panel of Fig. 1, the relationship of peak height to the number of carbon atoms is less simple. The presence of the derivatized interior tertiary nitrogen apparently alters the behavior of the molecule in this system. Also note that there is a 40-fold linear range in these standard curves.

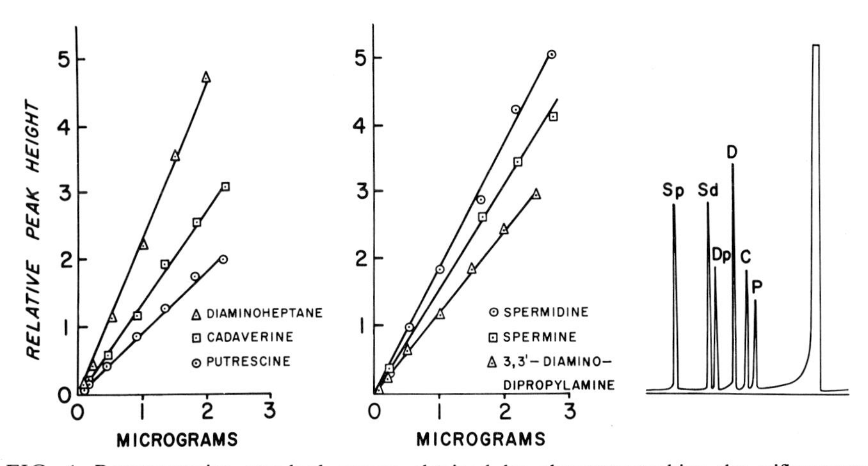

FIG. 1. Representative standard curves obtained by chromatographing the trifluroacetyl derivatives of individual aliquots of the amines are shown in the left and middle panels. On the right is shown a representative chromatogram in which each peak represents approximately 1 μg of the respective diamine or polyamine base. P, putrescine; C, cadaverine; D, diaminoheptane; Dp, 3,3′-diaminodipropylamine; Sd, spermidine; Sp, spermine.

We have insufficient data to present a complete statistical analysis of overall recovery at the present time. Preliminary examination of our recovery data using diaminoheptane as an internal standard indicates that quantitation of the overall procedure is also quite satisfactory. Recovery experiments in which exogenous polyamine and diamine standards were added to urine samples prior to any extraction or hydrolysis have been performed. Hydrolysis overnight in 6 N hydrochloric acid at 120°C was used to release polyamines from any bound or conjugated state. Determinations made after hydrolysis are referred to as "total" polyamine or diamine content. Comparable recovery data have been obtained using the methods for both hydrolyzed and nonhydrolyzed procedures.

B. Urinary polyamines

In Fig. 2 are shown chromatographs of the polyamines from the hydrolyzed urines of a non-cancer patient and a cancer patient. The urines were collected on the same day, and both patients were hospitalized at the Gen-

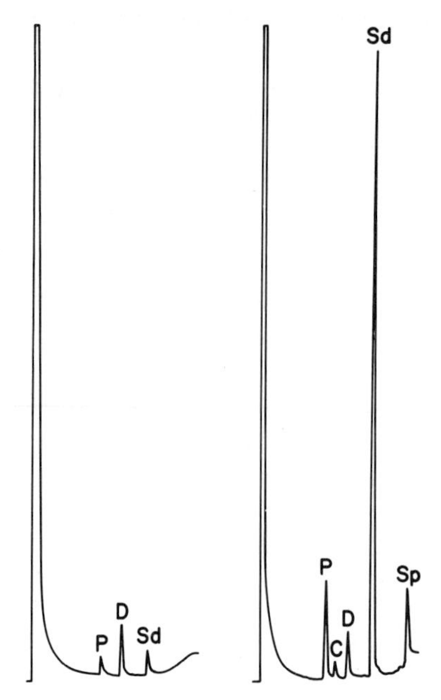

FIG. 2. Chromatograms of the urinary polyamines and diamines from a non-cancer patient on the left and a cancer patient on the right. Eight-ml aliquots were used in each case, and the urine was hydrolyzed overnight at 120°C in 6 N HCl. Diaminoheptane is the internal standard added before hydrolysis.

eral Clinical Research Center at the University of Cincinnati Medical Center. The heights of the peaks representing the polyamines spermidine and spermine were much higher in the sample extracted from the cancer patient's urine. The urine volumes were similar and equal amounts were taken for analysis. Since the recovery of the internal standard added to both urines is similar, these data indicate that the cancer patient excreted much greater amounts of spermidine and spermine than did the non-cancer patient.

The next area of investigation involved an examination of unhydrolyzed urine from the cancer patient who in this case was a patient with acute myelocytic leukemia. In Fig. 3 are shown chromatograms of extracts of both hydrolyzed and nonhydrolyzed samples of urine from a cancer patient. In the hydrolyzed urine shown on the right, a very large spermidine peak and a smaller spermine peak are present. In the nonhydrolyzed urine shown on the left of Fig. 3, no spermine peak is seen and the spermidine peak is much smaller, despite the fact that nearly twice as much urine was taken for the analysis. There is a large peak in the nonhydrolyzed urine which is due to the presence of acetylspermidine. This peak was identified as dis-

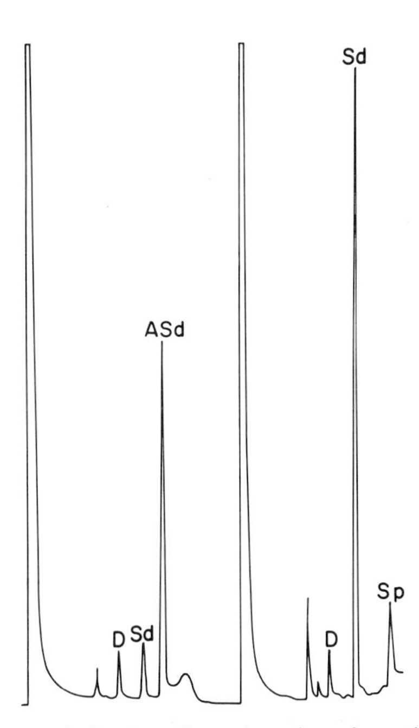

FIG. 3. Urinary polyamines and diamines from the urine of a patient with acute myelocytic leukemia. The chromatogram on the left is from a 15-ml aliquot extracted without hydrolysis. ASd, acetylspermidine. The chromatogram on the right is from an 8-ml aliquot extracted after hydrolysis overnight at 120°C in 6 N HCl.

cussed by Walle in the preceding chapter (6). Acetylspermidine has been previously detected by Nakajima et al. (7) in the urine of children using qualitative paper chromatographic and electrophoretic techniques.

C. Daily excretion of polyamines

The daily total excretion of spermidine in a patient with acute myelocytic leukemia during treatment with vincristine and prednisone is shown in Fig. 4.

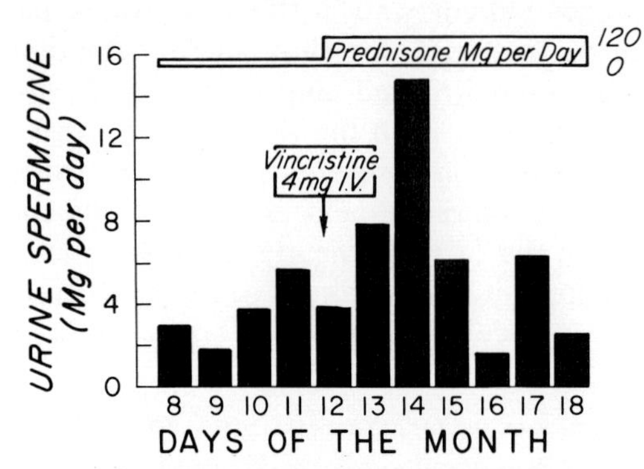

FIG. 4. Daily total spermidine excretion in a patient with acute myelocytic leukemia during a course of therapy with vincristine and prednisone. The therapy was associated with a transitory fall in the number of blasts in the peripheral blood, but on the 19th day of the month the patient had a return of blasts into the peripheral blood, became septic and hypotensive, and expired shortly thereafter.

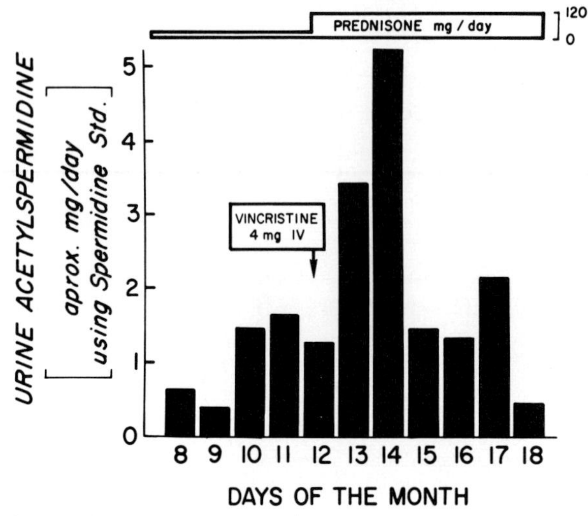

FIG. 5. Approximate daily acetylspermidine excretion in the same samples as shown in Fig. 4.

Within 48 hr after therapy was given, there was a marked increase in spermidine excretion from a level of 2 to 6 mg per day to a maximum of 15 mg per day. This maximum corresponds to the time of maximum cell dissolution reported for the use of vincristine in patients with leukemia (8).

The daily excretion of acetylspermidine follows a qualitatively similar pattern to the total spermidine excretion. The data shown in Fig. 5 are expressed on the basis of a spermidine standard. Exact quantitation of the acetylspermidine using flame ionization detector gas chromatography has not been done because we do not have a pure acetylspermidine standard at the present time. This may lead to an absolute error but the relative daily variation pattern should be the same.

III. DISCUSSION

The study of polyamines in the biological fluids of patients has been limited primarily by the lack of quantitative analytical procedures of sufficient sensitivity, accuracy, and specificity such that reliable and reproducible measurements of the polyamines can be made in large numbers of samples. We have presented today a quantitative gas chromatographic procedure which is highly specific and reproducible. The content of the peaks has been confirmed by gas chromatography–mass spectroscopy as discussed by Walle in the preceding chapter (6). This procedure allows one to measure the total polyamine and diamine content of urine after hydrolysis in hydrochloric acid has released the polyamines from any bound or conjugated state. This provides a measure of the complete extent of polyamine excretion by patients. The analytical procedure described here also allows the quantitative estimation of the free polyamines as they are excreted into the urine. This capability will prove exceedingly useful in future studies of the metabolism of these compounds, as shown by our description of acetylspermidine as a major fraction of the *in vivo* urinary form of spermidine in this patient.

The similarity of the pattern of daily total spermidine and acetylspermidine excretion suggests that a constant percent of the spermidine is excreted in the form of acetylspermidine. Since other quantitative measurements in our laboratory indicate that only about one-thirtieth of the spermidine was excreted in a free form (data not presented), either there is some other spermidine derivative excreted which we have not yet detected or the quantitation of acetylspermidine on the basis of peak height has a much different response factor than does spermidine itself. This quantitative aspect is currently under investigation.

In summary, we have developed a gas chromatographic procedure for the quantitative determination of polyamines in biological samples and have presented some of our data regarding polyamine excretion in patients. This new procedure not only allows very sensitive and specific measurement of total polyamine content but also allows mass spectroscopic confirmation of the identity and purity of the peaks obtained. In addition, quantitation and identification of free and conjugated derivatives as they are excreted into the urine are possible. Our preliminary data suggest that chromatographic analysis of unhydrolyzed urine samples may be even more useful and interesting both clinically and scientifically than measurements made after extensive hydrolysis. Not only is there a possible diagnostic role for polyamines but they also may serve as indicators of the responsiveness of neoplasms to therapy as previously postulated by Russell and Levy (9). With the capabilities afforded by these new analytical procedures, these questions can be answered in clinical studies of patients with precision, reliability, and sensitivity not previously possible.

ACKNOWLEDGMENTS

The excellent assistance of the following persons in this work is gratefully acknowledged: Miss Elizabeth Taylor assisted in the early phase of development of the method of polyamine analysis; Mrs. Edie Kreuter assisted in the preparation of this manuscript; and Mrs. Alice Bullock and the Staff of the General Clinical Research Center at the University of Cincinnati Medical Center assisted in the collection and storage of the patient samples.

The care of the patients and collection of specimens were supported by grant RR0068 to the General Clinical Research Center of the University of Cincinnati Medical Center.

REFERENCES

1. Tabor, H., Tabor, C. W., and Rosenthal, S. M., *Ann. Rev. Biochem.*, **30,** 579 (1961).
2. Tabor, H., and Tabor, C. W., *Pharmacol. Rev.*, **16,** 245 (1964).
3. Jänne, J., Raina, A., and Siimes, M., Acta Physiol. Scand., **62,** 352 (1964).
4. Snyder, S. H., and Russell, D. H., *Fed. Proc.*, **29,** 1575 (1970).
5. Russell, D. H., *Nature New Biol.* **233,** 144 (1971).
6. Walle, T., this volume.
7. Nakajima, T., Zack, J. F., Jr., and Wolfgram, F., *Biochim. Biophys. Acta*, **184,** 651 (1969).
8. Lampkin, B. C., Nagao, T., and Mauer, A. M., *J. Clin. Invest.*, **48,** 1124 (1969).
9. Russell, D. H., and Levy, C. C., *Cancer Res.*, **31,** 248 (1971).

Polyamines in Normal and Neoplastic Growth. Edited by D. H. Russell. Raven Press, New York © 1973.

A Quantitative Model for Relating Tumor Cell Number to Polyamine Concentrations

Kwang B. Woo and Richard M. Simon

Division of Cancer Treatment, National Cancer Institute, Bethesda, Maryland 20014

I. INTRODUCTION

In the clinical management of patients with cancer, there is a need for quantitative and specific methods for assaying the effectiveness of various treatments and evaluating disease status. Many types of tumor cells have identifiable metabolites, or biomarkers, whose kinetic parameters are indicative of the presence of a malignancy and reflect the total number of neoplastic cells involved and any change occurring due to treatment. The establishment of a quantitative relationship between tumor cells and a marker is, therefore, of considerable clinical interest as it affords the possibility of estimating the total number of tumor cells as well as any variation in number due to therapy. Recently, Sullivan and Salmon (1) carried out a kinetic study of tumor growth and regression in I_gG multiple myeloma based on the measurement of a marker immunoglobulin (M-component) whose metabolism has been well characterized.

Various patterns of tumor growth have been represented successfully by a number of mathematical equations, notably the Gompertz equation (2), the biological implication of which is based on the hypothesis that the major part of the growth is the result of two genetically determined processes whose magnitudes are defined by exponential coefficients: the initial exponential proliferation of the system and the exponential decay of this primary exponential growth rate. These growth equations do not, however, incorporate the characteristics of biomarkers for tumor cells. Several models of cellular proliferation using positive and negative feedback control loops have been developed (3, 4). That of Weiss and Kavanau (3) is based on a concept of growth regulation by negative feedback involving chemical substances which are excreted from the fully differentiated cells. Bronk et al. (4) developed a theoretical model for the cooperative control of cellular proliferation kinetics in which a critical substance is produced by the cells whose concentration in a given cell determines whether that cell can divide; it is assumed that the substance can leak out of the cells into the surrounding medium and be reabsorbed by the cells.

The biological properties of the polyamines and their clinical implications have been discussed extensively and reviewed in this volume and elsewhere (5–9). Some of the properties are that these compounds promote growth in a number of microorganisms and mammalian cell lines and that they might serve as indicators of rapidly proliferating tumors (10). A recent finding strongly suggests that the growth factor released from adult human skin fibroblasts is putrescine (11).

Andersson and Agrell (12) studied growth pattern, distribution of cell size, and other characteristics of the Ehrlich ascites carcinoma cells of tumor-bearing mice with an inoculum dose of 4×10^6 ascites tumor cells. Changes in the mean concentration of putrescine, spermidine, and spermine in this tumor system were measured at various intervals after inoculation of the tumor (13).

In this chapter, a quantitative model for tumor-polyamine interactions in the Ehrlich ascites carcinoma cells is developed which allows for the estimation of the total number as well as the size or rate of proliferation of tumor cells by the measurement of polyamine concentrations. With information available on cell-cycle and proliferation kinetics of the Ehrlich ascites tumor (14–16), and with the assumption (or a positive speculation) that putrescine may be the growth factor, the kinetic relationship between the tumor cells and polyamines is characterized by a system of nonlinear differential equations. Various kinetic characteristics of tumor cell proliferation and polyamine synthesis are studied by computer simulation and subsequently evaluated with the available experimental data.

II. STRUCTURE OF THE KINETIC MODEL

A kinetic model for the dynamics of tumor-polyamine interactions is schematically illustrated in Fig. 1. The tumor system is classified into two compartments: (a) proliferating cells, x_1, and (b) nonproliferating cells, x_2. All proliferating cells have the same generation time, τ. A fraction, δ, of the daughter cells are returned. The remaining fraction, $1-\delta$, developed into nonproliferating cells. These cells have lost, irrevocably, the ability to divide. Cell loss which represents cell death or removal by exfoliation is also included in both compartments. The compartment "Rate of Proliferating Tumor Cells Formed" stresses its significance to the excess amount of polyamine synthesis and accumulation; this relationship was

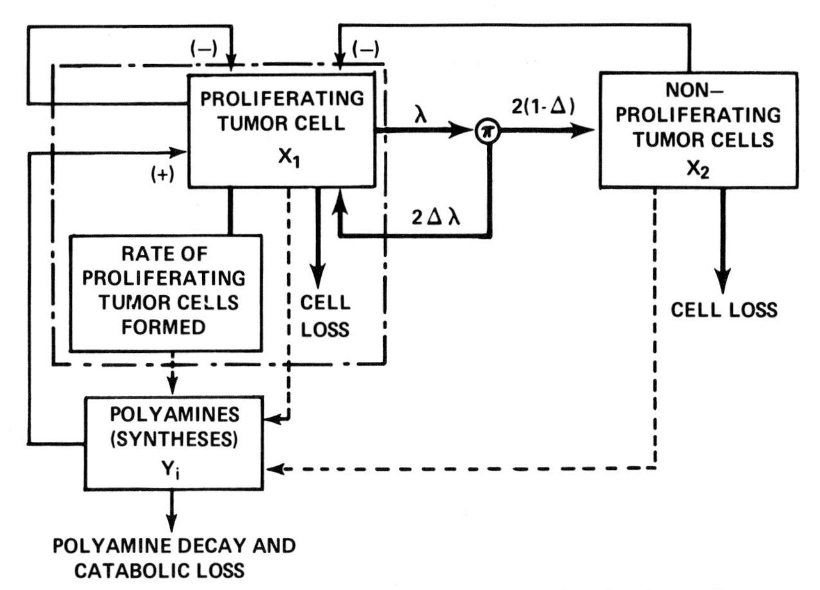

FIG. 1. A schematic diagram of the dynamics of tumor-polyamine interactions: ———, pathways of tumor cells; ———, control polyamines exert upon tumor cell proliferation with (+) stimulation and (−) inhibition; – – –, control tumor cells exert on polyamine synthesis.

previously elucidated by experiments (6, 10, 17, 18). The heavy continuous lines that connect these compartments represent fluxes of cells between them. Because the kinetics of polyamine synthesis and the control of this synthesis have not been fully characterized in the Ehrlich ascites tumor system, an attempt was made in this study to show that the rate of polyamine synthesis is specifically and quantitatively determined by density distribution of tumor cells, especially the rate of its variation in various compartments.

The interactions of all the compartments are integrated by implementing two modes of control, which are believed to function in the proliferation of tumor cells; they are (a) the self-inhibition by the tumor cell population itself and (b) the stimulation (activation) due to the critical level concentrations of polyamines (see Fig. 1). In fact, a theory on the self-inhibition of growth hypothesis was proposed earlier by Rose (19) in which self-limitation may in general result from the accumulation of specific products, and possibly the failure of malignant tumors to achieve self-limitation because of excessive deaths could be compensated for by the addition of sufficient quantities of their aberrant products. Recently, Burns (20) further elaborated this theory in reference to the Ehrlich ascites tumor cells, and suggested that these tumor cells had stopped growing because of the self-inhibition of a growth mechanism that is dependent on the attainment and maintenance of a critical viable tumor cell protoplasmic mass and most probably mediated through the ascitic fluid. The stimulation of tumor growth is implemented by a positive feedback of the level of putrescine concentration; no quantitative relationship has experimentally been established but it is assumed that its characteristics are determined essentially by the exhibition of cooperative, sigmoid response to the concentration of putrescine.

A kinetic model for tumor-polyamine interactions can be expressed by writing the material balance for the tumor cells and the polyamine concentrations.

Tumor Cell Growth:

$$\begin{bmatrix}\text{Total rate of change of}\\ \text{proliferating tumor cell}\end{bmatrix} = \begin{bmatrix}\text{Rate of reproduction of}\\ \text{proliferating tumor cell}\end{bmatrix}\begin{bmatrix}\text{Control}\\ \text{factors}\end{bmatrix} - \begin{bmatrix}\text{Rate of proliferating}\\ \text{tumor cell loss}\end{bmatrix} \quad (1)$$

$$\begin{bmatrix}\text{Total rate of change of}\\ \text{nonproliferating tumor cell}\end{bmatrix} = \begin{bmatrix}\text{Rate of reproduction of}\\ \text{nonproliferating tumor cell}\end{bmatrix}\begin{bmatrix}\text{Control}\\ \text{factors}\end{bmatrix} - \begin{bmatrix}\text{Rate of nonproliferating}\\ \text{tumor cell loss}\end{bmatrix} \quad (2)$$

Polyamine Synthesis:

$$\begin{bmatrix}\text{Total rate of change of}\\ \text{polyamine concentration}\end{bmatrix} = \begin{bmatrix}\text{Rate of production of}\\ \text{polyamine due to}\\ \text{proliferating tumor cell}\end{bmatrix} - \begin{bmatrix}\text{Rate of polyamine}\\ \text{metabolized for mitosis of}\\ \text{proliferating tumor cell}\end{bmatrix} + \begin{bmatrix}\text{Rate of production of}\\ \text{polyamine due to non-}\\ \text{proliferating tumor cell}\end{bmatrix} - \begin{bmatrix}\text{Rate of polyamine}\\ \text{decay}\end{bmatrix} \quad (3)$$

Tumor-polyamine dynamics, characterized by these equations, are cross-coupled by cooperative components [the kinetic parameters expressed as the control factors in Eqs. (1) and (2)] that represent communication between the cells as well as regulation of the cellular proliferation. These components are nonlinear functions of the tumor cells and the polyamine concentrations. An attempt is made to investigate quantitatively the behavior of these cross-coupled systems of various functional forms of cooperative components in reference to tumor growth and its relationship to polyamine synthesis.

III. KINETIC EQUATIONS FOR TUMOR-POLYAMINE INTERACTIONS

The sigmoidal pattern of growth observed in some tumor systems is empirically described by a Gompertz equation in the following form (2):

$$N(t) = N_0 \exp \left\{ \frac{A_0}{\alpha_1} [1 - \exp(-\alpha_1 t)] \right\} \tag{4}$$

In this equation, $N(t)$ and N_0 are measures of tumor size, in appropriate units, at general times (t) and zero time, respectively. A_0 is related to the initial slope of the growth curve and α_1 is a retardation factor; they are adjustable constants. Although the Gompertz equation can empirically describe a variety of tumor growth data, its usefulness appears limited because the constants A and α_1 have no discernible connection with experimentally measurable tumor cell characteristics.

Several hypotheses have been advanced to account for the regulation of growth: the availability of diffusion-limited nutrient supplies (21), the growth control via changes in the generation time of cells including a reversible regulation in both the G_1 and the G_2 phases (2, 14, 15, 22), the proliferating compartment of cells as the control site (23), and the existence of intrinsic tissue specific inhibitors (chalone) for tissue growth control (22, 24, 25). Recently, a few theoretical studies based on certain hypotheses have been carried out on growth control involving specific metabolites or substances excreted from cells, which mainly act as feedback regulators (3, 4, 26). In the present study, the control of tumor growth is implemented by two kinds of control factors, stimulation and inhibition, which are nonlinear functions of polyamine concentrations and tumor cell numbers. As illustrated by Eqs. (1)–(3), the dynamics of tumor-polyamine interactions requires two sets of kinetic equations. If x_1 and x_2 are the proliferating and the nonproliferating tumor cells, respectively, the growth of tumor cells is described by the following differential equations:

$$\frac{dx_1}{dt} = \lambda\, (2\delta - 1)\, \rho\left(\frac{y_1}{x_1}\right) \alpha(x_1, x_2)\, x_1 - \beta_1\, x_1 \tag{5}$$

$$\frac{dx_2}{dt} = \lambda\, 2\, (1 - \delta)\, \rho\left(\frac{y_1}{x_1}\right) \alpha(x_1, x_2)\, x_1 - \beta_2\, x_2 \tag{6}$$

where λ is the inverse of the generation time ($\tau = 1/\lambda$) for all proliferating cells, δ the fraction of the daughter cells returning to the proliferating cell compartment, β_1 the loss rate constant of the proliferating tumor cell, and β_2 the loss rate constant of the nonproliferating tumor cells.

Polyamines are synthesized by a well-known pathway (5, 6, 8). In mammals, L-ornithine decarboxylase (ODC) catalyzes the formation of putrescine from the substrate L-ornithine. Additionally, spermidine is formed from the concentration of putrescine and a polyamine moiety derived from S-adenosyl-L-methionine. The spermine synthetase S-adenosyl-L-methionine decarboxylase (SAMD) catalyzes this reaction. The evolution of $^{14}CO_2$ from S-adenosyl-L-methionine-$^{14}COOH$ is a valid measure of spermidine formation. Mechanistic and regulatory features of the enzymes involved in polyamine biosynthesis have been reviewed (9), and these enzymes exhibit dramatic elevations in activity early in the sequence of events that lead to the proliferation of cells. No attempt has been made in this chapter to analyze the specific enzymic kinetics of polyamine synthesis and its control with respect to the growth of Ehrlich ascites tumor cells. Enzyme kinetic study in relation to L1210 mouse leukemia cells and their response to antileukemic agents is under way.

With an inoculum dose of 4×10^6 ascites tumor cells, the growth pattern, distribution of cell size, and other characteristics of the Ehrlich ascites tumors were studied (12), and the patterns of change in the intracellular putrescine, spermidine, and spermine concentrations were presented, which enabled us to determine polyamine concentrations at various intervals after inoculation of the tumor cells (13). It is believed that the rate of polyamine synthesis is related to the density distribution of tumor cells in various compartments. Assuming that the decay loss of polyamines is proportional to the cell number present at any given time, and the amount consumed in mitosis is specified by the rate constant θ, the kinetics of polyamine synthesis as a function of tumor cells is described by

$$\frac{dy_i}{dt} = (K_{1i} - \theta)\,\lambda\,\rho\left(\frac{y_1}{x_1}\right)\alpha(x_1, x_2)\,x_1 + \lambda\,K_{2i}\left\{\left[1 - \rho\left(\frac{y_1}{x_1}\right)\alpha(x_1, x_2)\right]x_1 + x_2\right\} - \beta_{3i}\,y_i \qquad i = 1, 2, 3 \quad (7)$$

where y_i is the concentration of polyamine in μ mole ($i = 1$, putrescine; $i = 2$, spermidine; and $i = 3$, spermine), K_{1i} the rate constant for the production of polyamine by the proliferating tumor cells, K_{2i} the rate constant for the production of polyamine by the nonproliferating tumor cells, and β_{3i} the decay constant of polyamine concentration.

It is assumed that the general characteristics of the control factors denoted by parameters ρ and α used in Eqs. (5)–(7) basically follow strong sigmoid dependence on the concentrations of stimulator and inhibitor, as fully established in allosteric control of regulatory enzymes (26, 28) and also a number of cellular control processes (26, 29). The stimulation factor, $\rho\left(\frac{y_1}{x_1}\right)$, is a function of the concentration of polyamine related to the total number of proliferating cells. As a growth factor released from adult human skin, fibroblasts are believed to be putrescine; it is assumed that the level of putrescine concentration determines the degree of activation of tumor growth with a sigmoid relationship, which functions at threshold value of putrescine concentration, c^*:

$$\rho\left(\frac{y_1}{x_1}\right) = \begin{cases} 0, & \frac{y_1}{x_1} < c^* \\ 1, & \frac{y_1}{x_1} \geqslant c^* \end{cases} \tag{8}$$

A Gompertz type of the sigmoid function is also investigated for simulation study as given by

$$\rho\left(\frac{y_1}{x_1}\right) = \exp\left\{ ln\ [\rho(0)]\ \exp\left(-\frac{ry_1}{x_1}\right)\right\} \tag{9}$$

where r denotes a time constant that determines the rate at which the function approaches its asymptotic level of unity, and $\rho(0)$ the initial value of the function. The inhibition of tumor growth given by the inhibitory factor, $\alpha(x_1, x_2)$, is manifested by a retardation of the growth at high density of the tumor cell population. Analyzing the Gompertz equation of tumor growth and expressing the retardation as a function of tumor cells, we have

$$\alpha(x_1, x_2) = A_0 - \alpha_1\ ln \left\{\frac{x_1{}^t + x_2 t}{x_1(0) + x_2(0)}\right\} \tag{10}$$

where $x_1(0)$ and $x_2(0)$ are the initial densities of the proliferating and the nonproliferating tumor cells, respectively.

The dynamics of a progressive transition of proliferating tumor cells into a nonproliferating state, including the loss of cells from each compartment, is illustrated in Fig. 1. Cell loss is known to be an age-specific elimination of noncycling and a small amount of mitotic death (15). In the anatomical sense, cell loss may be represented by cell death, removal by exfoliation, or migration by metastases. In most tumors, cell death seems to represent the major pathway. The rate of cell loss from the proliferating compartment β_1 may be estimated by measuring the cell production rate and comparing this with the rate at which cells are observed to be added to the compartment, and it can be calculated by (16)

$$\beta_1 = ln\frac{2}{T} \cdot \left(1 - \frac{T}{T_d}\right) \tag{11}$$

where T is the potential doubling time and T_d the observed population doubling time. The equation is also used for the one from the nonproliferating compartment β_2.

After being produced, polyamines will enter the cell fluid and undergo numerous conversions as they diffuse and/or are transported throughout the cell, and also can leak out of the cells into the surrounding medium as well as be reabsorbed and consumed in mitosis. Since emphasis is placed primarily on the behavior of cell proliferation and its control, however, the details of these conversions are ignored, and the polyamines being leaked and metabolized by the cells are assumed to undergo first-order decay as shown in Eq. (7). The catabolic loss occurring in mitosis is specific for the proliferating cells, and its rate constant θ relates to this compartment.

IV. CHARACTERISTICS OF TUMOR GROWTH AND ITS RELATIONSHIP TO POLYAMINE SYNTHESIS: SIMULATION STUDY

The dynamics of tumor-polyamine interactions as represented by Eqs. (5)–(7) are characterized by computer simulation utilizing an IBM System/360 program for the simulation of continuous system, CSMP (Continuous System Modeling Program). This computer program is very useful and effective for manipulating the key parameters of stimulation, inhibition, and cell loss, and for evaluating their effects on tumor growth and polyamine synthesis.

Equation (10) for growth inhibition, which is derived from the Gompertz equation, contains the same Gompertz constants A_0 and α_1. Specific values or the general ranges of values of these constants may be estimated by the method for tumor growth analysis developed by Laird (2). Laird's analysis showed that the growth of Ehrlich ascites tumor with an inoculum dose of 4.3×10^5 ascites tumor cells (N_0) gives rise to the maximum cell number 1.6×10^9 (N_∞), which gives the ratio of Gompertz constants $A_0/\alpha_1 = 8.22$. Andersson and Agrell (12) suggested that an inoculum dose of 4×10^6 ascites tumor cells reaches to the cell number 1.6×10^9 with $A/\alpha_1 = 5.98$. It should be noted that the constants A_0 and α_1 are population-density dependent and may be so adjusted that the Gompertz equation fits to the growth curves of tumors. The rate of cell loss denoted by β_1 or β_2 is also population dependent as shown by Lala and Patt (14); in Ehrlich ascites tumor, it is 0.012/hr at $N = 4.2 \times 10^6$ and 0.014/hr at $N = 6 \times 10^8$. Steel (16) also calculated that the rate of cell loss in this tumor is in the range of 0.0126 to 0.0173/hr or 0.3 to 0.4/day. The constants A_0, α_1, and β_1 are adjusted within these ranges so that the number of tumor cells increases from 10^6 to an asymptotic level of about 1.2×10^9 in 14 days as has been reported by Andersson and Agrell (12). The time at which the asymptotic level is reached is a decreasing function of α_1. The constants A_0, β_1, and β_2 affect the asymptotic levels of $x_1(t)$ and $x_2(t)$. The parameter δ plays an important role in determining the growth rate at which these asymptotic levels are reached. For $\delta < 0.5$, $x_1(t)$ decreases exponentially from its initial value to 0, whereas if $\beta_2 > 0$, then $x_2(t)$ also approaches 0 exponentially after an initial increase. For

$0.5 < \delta < 1.0$, $x_1(t)$ and $x_2(t)$ increase from their initial values to an asymptotic level if the decay constants β_1 and β_2 are not too large. Under these conditions, tumor cell growth is of a sigmoid form and the rate at which $x_1(t) + x_2(t)$ approaches its asymptotic level is an increasing function of δ. The activation factor demonstrates that when the stimulant concentration y_1/x_1 falls below a threshold value, cellular proliferation ceases completely and the number of tumor cells decays exponentially. It appears that if the rate of decay of tumor cells is sufficiently greater than the rate of decay of stimulant, then y_1/x_1 is increased and reaches above the threshold value by which cellular proliferation resumes and results in an oscillatory pattern of tumor growth. This effect was not clearly observed in the present study; however, a further search for it and its required conditions is underway. Figure 2 shows simulated results of tumor growth and putrescine synthesis.

With both the inhibitory and the stimulatory factors fixed, the forms of the growth curve of tumor and the corresponding response variation of putrescine concentration per cell are influenced by a number of kinetic constants, including the rates of polyamine production by proliferating and nonproliferating cells, and the rates of polyamine utilization, catabolism, and removal. These are also a function of the rate of cell degradation and removal, and of the initial conditions of tumor cells. The effect of the number of nonproliferating cells on putrescine synthesis is shown by two different values of the fraction of daughter cells returning to the proliferating compartment, $\delta = 1.0$ and 0.8 (Figs. 3a and b). In Fig. 4, several curves of the

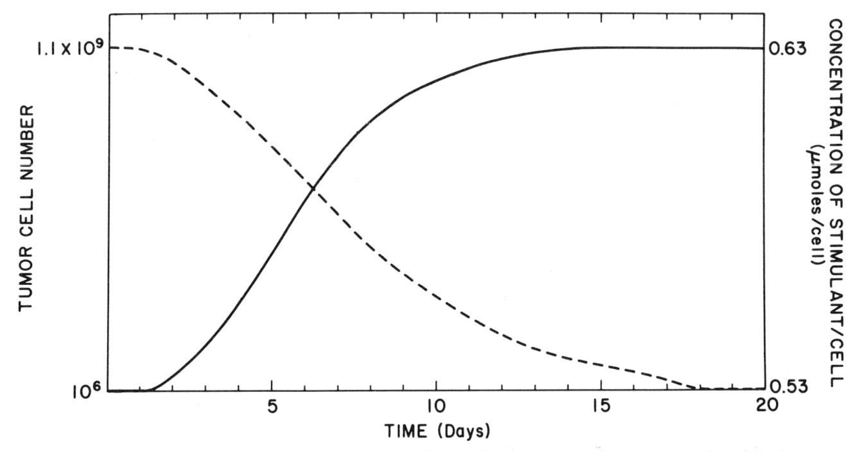

FIG. 2. Time courses of tumor cell growth and stimulant (putrescine) synthesis with the values of kinetic constants, $\delta = 1.0$, $\lambda = 1.0$, $\beta_1 = 0.25$, $K_{1i} - \theta = 0.70$, $K_{2i} = 0.05$, $\beta_3 = 0.40$, $A_0 = 3.2$, $\alpha_1 = 0.42$, $\rho(0) = 0.1$, and $r = 0.83$: ———, tumor cell number; – – –, stimulant (putrescine) concentration/cell.

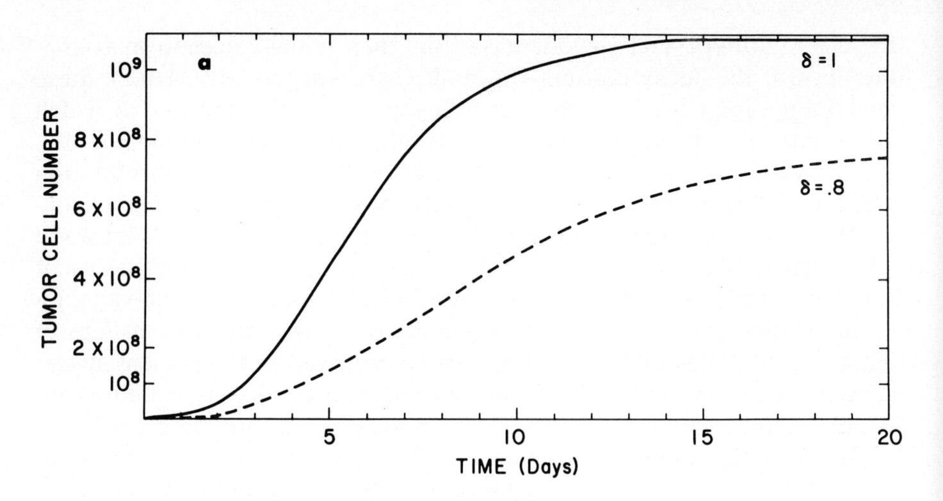

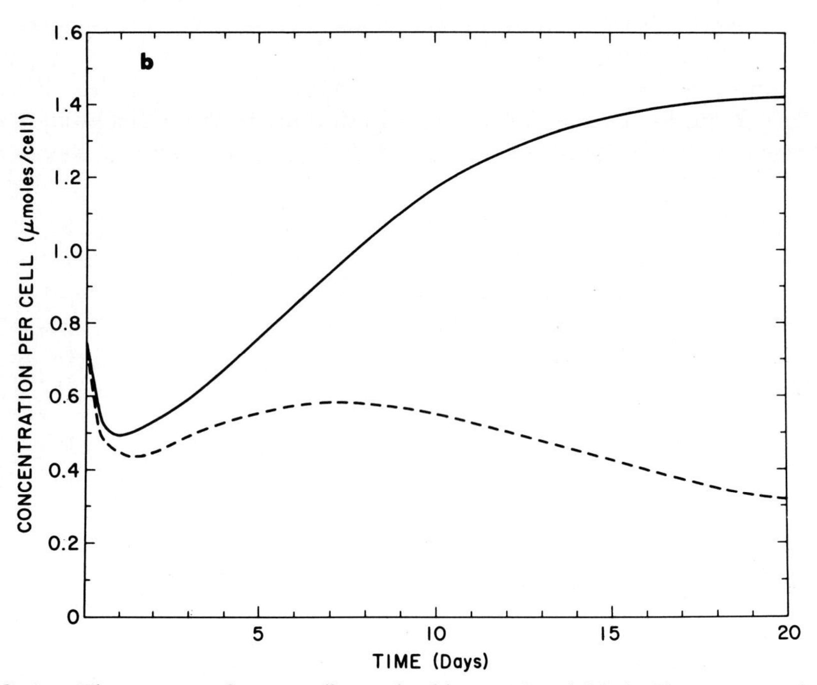

FIG. 3. a: Time courses of tumor cell growth with $\delta = 1.0$ and 0.8. b: Time courses of the concentrations of stimulant for tumor growth corresponding to the growth curves of Fig. 3a: ———, for $\delta = 1.0$, and – – –, for $\delta = 0.8$.

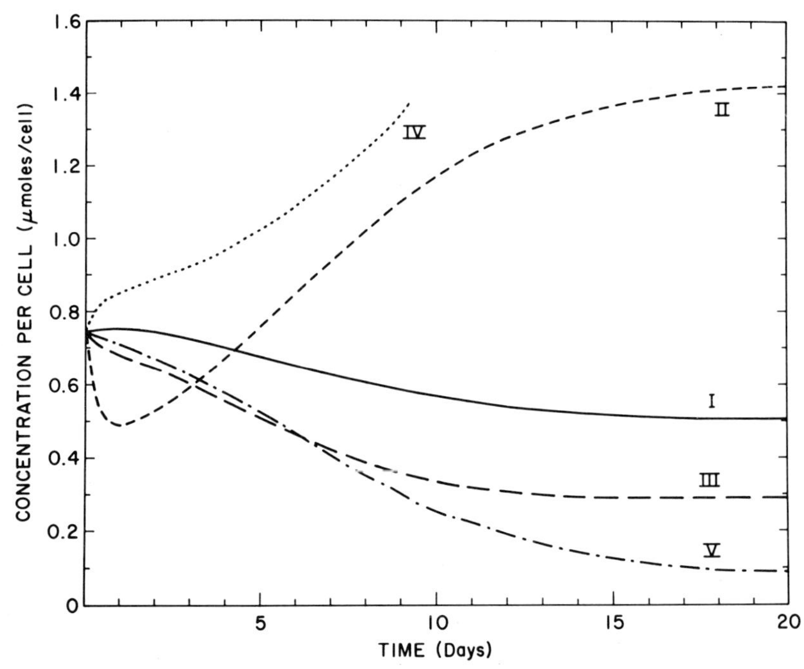

FIG. 4. Effects of various sets of kinetic constants summarized in Table 1 on the synthesis of stimulant (putrescine) for tumor growth.

concentration of putrescine are illustrated with the constant values of $A_0 = 3.2$, $\lambda = 0.42$, $\rho(0) = 0.1$, $r = 8.33$, and $\delta = 1.0$; the remaining constants are given in Table 1. As the rate of polyamine production or of tumor cell removal increases, the curve of polyamine concentration per cell increases with time. When the rate of polyamine utilization or removal increases, the curve decreases with time.

TABLE 1. *Values of kinetic constants* K_{2i}, β_1, *and* β_3 *used for simulation study of polyamine (putrescine) synthesis*

Curve	K_{2i}	β_1	β_3
I	0	0.4	0.25
II	0.5	0.4	0.25
III	0	0.7	0.25
IV	0	0.05	0.25
V	0	0.4	0.05

V. SUMMARY

A model based on three compartments—the proliferation and the nonproliferation of tumor cells and the synthesis of polyamines—was developed to investigate the dynamics of tumor-polyamine interactions. A system of nonlinear differential equations derived for the model includes two kinetic parameters for the control of tumor cell proliferation, self-inhibition, and stimulation. The main characteristics of these parameters are basically a cooperative control in nature which displays strong sigmoid response to the number of tumor cells and the concentration of polyamines. Implementing the experimental data obtained from cellular kinetic studies, several functional forms of the kinetic parameters for control are investigated and their effects on tumor growth and polyamine synthesis are analyzed. A quantitative relationship between tumor cells (Ehrlich ascites tumor) and corresponding markers (polyamines) is reasonably well demonstrated by the proposed model.

The important features derived from the analysis of the model are the nature of controls (inhibition and stimulation) in tumor cell proliferation, which are of a cooperative and sigmoid type, and the quantification of tumor cell numbers by the measurements of biological markers for the tumor. Further improvement and evaluation of the model are necessary to characterize the specific enzymic kinetics of polyamine synthesis and their control in relation to tumor growth; a study currently underway on this aspect with respect to L1210 leukemic mice cells is based on the experimental data available (30, 31). The clinical implications of the model would be greatly enhanced when the size or proliferation rate of tumor is estimated by this model, using urinary polyamine in patients with various types of solid tumors and leukemias.

ACKNOWLEDGMENTS

The authors wish to express their gratitude to Dr. S. Perry for early comments and encouragement and to Dr. M. Omine for his helpful comments and discussions on the subject of cell-cycle kinetics.

REFERENCES

1. Sullivan, P. W., and Salmon, S. E., *J. Clin. Invest.*, **51,** 1697 (1972).
2. Laird, A. K., *Brit. J. Cancer,* **18,** 490 (1964).
3. Weiss, P., and Kavanau, J. C., *J. Gen. Physiol.*, **41,** 1 (1957).
4. Bronk, B. V., Bienes, G. J., and Johnson, R. A., *Biophys. J.*, **10,** 487 (1970).

5. Tabor, H., and Tabor, C. W., *Pharmacol. Rev.,* **16,** 245 (1964).
6. Cohen, S. S., *Introduction to the polyamines,* Prentice-Hall, Englewood Cliffs, N.J., 1971.
7. Herbst, E. J., and Tangway, R. B., *Prog. Mol. Subcell. Biol.,* **2,** 166 (1971).
8. Williams-Ashman, H. G., Pegg, A. E., and Lockwood, D. H., *Adv. Enzyme Regul.,* **7,** 291 (1969).
9. Williams-Ashman, H. G., Jänne, J., Coppoc, G. L., Geroch, M. E., and Schenone, A., *Adv. Enzyme Regul.,* **10,** 225 (1972).
10. Russell, D. H., and Levy, C. C., *Cancer Res.,* **31,** 248 (1971).
11. Pohjanpelts, P., and Raina, A., *Nature,* **235,** 247 (1972).
12. Andersson, G., and Agrell, I., *Virchows Arch. Abt. B. Zellpath.,* **11,** 1 (1972).
13. Andersson, G., and Heby, O., *J. Nat. Cancer Inst.,* **48,** 165 (1972).
14. Lala, P. K., and Patt, H. M., *Proc. Nat. Acad. Sci. U.S.A.,* **56,** 1735 (1966).
15. Lala, P. K., *Cancer,* **29,** 261 (1972).
16. Steel, G. G., *Cell Tissue Kinet.,* **1,** 193 (1968).
17. Snyder, S. H., and Russell, D. H., *Fed. Proc.,* **29,** 1575 (1970).
18. Russell, D. H., Medina, V. J., and Snyder, S. H., *J. Biol. Chem.,* **245,** 6732 (1970).
19. Rose, S. M., *J. Nat. Cancer Inst.,* **20,** 653 (1958).
20. Burns, E. R., *Growth,* **33,** 25 (1969).
21. Burton, A. C., *Growth,* **30,** 157 (1966).
22. Bichel, P., *Eur. J. Cancer,* **7,** 349 (1971).
23. Baserga, R., *Cell Tissue Kinet.,* **1,** 167 (1968).
24. Bullough, W. S., and Deol, J. U. R., *Symp. Soc. Exper. Biol.,* **25,** 255 (1971).
25. Bichel, P., *Nature,* **231,** 449 (1971).
26. Glass, L., and Kauffman, S. A., *J. Theor. Biol.,* **34,** 219 (1972).
27. Good, P., *J. Theor. Biol.,* **34,** 99 (1972).
28. Monod, J., Wyman, J., and Changeux, J., *J. Mol. Biol.,* **12,** 88 (1965).
29. Goodwin, B., *Temporal organization in cells,* Academic Press, London, 1963.
30. Russell, D. H., *Cancer Res.,* **32,** 2459 (1972).
31. Heby, O., and Russell, D. H., *Cancer Res.,* **33,** 159 (1973).

Polyamines in Normal and Neoplastic Growth. Edited by D. H. Russell. Raven Press, New York © 1973.

Polyamines—Potential Roles in the Diagnosis, Prognosis, and Therapy of Patients with Cancer

Stephen C. Schimpff, Carl C. Levy, Inez A. Hawk, and Diane H. Russell

Baltimore Cancer Research Center, National Cancer Institute, Baltimore, Maryland 21211

I. INTRODUCTION

The physician faced with the problem of diagnosing and treating patients with suspected or known cancer frequently faces a multitude of problems, often including the need to do major surgical procedures to differentiate benign from malignant disorders. Another problem involves knowing whether surgery, radiation therapy, or drug therapy succeeded in eliminating the tumor present; if not eliminated, recurrence often cannot be detected until the total tumor mass is extensive. In such circumstances, the availability of a simple and reliable biochemical test that would estimate the presence and growth of tumor would make the clinician's task not only easier but more accurate.

Investigations from this Center first suggested that the assay of polyamines might be useful as such a marker of tumor presence (1, 2). Those findings will be reviewed, followed by our suggestions for further clinical evaluation of the polyamine assays.

II. URINARY POLYAMINES AS POTENTIAL CANCER MARKERS

Approximately 2 yr ago a 24-hr urine sample was collected from a teenager who presented with a massive, recurrent, abdominal ovarian teratoma (Table 1). The polyamine levels proved to be elevated far above normal (1). At surgery a few days later, the bulk of a grapefruit-sized mass was resected leaving considerable residual tumor. Ten days following surgery, the polyamine level was markedly reduced but still well above normal. The patient then received intensive cytotoxic drug therapy, and 3 weeks after the last dose of drug, her physical examination and urinary polyamine levels were normal. Monthly urines were collected for polyamines, and they remained at normal levels until some months later when tumor recurrence was detected.

These were most interesting findings: not only were the polyamine levels elevated at a time of massive tumor involvement but the subsequent decline and rise in values corresponded with the patient's known clinical tumor status.

Meanwhile, the following preliminary investigation was designed to determine if polyamines were generally elevated in patients with cancer and if polyamines fluctuated with known changes in tumor status (2).

TABLE 1. *Urinary polyamine excretion in a patient with ovarian teratoma*

	Polyamines (mg/24 hr)		
Status	Putrescine	Spermidine	Spermine + unknown
Pre-therapy	72.0	84.0	64.0
Post-surgery	10.0	28.0	<2.5
Post-chemotherapy	<2.5	5.6	<2.5

Twenty-four-hr urine samples were collected from 50 normal volunteers for polyamine analysis. The resulting normal values are in Table 2. Because polyamines are present in most rapidly dividing growing cell systems, urines from a group of normal pregnant women were analyzed. Not surprisingly, these values proved to be somewhat higher than those of the normal volunteers. Urines from a small group of patients with severely impaired liver function were evaluated to determine if hepatic dysfunction would affect urinary levels of polyamines; these levels were not significantly elevated above normal. As a final control group, urines from 35 hospitalized patients who had no evidence of cancer were analyzed. Included were patients with infection, chronic obstructive pulmonary disease, myocardial infarction, diabetes, and hypertension. None of these patients had elevated polyamine levels in their 24-hr urine samples.

The next group of patients in Table 2, all with histologically proven cancer, submitted urine samples *before* receiving anti-tumor therapy. As a group, the patients with solid tumors (e.g., lung, breast, bone, brain) had the highest polyamine levels, but the range was wide and the resulting standard errors are relatively high. Patients with lymphoma (Hodgkin's, lymphosarcoma, and reticulum cell sarcoma) all had elevated levels of polyamines. Likewise, all of the patients tested who had acute nonlymphocytic leukemia (ANLL) had elevated polyamines before therapy. Overall, every patient studied who had known cancer (with the single exception of one patient with multiple myeloma) had elevated urinary polyamines before the initiation of anticancer therapy (2).

Care must be taken in the handling and storage of the urine specimens. Collection under toluene may not be a sufficient precaution. Storage of the aliquot of urine to be assayed for polyamines in 6 N HCl will ensure an accurate measurement. To obtain polyamines in a nonconjugated form, it is necessary to subject the urine to acid hydrolysis prior to butanol extraction.

TABLE 2. *Urinary polyamine excretion in a variety of subjects*

		Polyamines (mg/24 hr)		
Subject	Number tested	Putrescine	Spermidine	Spermine + unknown
Normal volunteers	50	2.7 (0.5)	3.1 (0.6)	3.4 (0.7)
Pregnant patients	8	3.7 (0.6)	7.7 (0.8)	10.5 (1.8)
Cirrhotic patients	5	4.1 (0.7)	5.1 (0.7)	4.0 (0.6)
Nontumor hospital patients	35	2.9 (0.5)	4.7 (0.6)	3.5 (0.7)
Tumor patients (pre-therapy)				
Solid tumor	58	7.3 (0.9)	22.8 (10.1)	41.3 (16.3)
Lymphoma	48	6.1 (2.5)	15.5 (7.6)	26.2 (9.9)
Acute lymphocytic leukemia	7	7.2 (1.3)	39.2 (6.2)	68.3 (7.6)
Acute nonlymphocytic Leukemia	7	7.3 (1.2)	20.3 (9.8)	29.1 (11.5)

S.E. given in parentheses.

An accurate determination of the spermine level in urine is not possible by the high-voltage electrophoresis method because of the excretion of an unknown which has a migratory pattern identical to spermine. Whatever this compound is, however, it is not excreted in detectable amounts by normal individuals.

III. VARIATION IN POLYAMINE LEVELS FOLLOWING TUMOR THERAPY

The seven patients with ANLL were studied with serial urine samples. Table 3 indicates that, on the average, the polyamine levels rose during drug therapy. Most interesting, however, is the polyamine decline toward normal levels in the four patients who developed a remission of their leukemia. For

TABLE 3. *Urinary polyamine excretion in patients with acute nonlymphocytic leukemia*

Patient status	Number tested	Polyamines (mg/24 hr)		
		Putrescine	Spermidine	Spermine + unknown
Relapse	7	7.3 (1.2)	20.3 (9.8)	29.2 (11.5)
During therapy	7	3.1 (0.6)	31.6 (13.1)	45.6 (15.3)
Remission	4	<2.5	12.1 (5.4)	17.4 (8.7)

S.E. given in parentheses.

each patient, the total polyamine urinary excretion declined. In two of the four patients, the levels became undetectable and in two the levels were lower but still within the abnormally elevated range. Clearly, four patients represent too small a sample to hazard any estimation of whether the polyamine level was indicative of the absolute number of tumor cells remaining during remission or any correlation with subsequent length of remission. Likewise, the number of patients is too small to evaluate whether the individual level of polyamines at admission would suggest which patients may have the highest percent of cells dividing and hence the greatest chance to have a high proportion of cells affected by cell cycle-specific drugs. These are questions, however, which deserve consideration in future, more extensive investigations.

Table 4 lists the polyamines for three patients with cancer who had 24-hr urinary polyamine levels determined before and after surgery. In each of these three patients (and also in the patient presented in Table 1), the urinary polyamines declined to or toward normal following surgical removal of a portion of the tumor.

TABLE 4. *Urinary polyamine excretion following tumor surgery*

Patient	Status	Polyamines (mg/24 hr)		
		Putrescine	Spermidine	Spermine + unknown
Brain tumor	pre	<2.5	8.5	14.7
	post	<2.5	<2.5	<2.5
Brain tumor	pre	<2.5	20.3	<2.5
	post	<2.5	13.7	3.8
Testicular tumor	pre	7.8	21.9	32.1
	post	5.4	6.1	5.0

To summarize this initial clinical study of urinary polyamines in patients with cancer: (a) a relatively narrow range of normal was identified; (b) pregnancy led to some elevation of polyamines; (c) a small but diversified group of hospitalized nontumor patients had normal polyamines; (d) all but one patient with histologically proven advanced cancer had elevated urinary polyamines; and (e) polyamine levels in certain selected patients fluctuated with the known clinical status of the patient.

These preliminary clinical observations raise a number of interesting possibilities. Are urinary polyamines elevated in all cancers? Does the absolute polyamine level correspond to the total cell burden of the individual patient? Does the polyamine level mirror the degree of tumor cell proliferation? Do serial changes in urinary polyamines accurately reflect changes in the patient's clinical course? In other words, can the urinary polyamine determination be useful in the diagnosis, management, and follow-up of patients with cancer?

This initial clinical evaluation had some exciting and provocative results, but it was a relatively small investigation and will need considerable follow-up. Before continuing with specific suggestions for future evaluation, it might be appropriate to consider certain other known biochemical markers of cancer and the problems encountered in their evaluation.

IV. LABORATORY MARKERS OF CANCER

Table 5 presents a list of some of the known markers. Despite multiple markers, none has proven to be as specific as was first supposed.

For example, the carcino-embryonic antigen (CEA) was first found to be present in 35 of 36 patients with intestinal malignancy and not present in

TABLE 5. *Biochemical markers of cancer*

Marker	Tumor
Carcinoembryonic antigen (CEA)	Gastrointestinal
Alpha-feto protein	Hepatoma
Placental alkaline phosphotase (Regan)	Lung
Muramidase	Leukemia
Acid phosphotase	Prostate
Chorionic gonadotropin (HCG)	Trophoblastic
Adrenocorticotrophic hormone (ACTH)	Lung, thymus, pancreas
Erythropoietin	Renal
Parathormone	Lung, kidney, lymphoma
Antidiuretic hormone (ADH)	Lung
Serotonin	Carcinoid

any other condition tested (3). More recently, it has been found present in only about 70 to 75% of intestinal tumors but in a variety of other tumor and nontumor conditions (4).

The alpha-fetoprotein was initially believed present only in patients with hepatoma. Further investigations have shown it present in 50 to 80% of hepatoma patients plus occasional patients with nonhepatic cancer (5, 6).

Muramidase is an enzyme useful in differentiating various types of acute leukemia and has been found to fluctuate with the status of the patient; e.g., it returns toward normal during remission. Nevertheless, patients with morphologically distinct leukemia occasionally have unexpected muramidase values, and variations in known tumor status are not always accompanied by muramidase changes (7). Similar statements could be made for each of the markers listed in the table.

V. AREAS FOR FURTHER INVESTIGATION

A. Methodology

To a clinician there are certain aspects of any clinical test that make it of greater or lesser value. Of primary importance is rapidity—methodology presently under evaluation may make the analysis both rapid and inexpensive. A second problem is the present need for a 24-hr urine collection. It is well known that collecting a 24-hr urine sample is not only difficult but almost impossible on a busy general hospital ward. The same problem exists for outpatient urine collection. As a result, tests that utilize 24-hr urine collections are generally avoided by physicians whenever possible. A real methodologic improvement from a practical standpoint would be to determine if "spot" samples (e.g., first morning specimen) yield results comparable to a 24-hr specimen. Another direction being explored is blood sampling; blood collection is often more accurate and convenient than urine collection. Blood samples avoid contamination, are immediately available, and avoid many of the problems of a 24-hr urine collection. There are some preliminary data that blood evaluation may be possible. Marton et al. (this volume), utilizing an automated amino acid analyzer technique, have detected elevated spermidine and sometimes putrescine in the sera of a group of patients with cancer.

B. Normal variations

Considering the data already available regarding polyamines and cancer, it is now pertinent to consider an overall approach to further human polyamine investigation. Table 5 presents suggestions for investigation of the normal variations in patients with no evidence of cancer.

First, a large number of multiple samples from normal volunteers would enable investigators to establish a normal range and to assess intrapatient variations. Second, a large number of hospitalized patients with no evidence of tumor should be screened. It is important to obtain samples from a large variety of disease states so that conditions which may lead to elevated polyamines are detected. Third, a large group of patients should be screened who have conditions that might be expected to cause polyamine elevations. Obvious examples are pregnant women, patients with resolving hepatitis where liver cell regeneration is probably maximal, patients in the healing phases after major surgery, patients receiving anabolic steroid preparations, and growing children. This aspect of future evaluation is essential in establishing the areas for "false-positive" results in cancer detection.

After these investigations are complcte—or perhaps concurrently—there should be further evaluation of the polyamine assay in patients with cancer. This aspect is divided into four sections: universality, diagnosis, prognosis, and status (Table 6).

TABLE 6. *Further evaluation to assess clinical role of polyamines*

- Clinical trials—Cancer
  - Universality
    - Multiple patients, various centers, various tumor types, various tumor burdens
  - Diagnosis
    - Blinded assay prehistologic confirmation (e.g., breast mass, lung mass)
  - Prognosis
    - Blinded assay pre-therapy (surgery, drug)
  - Status
    - Variations in levels during course of disease

Polyamine assays should be done on a large group of patients with histologically proven cancer. For the evaluation to be of the most value, it should include only patients who have had careful staging to determine the degree of tumor dissemination. For example, it may be that the polyamines are elevated only in patients with massive local involvement or metastatic spread but not in those patients with relatively small or nonmetastatic cancers. Obviously, it is important to obtain samples from patients with a wide variety of tumor types, all collected before any therapy. The results of such a survey should establish if this assay has universality.

The next area, diagnosis, is one of great hope but for which there is presently little or no information. The patients we tested in Baltimore all had obvious and widely spread cancer—the important question now is whether patients with small tumors or nonmetastatic tumors also would have elevated polyamines. Three clinical situations for which adequate patients and appropriate information should be readily available are the initial diagnostic surgery for breast masses, pulmonary densities, and X-ray detected colonic masses.

Women with a benign or malignant breast mass but no evidence of distant tumor spread could have polyamines assayed before biopsy and correlated with the histologic results obtained after the subsequent surgery. This would establish whether the polyamines are elevated in benign and/or localized malignant masses and thus give some idea of their diagnostic value.

A similar investigation in patients with pulmonary lesions of unknown etiology before planned surgery would establish whether a polyamine assay would differentiate benign from malignant pulmonary masses. Similarly, patients with colonic lesions detected by X-ray could have polyamines determined before surgery.

These three lesions were chosen because they are fairly common, and because, in each case, biopsy is the only acceptable method for diagnosis between malignancy and other more benign conditions. Therefore, adequate numbers of patients should be available in association with histologic confirmation. This aspect of polyamine investigation should be launched after it is firmly established that these three tumors are consistently associated with elevated polyamines.

The next area to consider is prognosis: will the level of polyamines help the physician classify his patient? For example, do higher levels equate with metastasis? Or, does a high level indicate rapid cell turnover and thereby indicate a tumor which would be likely to respond to cell cycle-specific therapy?

Finally, will an assay of polyamines be useful in following the status of the patient and his tumor over time as he undergoes surgery, radiation, and/or drug therapy? The patient with the ovarian tumor had distinct polyamine changes as her tumor status varied. The four patients with acute nonlymphocytic leukemia who developed a remission had a drop in their polyamine levels. Two patients with brain tumors and one with a testicular tumor all had a reduction of 24-hr urinary polyamines following surgery. Investigations into this question can be done in conjunction with the universality and diagnosis sections.

VI. SUMMARY

A preliminary clinical evaluation at the Baltimore Cancer Research Center has suggested that a 24-hr urinary polyamine assay is positive in the great majority of patients with advanced cancer and that the polyamine levels often vary as the tumor mass varies with therapy. These initial findings suggest that further investigation is appropriate to determine if there is a potential clinical applicability of the polyamine assay to human cancer. Specific suggestions for further investigation include establishing the normal range, biologic states that elevate polyamines other than cancer, and the role of the polyamine assay in the diagnosis, prognosis, and clinical evaluation of patients with cancer.

REFERENCES

1. Russell, D. H., *Nature,* **233,** 144 (1971).
2. Russell, D. H., Levy, C., Schimpff, S. C., and Hawk, I. A., *Cancer Res.,* **31,** 1555 (1971).
3. Gold, P., and Freedman, S. O., *J. Exp. Med.,* **122,** 467 (1965).
4. LeBel, S. S., Deodhan, S. D., and Brown, C. H., *Dis. Col. Rect.,* **15,** 111 (1972).
5. Alpert, M. E., Uriel, J., and deNuchand, B., *New Eng. J. Med.,* **278,** 984 (1968).
6. Mehlman, D. J., Bulkley, B. H., and Wiernik, P., *New Eng. J. Med.,* **285,** 1060 (1971).
7. Wiernik, P., and Serpick, A. A., *Amer. J. Med.,* **46,** 330 (1969).

Polyamines in Normal and Neoplastic Growth. Edited by D. H. Russell. Raven Press, New York © 1973.

Some Recent Observations on the Molecular Biology of RNA Tumor Viruses and Attempts at Application to Human Leukemia

Robert C. Gallo

Laboratory of Tumor Cell Biology, National Cancer Institute, National Institutes of Health, Bethesda, Maryland 20014

I. BACKGROUND

The mechanism of leukemogenesis in man appears to involve a block or an aberration in the normal process of bone marrow leukocyte differentiation, rather than a primary proliferative disorder. This has been discussed and reviewed in more detail elsewhere (1–3). Our initial prejudices were that this process was irreversible. However, a recent important report by Paran et al. in Sachs' laboratory (4) indicates that the leukemic process can be reversed *in vitro.* Sachs and his colleagues had previously developed a technique of growing leukocytes on soft agar plates (5). If the feeder layer of normal cells is provided below the soft agar layer, leukemic blast cells from bone marrow peripheral blood can apparently be induced to mature (differentiate). This is clearly one of the more important areas of research at the moment in human acute leukemia. The questions of the origin(s) and molecular mechanism of action of this factor(s) and the demonstration that these *in vitro* factors play physiological roles *in vivo* will be the areas of greatest interest.

What is the etiology of this defect in maturation? There is as yet no definitive answer to this question, but for many reasons information related to RNA tumor viruses is strongly suspected as playing some role in the development of this disorder. The main reason for suspecting that information from these viruses is involved in human leukemia is still the experience with animal leukemias. However, it is important to stress that in man there is no clear electron-microscopic evidence of the presence of complete RNA tumor viruses. This might be because too few viruses are present to be detectable by microscopic techniques. Alternatively, mature viral particles may never appear. (There are, of course, examples of virally induced neoplasias in animals where intact viral particles are subsequently not observed.) In any case, other approaches were obviously needed. Biochemical observations made within the past few years have further suggested that virally related information is present in human leukemic cells. The impetus for these studies came from the discovery of the DNA polymerase of RNA tumor viruses, the so-called reverse transcriptase. This enzyme, as is now well known, catalyzes formation of a DNA intermediate in the replication cycle of an RNA tumor virus. Presumably this DNA (Temin's provirus) intermediate can be integrated into the host normal cell DNA where it may stay in repressed form or where it may be transcribed to produce viral RNA. The viral RNA which is associated with reverse transcriptase has certain characteristics, for example, its size (70S). Several biochemical characteristics of this enzyme have provided a means for distinguishing it from normal cellular DNA polymerases.

II. LOOKING FOR REVERSE TRANSCRIPTASE IN CELLS

During the past 2 years, we have examined leukemic cells for an enzyme with the biochemical characteristics of the viral reverse transcriptase. We have attempted to define criteria for distinguishing this enzyme from the major normal cellular DNA polymerase. In other words, the goal was to determine if we could utilize the finding of reverse transcriptase as a "footprint" for an RNA tumor virus. It was also of interest to look in cells for the presence of enzymes which could catalyze reverse transcription but which were not necessarily viral in origin. In fact, such enzymes are predicted in Temin's protovirus theory (6). Temin has speculated that the association of messenger RNAs and DNA polymerase with reverse transcriptase properties may play a role in normal differentiation. These nucleic acid polymerase systems under certain circumstances might "evolve" into RNA tumor viruses (6).

In order to define a reverse transcriptase in cellular systems, one must be familiar with the properties of reverse transcriptases from known RNA tumor viruses, as well as with the DNA polymerases present in normal proliferative cells. During the past 2 years, we have purified and characterized the biochemical properties of the major DNA polymerases of normal human blood lymphocytes stimulated to divide with PHA (7). There are at least two *major* DNA polymerases in these cells, an enzyme with a molecular weight of approximately 150,000 (which can dimerize to a form with a molecular weight of 300,000 daltons), apparently present primarily in the cytoplasm. The second enzyme appears to be located in both nucleus and cytoplasm and is of much lower weight (approximately 35,000 daltons). Drs. Graham Smith and Marjorie Robert in my laboratory have clearly demonstrated that these enzymes are capable of copying some DNA-RNA synthetic structures such as polythymidylic acid–polyriboadenylic acid (poly dT·poly rA) (7, 8). These enzymes can also catalyze the transcription of poly A stretches in messenger RNA or in other RNAs when supplied with an appropriate primer, i.e., oligo dT. Therefore, in a limited sense these enzymes are capable of transcribing RNA. The observation that they were able to utilize poly dT·poly rA as a template produced some confusion regarding the question whether viral reverse transcriptase was present in normal proliferative cells. It appears that virtually every DNA polymerase can copy appropriately primed poly rA (8, 9). On the other hand, we have found these enzymes do not catalyze transcription of heteropolymeric portions of natural RNAs. Although it is apparent that some of the fine properties of reverse transcriptase from different species of RNA tumor viruses may differ, there are common features of every reverse transcriptase that have been studied to date. These common features are what we have utilized as criteria for defining the presence of reverse transcriptase in leukemic cells. These properties are listed below and have been discussed in detail elsewhere (9–12).

(a) The DNA polymerase must be isolated in a particulate fraction.

(b) Particulate DNA polymerase should catalyze *endogenous* ribonuclease-sensitive DNA synthesis, i.e., DNA polymerase activity without an added template, presumably utilizing a nucleic acid in the same particulate fraction. The ribonuclease sensitivity would demonstrate that the reaction is indeed dependent on RNA but does not in itself prove that RNA *directs* the reaction.

(c) The product of the endogenous DNA polymerase reaction exists at least in part as a hybrid (i.e., DNA is joined to a larger RNA molecule).

(d) The DNA polymerase purified from the particulate fraction, to a point where it is free of its endogenous nucleic acid, and from the bulk of contaminating proteins, should be able to utilize natural RNA, DNA-RNA hybrids, and DNA as template primers.

(e) At the present time, the most specific template would appear to be 70S RNA from a known RNA tumor virus. The DNA product of this reaction should be shown to exist at least in part as a hybrid structure.

(f) DNA product of the reaction with viral 70S RNA as template primer should be able to "back-hybridize" specifically to the template RNA and not to heterologous RNA.

(g) The purified reverse transcriptases from every RNA tumor virus we have looked at respond in a characteristic way to certain synthetic templates. They respond to oligo dT·poly rA $\gg$ oligo dT·poly dA. In the presence of Mg^{++}, we have not found any normal cellular DNA polymerase that showed this pattern of response to synthetic templates. In the presence of Mg^{++}, normal cellular DNA polymerase in our hands has always shown a much greater affinity for oligo dT·poly dA than for oligo dT·poly rA.

In 1970 we published a preliminary report of an enzyme in human leukemic cells that could copy, at least partially, some natural RNAs, particularly ribosomal RNA (13). During the last year more definitive characterization of the enzyme has been carried out (11). We have been able to show that this enzyme has all the biochemical characteristics associated with the reverse transcriptase of RNA tumor viruses. We have isolated the enzyme in the cytoplasmic "pellet" fraction from both acute lymphoblastic leukemic cells and acute myeloblastic leukemic cells. Within this particulate fraction the enzyme catalyzes an endogenous, completely ribonuclease-sensitive synthesis of DNA (11). We have found this polymerase in approximately 40% of patients with acute leukemia (12). The product of early time points of this endogenous reaction can be shown to exist as hybrid structures. We have purified the enzyme from the particulate fraction and it has the following characteristics: (a) its molecular weight is approximately 120,000 daltons; (b) it shows a three- to sixfold preference for oligo dT·poly rA compared to oligo dT·poly dA as a template primer; (c) it transcribes Rauscher leukemia virus and avian myeloblastosis virus 70S RNA, and the DNA product was shown specifically to hybridize back to the template RNA and not to heterologous RNA. The details of the purification and properties of the enzyme have been published elsewhere (9, 11, 12). The enzyme is also clearly distinguishable from a terminal addition deoxynucleotidyl transferase. Immunological studies are now in progress to obtain information that might allow us a clear distinction on its origin (viral vs. cellular). Antibodies prepared against reverse transcriptase RNA tumor viruses will be assayed for their inhibitory effect on this leukemic DNA polymerase.

III. RNA OF RNA TUMOR VIRUSES

Recently 70S viral RNA molecules were shown to contain large tracts of poly A (14–16). We used an assay of hybridization of the 70S RNA to ^{3}H-poly U followed by hydrolysis of single-stranded regions. The poly A tracts, of course, are hydrogen bonded to poly U, and these hybrid structures are collected on Millipore filters. With this assay we can determine the content of poly A in the 70S RNA of various RNA tumor viruses. In addition, we can analyze the size of the poly A tract and its homogeneity by gel electrophoresis. Taken together, the amount of poly A, its size, the fact that the peak of poly A is homogeneous, and its association with a 70S RNA are criteria which we can use to distinguish the poly A of 70S RNA from the poly A of message RNA or the poly A from RNA viruses which are non-tumorigenic (15). This assay should prove to be an extremely useful biochemical tool (a) for determining if a particular RNA molecule is of viral origin, (b) for quantitation of virus, and (c) for detection of virus mutants lacking infectivity and other normally "testable" components. This assay also offers a new, additional criterion for establishing that RNA isolated from questionable particles or from cells is virally related.

We have recently identified high molecular weight poly A containing RNA in the same cytoplasmic particulate fraction from human leukemic cells that contains the polymerase with reverse transcriptase properties (M. Robert, D. Gillespie, and R. Gallo, *unpublished results*). We find that the poly A tracts have properties similar to those we have found in the 70S RNA of RNA tumor viruses. However, the size of the RNA isolated in these particulate fractions has varied from approximately 60S to over 100S. It remains to be determined whether these particulate fractions represent immature virions, complete virions which are present in too small a number to be visibly detectable, or biological systems independent of RNA tumor viruses.

Are reverse transcriptase systems present in normal cells? Three DNA polymerases which may function in transcribing cellular RNA have been described. These consist of the developing frog oocyte where it has been postulated that reversal of information flow may be responsible for the amplification of the genes for ribosomal RNA during oocyte development (17). The second system is the chick embryo where Kang and Temin (18) have found an enzyme that appears to have properties similar to viral reverse transcriptase. They have been able to differentiate this enzyme from the viral enzyme immunologically and to postulate a role for this polymerase in cell differentiation. There is, as yet, no published definitive evidence, however, that this enzyme can catalyze transcription of viral RNA. The third finding is in PHA-stimulated lymphocytes (19). We found that stimulated but not resting lymphocytes contain particulate, endogenous, ribonuclease-sensitive DNA polymerase activity. If this enzyme was purified, however, it could be distinguished from the enzyme isolated from leukemic cells. We have been unable to show that this enzyme is capable of catalyzing transcription of heteropolymeric portions of viral 70S RNA. It remains to be determined whether this endogenous reaction functions in reversal of information flow in cells or whether it is RNA primed and DNA directed (rather than RNA directed).

IV. INHIBITORS OF VIRAL REPLICATION

If RNA tumor viruses could be shown definitively to be involved in human neoplasia, would there be rationale for inhibiting viral replication? This is a question to which perhaps too little attention has been directed. There is no reason to think that inhibiting viral replication would have any effect on cells after they have been transformed. There does seem to me, however, to be good rationale for using inhibitors of viral replication for remission maintenance or possibly in the future for prophylaxis, *if* one could convincingly demonstrate replicating infectious viruses in these cells. I say this because there is evidence in some viral-induced animal tumors for what might be referred to as reinduction or reinfection. In other words, some tumors in animals may not be monoclonal—there may be additional transformations in normal cells after the initial transforming event. There is also suggestive evidence that something like this may occur in particular situations in man. It is almost certain that, after a patient experiences remission with standard cytotoxic chemotherapeutic agents, relapse in most instances is due to the physician's failure to destroy the last tumor cell, especially when relapse occurs after a short remission. In some instances, however, relapse may be caused by an additional, new transformation event, particularly when relapses occur after long remissions. The suggestive clinical evidence to support this notion is summarized below.

(a) There is a high frequency of additional primary neoplasias in patients with breast cancer.

(b) We are now observing very late relapses in childhood leukemia, some occurring after as much as 10 years of remission. It seems unlikely that this is due to a small residual population of leukemic cells which stay quiescent for a decade and then begin to proliferate.

(c) If new transformation events occur, one would predict that cytogenetic analysis should reveal an increase in the appearance of new chromosomal markers with increase in time of survival. For example, in one disease where enough cytogenetic studies have been performed over a sufficient interval we find that this prediction holds true: with increase in time of survival, there is an increased incidence in new chromosomal markers in chronic myelocytic leukemia (20).

(d) In patients with Burkitt's lymphoma who have been put into remission with cytotoxin, two types of relapses have been described, early and late. Apparently in the late relapses cytogenetic analysis reveals new chromosomal markers which would also suggest the possibility of additional transformation events (J. Ziegler, *personal communication*).

(e) Finally and most interesting are the results of bone marrow transplantation studies in leukemic patients. There are now at least three incidents where evidence of phenotypic leukemic transformation of normal donor cells in recipient leukemic patients has occurred. I think the most likely interpretation of these findings is that an initial inciting agent is present and is somehow activated, perhaps by the treatment (radiation or chemotherapy) the leukemic patient receives in preparation for the bone marrow transplantation. This "agent" would then induce the leukemogenic change in the donor cells. It must be stressed, however, that there are alternative explanations for the above clinical phenomena.

We have been studying two approaches to inhibit viral replication. First of these is to block formation of the so-called provirus by inhibiting reverse transcriptase. We recently analyzed over 200 derivatives of the antibiotic rifamycin SV for its ability to inhibit reverse transcriptase from various RNA tumor viruses and for its activity on normal cellular DNA polymerase (21). We showed that rifamycin SV derivatives which have relatively strong inhibitory effects on reverse transcriptase completely inactivate the virus biological activity (as measured by blocking plaque formation and focus formation) (22). Rifamycin SV derivatives which have moderate effects on reverse transcriptase also have intermediate effects on the biological activity of the virions, whereas derivatives which have little or no effect on the enzyme have no effect on biological activity of the virions (22). These observations have recently been extended to *in vivo* leukemogenesis in mice. Rifamycin derivatives which are relatively potent inhibitors of reverse transcriptase completely inactivate the ability of Rauscher leukemia virus to induce leukemia in normal mice (23).

The second approach we have been using to inhibit viral replication is an attempt to block the expression of the provirus. In other words, we are attempting to block the process of transcription of the viral DNA to viral RNA after integration of the viral DNA into the host cell DNA. The system we have been utilizing to study the transcription of the provirus DNA to viral RNA is the one developed by Lowy et al. (24). They found that treating certain normal tissues in short-term cultures with iododeoxyuridine (IDU) could induce these cells to produce C-type virus. These were, of course, cells not previously producing detectable virus. For example, when one treats Balb/3T3 with IDU, a "pulse" of C-type virus is released into the medium, reaching a peak in about 3 to 4 days after treatment with IDU (25). We have studied the effects of cordycepin (3′-deoxyadenosine), an analogue of deoxyadenosine and a known inhibitor of poly A formation, on the inducibility of virus from these cells. We find that if cordycepin is present specifically between 8 and 12 hr after treatment with IDU, the induction of the virus is completely blocked (26). Although we are uncertain of the mechanism, the most likely interpretation is that the poly A is necessary for the proper formation for viral RNA and cordycepin is interfering with poly A synthesis. This does not appear to be due to a nonspecific cell toxicity for several reasons: (a) the effect can be obtained when cell growth rate is unaffected; (b) pretreatment of the cells with cordycepin for 24 hr, followed by cell washing (to remove cordycepin), and then by the addition of IDU, does not affect IDU induction of virus (26); (c) virus induction can be blocked with cordycepin added for a short interval (8 to 12 hr after IDU), and under these conditions total cell RNA synthesis is not detectably reduced (D. Gillespie, A. Wu, and R. C. Gallo, *unpublished results*). Experiments are now in progress to determine if viral RNA, deficient in poly A, accumulates in the cell with cordycepin treatment.

REFERENCES

1. Gallo, R. C., *Cancer Res.,* **31,** 621 (1971).
2. Gallo, R. C., *Acta Haematol.,* **45,** 136 (1971).
3. Perry, S., and Gallo, R. C., in A. S. Gordon (Editor), *Regulation of Hematopoiesis,* Vol. II, Appleton-Century-Crofts, New York, 1970, p. 1221.
4. Paran, M., Sachs, L., Barak, Y., and Resnitzky, P., *Proc. Nat. Acad. Sci. U.S.A.,* **67,** 1542 (1970).
5. Pluznik, D. H., and Sachs, L., *J. Cell. Comp. Physiol.,* **66,** 319 (1965).
6. Temin, H. M., *J. Nat. Cancer Inst.,* **46:** iii (1971).
7. Smith, R. G., and Gallo, R. C., *Proc. Nat. Acad. Sci. U.S.A.,* **69,** 2879 (1972).
8. Robert, M. S., Smith, R. G., Gallo, R. C., Sarin, P. S., and Abrell, J. W., *Science,* **176,** 798 (1972).
9. Gallo, R. C., Sarin, P. S., Smith, R. G., Bobrow, S. N., Sarngadharan, M. G., Reitz, M. S., Jr., and Abrell, J. W., in R. Wells and R. Inman (Editors), *Proceedings of the second annual Steenbock symposium,* in press.
10. Gallo, R. C., *Nature,* **234,** 194 (1971).
11. Sarngadharan, M. G., Sarin, P. S., Reitz, M. S., and Gallo, R. C., *Nature New Biology,* **240,** 67 (1972).
12. Gallo, R. C., Sarin, P. S., and Bhattacharyya, J., *Cancer,* in press.
13. Gallo, R. C., Yang, S. S., and Ting, R. C., *Nature,* **228,** 927 (1970).
14. Lai, M. M. C., and Duesberg, P. H., *Nature,* **235**, 383 (1972).
15. Gillespie, D., Marshall, S., and Gallo, R. C., *Nature New Biology,* **236,** 227 (1972).
16. Green, M., and Cartas, M., *Proc. Nat. Acad. Sci. U.S.A.,* **69,** 791 (1972).
17. Brown, R. D., and Tocchini-Valentini, G. P., *Proc. Nat. Acad. Sci. U.S.A.,* **69,** 1746 (1972).
18. Kang, C.-Y., and Temin, H. M., *Proc. Nat. Acad. Sci. U.S.A.,* **69,** 1550 (1972).
19. Bobrow, S. N., Smith, R. G., Reitz, M. S., and Gallo, R. C., *Proc. Nat. Acad. Sci. U.S.A.,* **69,** 3228 (1972).
20. Gallo, R. C., Smith, R. G., Whang-Peng, J., Ting, R. C. Y., Yang, S. S., and Abrell, J. W., *Medicine,* **51,** 159 (1972).
21. Yang, S. S., Herrera, F., Smith, R. G., Reitz, M. S., Lancini, G., Ting, R., and Gallo, R., *J. Nat. Cancer Inst.,* **49,** 7 (1972).

22. Ting, R. C., Yang, S. S., and Gallo, R. C., *Nature New Biology,* **236,** 163 (1972).
23. Wu, A., Ting, R. C., and Gallo, R. C., *Proc. Nat. Acad. Sci. U.S.A.,* in press.
24. Lowy, D. R., Rowe, W. P., Teich, N., and Hartley, J. W., *Science,* **174,** 155 (1971).
25. Aaronson, S. A., Todaro, G. J., and Scolnick, E. M., *Science,* **174,** 157 (1971).
26. Wu, A. M., Ting, R. C., Paran, M., and Gallo, R. C., *Proc. Nat. Acad. Sci.,* **69,** 3820 (1972).

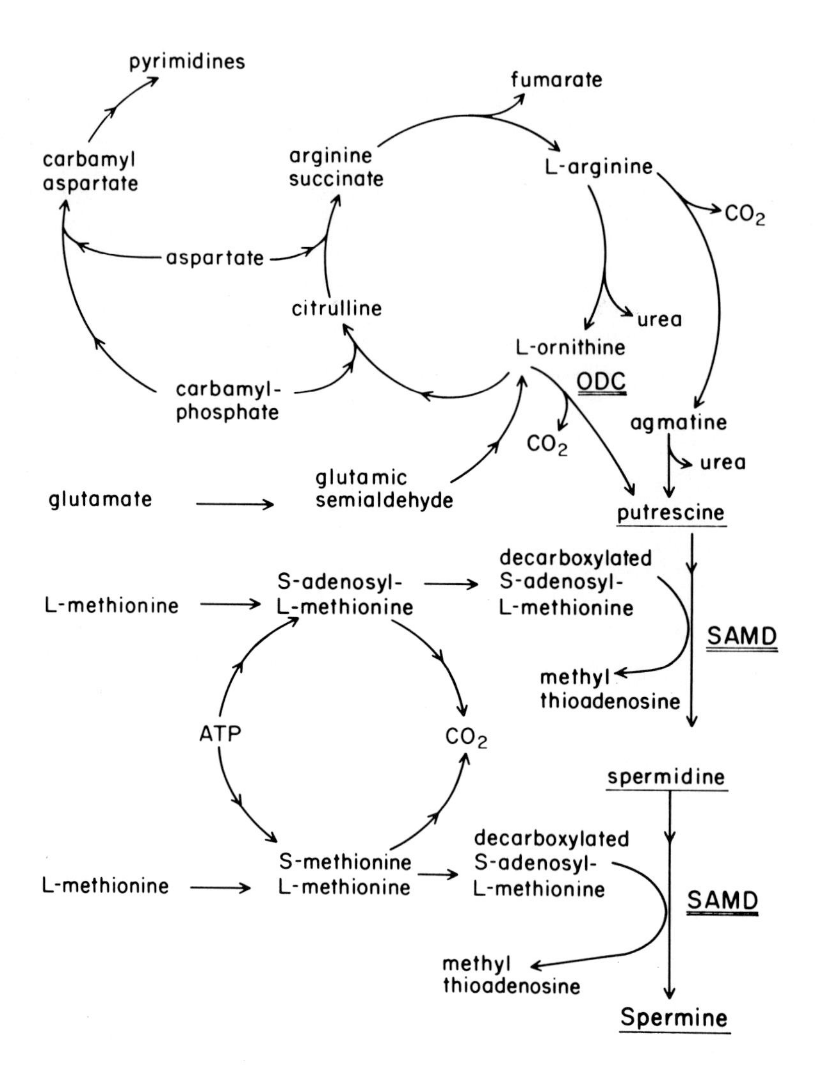

Metabolic pathways of polyamines

Index

8106257
3 1378 00810 6257